F²

Thermophiles

GENERAL, MOLECULAR, AND APPLIED MICROBIOLOGY

edited by

THOMAS D. BROCK
Department of Bacteriology
1550 Linden Drive
University of Wisconsin-Madison
Madison, Wisconsin 53705 USA

A WILEY-INTERSCIENCE PUBLICATION
JOHN WILEY & SONS
New York • Chichester • Brisbane • Toronto • Singapore

Library of Congress Cataloging-in-Publication Data

Thermophiles : general, molecular, and applied
  microbiology.

  (Wiley series in ecological and applied microbiology)
  "A Wiley-Interscience publication."
  Includes index.
  1. Micro-organisms, Thermophilic.   2. Bacteria,
Thermophilic.   3. Industrial microbiology.   I. Brock,
Thomas D.   II. Series.
QR84.T46   1986   576   86-7785
ISBN 0-471-82001-6

Printed in the United States of America

10 9 8 7 6 5 4 3 2 1

CONTRIBUTORS

SHUICHI AIBA Department of Fermentation Technology, Faculty of Engineering, Osaka University, Yamada-oka, Suita-shi, Osaka 565, Japan

PIERRE BÉGUIN Unit of Cellular Physiology, Department of Biochemistry and Molecular Genetics, Pasteur Institute, 28 rue du Dr. Roux, 75724 Paris Cedex 15, France

CORALE L. BRIERLEY Advanced Mineral Technologies, Inc., 5920 McIntyre, Golden, Colorado 80403 USA

JAMES A. BRIERLEY Advanced Mineral Technologies, Inc., 5920 McIntyre, Golden, Colorado 80403 USA

THOMAS D. BROCK Department of Bacteriology, 1550 Linden Drive, University of Wisconsin-Madison, Madison, Wisconsin 53706 USA

TADAYUKI IMANAKA Department of Fermentation Technology, Faculty of Engineering, Osaka University, Yamada-oka, Suita-shi, Osaka 565, Japan

WILLIAM R. KENEALY Central Research and Development Department, E.I. du Pont de Nemours and Company, Experimental Station, Wilmington, Delaware 19898 USA

THOMAS A. LANGWORTHY Department of Microbiology, School of Medicine, University of South Dakota, Vermillion, South Dakota 57069 USA

JACQUELINE MILLET Unit of Cellular Physiology, Department of Biochemistry and Molecular Genetics, Pasteur Institute, 28 rue du Dr. Roux, 75724 Paris Cedex 15, France

THOMAS K. NG Central Research and Development Department, E.I. du Pont de Nemours and Company, Experimental Station, Wilmington, Delaware 19898 USA

v

TAIRO OSHIMA   Department of Life Science, Tokyo Institute of Technology, Ookayama, Tokyo 152, Japan

JEAN L. POND   Department of Microbiology, School of Medicine, University of South Dakota, Vermillion, South Dakota 57069 USA

KARL O. STETTER   Institute for Microbiology, University of Regensburg, 8400 Regensburg, Universitätsstrasse 31, Federal Republic of Germany

T.K. SUNDARAM   Department of Biochemistry and Applied Molecular Biology, University of Manchester Institute of Science and Technology, Manchester M60 1QD, United Kingdom

PAUL J. WEIMER   Central Research and Development Department, Building 402, E.I. du Pont de Nemours and Company, Experimental Station, Wilmington, Delaware 19898 USA

JUERGEN WIEGEL   Department of Microbiology and Center for Biological Resource Recovery, University of Georgia, Athens, Georgia 30602 USA

STEPHEN H. ZINDER   Department of Microbiology, Stocking Hall, Cornell University, Ithaca, New York 14853 USA

SERIES PREFACE

The Ecological and Applied Microbiology series of monographs and edited volumes is being produced to facilitate the exchange of information relating to the microbiology of specific habitats, biochemical processes of importance in microbial ecology, and evolutionary microbiology. The series will also publish texts in applied microbiology, including biotechnology, medicine, and engineering, and will include such diverse subjects as the biology of anaerobes and thermophiles, paleomicrobiology, and the importance of biofilms in process engineering.

During the past decade we have seen dramatic advances in the study of microbial ecology. It is gratifying that today's microbial ecologists not only cooperate with colleagues in other disciplines but also study the comparative biology of different habitats. Modern microbial ecologists, investigating ecosystems, gain insights into previously unknown biochemical processes, comparative ecology, and evolutionary theory. They also isolate new microorganisms with application to medicine, industry, and agriculture.

Applied microbiology has also undergone a revolution in the past decade. The field of industrial microbiology has been transformed by new techniques in molecular genetics. Because of these advances, we now have the potential to utilize microorganisms for industrial processes in ways microbiologists could not have imagined 20 years ago. At the same time, we face the challenge of determining the consequences of releasing genetically engineered microorganisms into the natural environment.

New concepts and methods to study this extraordinary range of exciting problems in microbiology are now available. Young microbiologists are increasingly being trained in ecological theory, mathematics, biochemistry, and genetics. Barriers between the disciplines essential to the study of modern microbiology are disappearing. It is my hope that this series in Ecological and Applied Microbiology will facilitate the reintegration of microbiology and stimulate research in the tradition of Louis Pasteur.

In recent years interest in the biology of thermophiles has increased dramatically. Scientists working in widely divergent disciplines are studying this fascinating group of organisms. Questions are being raised about their mode of survival, ecological significance, and evolutionary strategies.

Thomas Brock's pioneering research on thermophiles has led to a new understanding of the adaptive behavior of microorganisms living in extreme environments. In this volume he has brought together chapters by leading authorities on the molecular, physiological, and genetic properties of thermophiles, as well as discussions of their importance in ecology and biotechnology.

RALPH MITCHELL

Cambridge, Massachusetts
April 1986

PREFACE

Thermophiles are a fascinating group of microorganisms which have received considerable interest in recent years because of their potential biotechnological applications. However, thermophiles are more than biotechnological wonders. Thermophiles are of considerable general interest because of the fascinating problems in basic biology which they present. Biologists are conditioned to expect high temperatures to be harmful to living organisms, yet here is a group of organisms that not only tolerates high temperatures, but actually thrives under these extreme conditions. The existence of thermophiles immediately raises evolutionary and molecular questions of considerable importance.

Fundamental and applied research on thermophiles is underway in a large number of laboratories throughout the world. It is the purpose of the present book to present the current status and future direction of thermophile research. The authors of the chapters in this book have made important research contributions on thermophiles, and they are also knowledgeable about the current research in other laboratories. We are thus fortunate in having such an expert collection of authors for this important and timely book.

The initial discussion on the content and authorship of this book was done in association with J. Gregory Zeikus, who was originally to have been a coeditor and coauthor. Subsequently, Dr. Zeikus' commitments became so extensive that he could not be an active participant in this project, but his advice and counsel are greatly appreciated.

Editorial/production supervision of this book was competently handled by Dr. Katherine M. Brock of Science Tech, Inc. and I greatly appreciate the extensive contributions which she made to make this book possible. My work as editor of this book was supported in part by the College of Agricultural and Life Sciences of the University of Wisconsin-Madison, the National Science Foundation, and the Wisconsin Alumni Research Foundation through a grant to the Graduate School of the University of Wisconsin-Madison. I hope that *Thermophiles: General, Molecular, and Applied*

Microbiology will be a useful book for all those with an interest in basic and applied microbiology.

THOMAS D. BROCK

Madison, Wisconsin, USA
April 1986

CONTENTS

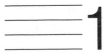

INTRODUCTION: AN OVERVIEW OF THE THERMOPHILES

THOMAS D. BROCK

Department of Bacteriology, University of Wisconsin, 1550 Linden Drive, Madison, WI 53706

Thermophiles are defined as organisms living at high temperatures. They have fascinated scientists for many years, but there has been a burst of interest in these organisms over the past generation, spurred on by the discoveries of bacteria living at or even above the boiling point of water.

1

Thermophiles are of interest both from fundamental and applied points of view. Basic research on thermophiles deals with molecular biology, genetics, biochemistry, evolution, taxonomy, ecology, and origins. From an applied or biotechnological point of view, thermophiles are of interest as sources of unique enzymes with unusual properties, as the active agents in high-temperature fermentations, in waste-treatment processes, and in mineral leaching. Although research in all of these areas is already extremely active, it seems clear from the articles in this book that the future holds vast promise for both basic and applied research on thermophilic microorganisms.

Because of the current excitement and extensive research on thermophilic microorganisms, the present time is apt for a book which provides a synthesis of current knowledge. In the present volume, some of the most active research workers have provided careful, considered overviews of thermophile research. This book can be viewed in two ways: 1) as a description of the current status of the field; 2) as a collection of guideposts to future work.

In the present book we deal almost exclusively with thermophilic bacteria (i.e., procaryotes), the microbial group capable of growth at the highest temperatures of any group of living organisms. Thermophilic bacteria have been of most biotechnological interest and have also been most extensively studied biochemically. Work on thermophilic algae is not dealt with to any extent in this book, but is well covered in Brock (1) and Castenholz (2). Thermophilic fungi, another major group of thermophiles, have been reviewed by Tansey and Brock (3).

1. THE DEFINITION OF A THERMOPHILE

The term *thermophile* has been used in a variety of ways, and there is no universally accepted definition. Simply put, a thermophile is an organism capable of growth at high temperature. How high a temperature? This will depend on the group of organisms under consideration, as illustrated in Table 1. In this table, the upper temperature limits for the growth of various groups of organisms are given. Note in this table that for any given group of organisms only a *few* species are capable of living close to the upper limit for the group; most organisms in the group live only at lower temperatures. Using Table 1 as a guide, we can construct the following definition: *a thermophile is an organism capable of living at temperatures at or near the maximum for the taxonomic group of which it is a part.* Thus, a thermophilic vertebrate would be one capable of living near 37°C, a thermophilic fungus one capable of living near 60°C, a thermophilic cyanobacterium one capable of living near 70°C, and so on. This definition has the advantage that it emphasizes the taxonomic distinctions of thermophily in different groups of organisms, but it has the disadvantage that it is imprecise and says nothing about the temperature range or optimum of the organism.

TABLE 1. Upper temperature limits for growth

Group	Approximate upper temperature (°C)
Animals	
Fish and other aquatic vertebrates	38
Insects	45–50
Ostracods (crustaceans)	49–50
Plants	
Vascular plants	45
Mosses	50
Eucaryotic microorganisms	
Protozoa	56
Algae	55–60
Fungi	60–62
Procaryotic microorganisms	
Cyanobacteria (blue-green algae)	70–73
Photosynthetic bacteria	70–73
Chemolithotrophic bacteria	>100
Heterotrophic bacteria	>100

From Brock (1)

2. TERMINOLOGY OF THERMOPHILIC BACTERIA

Unfortunately, terminology has proliferated for the thermophilic bacteria, and many of the terms have been ill defined. The problem can be discussed with reference to Figure 1. Some bacteria grow over a fairly wide range; such organisms are called *eurythermal*. Other bacteria, called *stenothermal*, grow only over a narrow temperature range. The concepts of *stenothermal* and *eurythermal* do not apply specifically to thermophiles but are broad terms describing the temperature/growth relationships of any organism. In general, organisms living in environments of constant temperature tend to be stenothermal, whereas organisms living in habitats of fluctuating temperatures tend to be eurythermal.

The *thermophile boundary* of 55 to 60°C defined in Figure 1 is arbitrary, but has some ecological and evolutionary basis. Temperatures lower than 50°C are widespread on earth, associated with sun-heated habitats, whereas temperatures greater than 55 to 60°C are much rarer in nature, being associated almost exclusively with geothermal habitats. Another reason for defining a thermophile boundary for bacteria at a temperature around 60°C is that 60°C is the upper temperature limit for eucaryotic life (Table 1), so that at temperatures above 60°C only procaryotes are found.

In Figure 1b, several terms frequently used with reference to thermophiles are illustrated. The terms described in Figure 1b may or may not be useful,

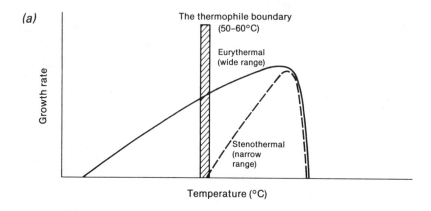

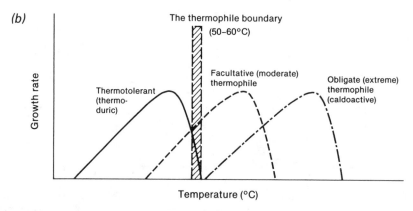

Figure 1. The relationship between growth rate and temperature for various bacteria, illustrating the thermophile boundary and the terminology which has been used to describe organisms exhibiting different temperature responses. For definitions of some of these terms, see Williams (4).

depending on circumstances. It is of some importance to emphasize that there are no real boundaries, but only a continuum. Thermophilic bacteria have been described, for instance, with temperature optima of 55, 60, 65, 70, 75, 80, 85, 90, 95, 100, and 105°C, and at many temperatures in between those listed. Furthermore, the temperature optimum of an organism may change a few degrees depending on the culture medium and conditions in which the organism is grown (see Chapter 4, this volume). For any careful description of a thermophilic bacterium, the precise temperature range over which it grows (and the culture conditions used) must be described (see Chapter 2).

Another important point is that the temperature responses listed in Figure 1 are for growth, *not* survival. Many bacteria, especially those capable of forming endospores, can *tolerate* temperatures much higher than the tem-

TABLE 2.   Genera of thermophilic bacteria

Genus	Number of species	Temperature range (°C)
Phototrophic bacteria		
Cyanobacteria	16	55–70
		(One strain, 74)
Purple bacteria	1	55–60
Green bacteria	1	70–73
Gram-positive bacteria		
Bacillus	15	50–70
Clostridium	11	50–75
Lactic acid bacteria	5	50–65
Actinomycetes	23	55–75
Other eubacteria		
Thiobacillus	3	50–60
Spirochete	1	54
Desulfotomaculum	7	37–55
Gram-negative aerobes	7	50–75
Gram-negative anaerobes	4	50–75
Archaebacteria		
Methanogens	4	55–95
Sulfur-dependent	10	55–110
Thermoplasma	1	37–55

For references see Tansey and Brock (3) and Brock (5)

peratures at which they grow. But evolutionarily, it is the temperature range over which a bacterium is able to *maintain a population* that is important.

3.   SPECIES DIVERSITY OF THE THERMOPHILIC BACTERIA

An extensive review of thermophilic microorganisms was published by Tansey and Brock (3) and a summary of organisms described since then has been presented by Brock (5). Although there are many uncertainties about the taxonomy of some species of thermophilic bacteria, an overview of the major genera is given in Table 2. It can be seen that a wide variety of genera have thermophilic representatives, but that bacteria growing at the highest temperatures, up to and above boiling, are all archaebacteria.

As a result of detailed analyses of 16S ribosomal RNA, Fox et al. (6) and Woese and Wolfe (7) have presented a phylogenetic scheme for the living world which is illustrated in Figure 2. Three kingdoms are recognized: the Archaebacteria, the Eubacteria, and the Eucaryotes. Thermophilic bacteria, defined here as organisms growing at temperatures of 55°C or higher, are

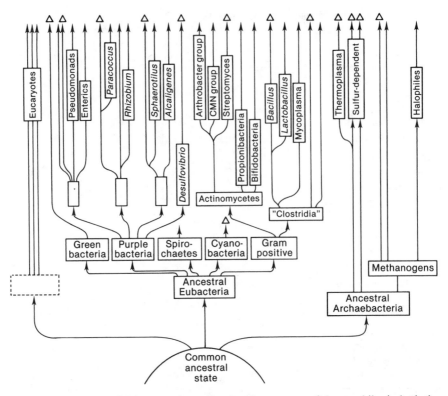

Figure 2. Phylogeny of living organisms, showing the presence of thermophiles in both the eubacteria and the archaebacteria lines. Lines with thermophilic members (defined by the ability to grow above 55°C) are indicated by the triangles. The CMN group contains the genera *Corynebacterium*, *Mycobacterium*, and *Nocardia*.

found in both the Archaebacteria and the Eubacteria kingdoms. As also illustrated in Figure 2, many of the lines of descent of the two kingdoms contain thermophilic members. From this fact it seems reasonable to conclude that thermophily is a polyphyletic character, having arisen a number of times during the course of evolution.

3.1. Thermophilic Bacteria Currently Under Study

In recent years, a number of new and interesting thermophilic bacteria have been isolated and characterized. Many of these bacteria are now available from the German Collection of Microorganisms, and are summarized in Table 3. The extreme thermophiles are described in detail in Chapter 3. These organisms can be cultured readily, even in large-scale fermenters using well-defined culture medium. Many of these organisms will prove of considerable interest to both the molecular biologist and the biotechnologist.

TABLE 3. Species and genera of thermophilic bacteria discovered in recent years

---

Archaebacteria
  Aerobic acidophiles, autotrophs
  Sulfolobus acidocaldarius (type species)
  Sulfolobus brierleyi
  Sulfolobus solfataricus
Grow at 70–90°C (optimum, 75–85°C); pH 1 to 4; use organic compounds and S^0 as energy source; use O_2 or Fe^{3+} as electron acceptor

Thermoproteales
  Thermoproteus tenax
  Desulfurococcus mobilis
  Desulfurococcus mucosus
  Thermophilum pendens
  Thermococcus celer
Grow at 70–85°C (optimum, 85°C); anaerobic; acidophilic to neutrophilic; use organic compounds as energy sources; use S^0 as electron acceptor

Methanothermaceae
  Methanothermus fervidus
Grows at 70–95°C (optimum, 85°C); anaerobic; uses H_2 and CO_2 exclusively

Methanococcales
  Methanococcus jannaschii
Grows at 60–90°C (optimum, 85°C); anaerobic; uses H_2 and CO_2 exclusively

Pyrodictium
  Pyrodictium brockii
  Pyrodictium occultum
Grow at 85–110°C (optimum, 105°C); anaerobic; use H_2 and S^0; autotrophic

Other thermophiles—all anaerobes
  Clostridium thermohydrosulfuricum—optimum, 65°C; produces ethanol
  Clostridium thermosulfurogenes—optimum, 60°C; ferments pectin; forms S^0
    from S_2O_3
  Thermoanaerobacter ethanolicus—optimum, 65°C; produces ethanol
  Thermoanaerobium brockii—optimum, 65°C; produces ethanol
  Thermobacteroides acetoethylicus—optimum, 65°C
  Thermodesulfobacterium commune—optimum, 70°C

From Brock (5)

4. HABITATS OF THERMOPHILIC MICROORGANISMS

Natural geothermal habitats are worldwide in distribution, associated primarily with tectonically active zones where major crustal movements of the earth occur. In such areas, deep-seated magmatic materials are thrust close to the earth's surface and serve as heat sources (Figure 3). Seawater or

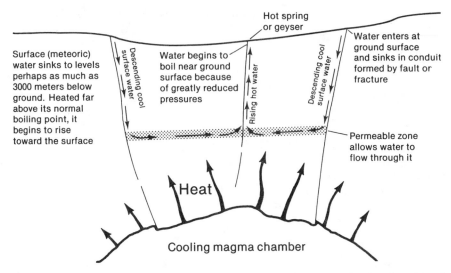

Figure 3. Origin of geothermal habitats. Heat flow and surface water, showing the origin of a geothermal system. Redrawn from Keefer (8).

groundwater percolating into the earth becomes intensely heated but does not boil because of lithostatic pressure. When the percolating fluid reaches a sufficiently high temperature, the pressure generated forces the fluid through pores and fissures back to the surface of the earth, where it can issue as *thermal springs and geysers.* As the hot fluid passes up through the earth's surface, minerals dissolve, thus accounting for the extensive mineralization of the thermal springs and geysers associated with geothermal sources. Microbiologically important constituents dissolved in the water may include hydrogen sulfide, carbon dioxide, low-molecular-weight organic carbon compounds, methane, hydrogen, ammonia, and trace elements. Chloride and bicarbonate are usually the dominant anions. The exact composition of the thermal waters will depend on the chemistry of the rocks through which the fluid passes and on the temperature of the water (1).

Because of the localized nature of geothermal sources, thermal habitats are generally concentrated in small areas, called *thermal basins.* If water supply to a deep-seated thermal source is low, steam rather than hot water may issue to the surface. Steam-dominated thermal sources are called *fumaroles* and are common in certain parts of the world (for instance, the famous area of Solfatara near Pozzuoli, Italy). In addition to such major fumarolic systems, small steam-dominated areas are often found in association with thermal springs, generally on hills above the thermal basin.

A common type of thermal basin will have geysers and/or flowing springs at the bottom of the hill, sulfur- and iron-rich nonflowing (or weakly flowing) springs on the flanks of the hill, and fumaroles on the highest ground. The geysers and flowing springs will generally be neutral to alkaline in pH; their waters will usually be highly mineralized. The sulfur- and iron-rich springs are usually acidic, due primarily to the presence of sulfuric acid, but except for the sulfate ion (and occasionally iron in soluble form), the mineralization of these springs is low. This is because the water of the sulfur-rich springs, if it exists, is primarily superficial groundwater which has never undergone deep circulation. The water of acid springs is derived primarily from the local region on the hills surrounding a basin. Such water, moving slowly down the slope and coming in contact with rising fumarolic steam, becomes heated. The sulfuric acid (H_2SO_4) is derived from the oxidation of hydrogen sulfide (H_2S) which rises with the steam from deep in the earth. When the hydrogen sulfide meets oxygen (O_2) derived from the atmosphere, elemental sulfur (S^0) is formed (usually nonbiologically) and becomes deposited in the soil. Sulfur-oxidizing bacteria (generally nonthermophilic) oxidize this elemental sulfur to sulfuric acid and the pH of the soil drops. Because of evaporation, the acidity of these fumarolic soils is often very high. Some of the acid formed in these soils then moves down the slope with rain-derived groundwater, and if a depression in the earth is present, an acid spring is formed. Such acid springs often have little or no surface-water flow, although subsurface flow usually occurs. Whether or not such a spring is hot will depend on whether it has steam sources beneath it.

There is a bimodal distribution of pH values of thermal springs, with many pH values around 1.8 to 2.2 and many values of 7.5 to 9.0 (9). These two pH ranges reflect the two main buffering components of natural thermal waters, sulfuric acid with a pK of 1.8, and the carbonate-bicarbonate system with pK values at 6.3 and 10.2.

In summary, boiling and superheated geysers found at the bottom of a thermal basin have extensive water flow, are highly mineralized, are neutral to alkaline in pH, and are of exceedingly constant temperature (at the source). The major anions are generally bicarbonate and chloride. Such geothermal habitats are often long-lived, on the order of hundreds of years. As we move upslope, the springs have progressively less flow, become less mineralized, more acidic, with the major anion becoming sulfate. The temperatures of the acidic springs are often highly variable, due to changes in steam sources beneath them. The springs themselves are often short-lived, on the order of years to tens of years.

Another type of geothermal habitat, that associated with *deep-sea geo-thermal vents* (10), has an origin similar to that of the neutral boiling spring, except the source of the groundwater is sea water rather than freshwater (Figure 4). Because of the saline water source, the hot water is highly reactive and the hydrothermal fluids of the deep-sea vents become highly mineralized.

Thermophilic bacteria have been found in association with all of the types of geothermal habitats described above: boiling springs of neutral pH, sulfur-rich acidic springs, and deep-sea vents. The most common bacterium

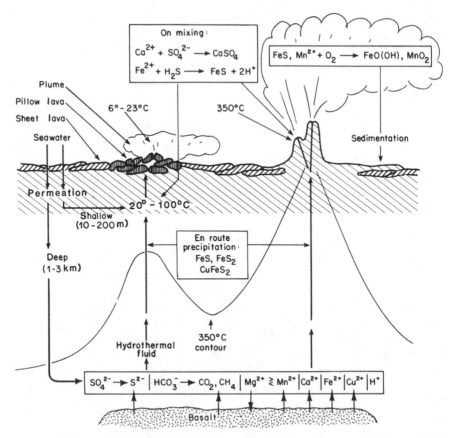

Figure 4. Diagram showing chemical processes at deep-sea vents. Seawater circulates to depth and becomes heated to 350 to 400°C, reacts with basaltic rocks, and leaches various minerals into solution. The hot water, under pressure, rises to the sea floor where various precipitation reactions occur. From Jannasch and Mottl (10). Copyright 1985 by the AAAS.

of the sulfur-rich acid springs is the archaebacterium *Sulfolobus*, although other archaebacteria have also been found. In the neutral pH boiling springs, *Pyrodictium*-like organisms dominate (11, 12, and Stetter, this book, Chapter 3). A thermophilic methanogen, *Methanococcus jannaschii*, has been isolated from a sample of material taken from a deep-sea vent (13).

5. ECOLOGY OF THERMOPHILIC MICROORGANISMS

Although it is absolutely essential to obtain pure cultures of dominant organisms in extreme environments (see Chapter 2), it is also necessary to study the activity of organisms directly in nature, because organisms often do different things in culture than in nature. If we are interested in evolution, we must admit that organisms do respond to environment by genetic change, and it is only by understanding the environmental relationships of organisms that we will have a full understanding of evolution. Extreme environments are very favorable for direct ecological study because of the restricted species diversity.

Although some thermal habitats are difficult to study, research on hot-spring habitats is simplified by the fact that hot springs have relatively constant chemical and physical properties and occur in nicely confined domains where experimentation is easily done. Measurements of temperature are easy to make with thermistor probes, and because of the high heat capacity of water, it can be assured that the temperature of bacteria present in the spring is the same as the measured temperature of the water (this is not true for thermal soils where bulk temperature and microenvironmental temperature may be quite different). Water samples for chemistry and microbiology are easy to obtain and repeated measurements can be made. In acid springs, the bacteria are found growing primarily free in the water and can be simply sampled by removing water samples. In neutral springs, the bacteria are primarily attached to siliceous walls of the spring and must be studied by the use of artificial substrates, such as glass slides (11).

Approaches to the study of the ecology of thermophilic microorganisms have been outlined in detail (1), and will only be summarized here. Several different types of approaches have been used: 1) Direct microscopy either with the light microscope or the electron microscope. For electron microscopy, Mylar film can be immersed in the environment and after microbial colonization the film fixed, embedded, sectioned, and examined under the transmission electron microscope. Light microscopy can be readily done using the phase-contrast microscope, preferably on organisms that have colonized microscope slides immersed in the habitat. 2) Growth rates can be measured, using microscopic techniques, on organisms colonizing substrates. An ultraviolet radiation technique was used to show that increases in cell numbers were really due to growth (14). 3) To determine temperature, pH, or salinity optima, radioisotope techniques can be used to measure the activity of natural populations and the effect of environmental change. It is most useful to carry out such measurements on microorganisms which have colonized specially prepared substrates (11). 4) Chemical changes in the natural environment can frequently be related to the activities of microorganisms (15). 5) One useful technique at the whole-system level is to measure turnover rates of microbes and chemicals by use of the addition of a tracer chemical to the water (16). This permits direct measurement of replication rates of nonattached organisms and determination of the connection between chemical change and microbial activity.

A major conclusion from studies of the activities of organisms directly in natural geothermal habitats is that in many cases the organisms present are optimally adapted to the temperatures at which they are living. Results of this kind suggest that the organisms living at high temperatures are probably *optimally* adapted to those temperatures. This conclusion has evolutionary implications, because it indicates that the organisms living at high temperatures are not just eurythermal organisms which have been able to extend their ranges to higher temperatures, but organisms uniquely adapted to these high temperatures. This conclusion also leads to a prediction that the temperature of the habitat is the best incubation temperature for cultural studies. Thus, as research moves to habitats of higher and higher temperature, the search for organisms by cultural methods should employ higher and higher incubation temperatures. However, there must be an upper temperqture limit beyond which this relationship should no longer hold: the upper temperature for life itself. Organisms living near the upper temperature limit for life may no longer be optimally adapted to their habitat temperatures. As yet, the upper temperature for life has not been defined, although there is some reason to believe that this temperature will be discerned in the near future (see Chapter 3).

6. MOLECULAR BASIS OF THERMOPHILY

It is now well established that the macromolecular structures of thermophiles are inherently thermostable. The basis for thermostability is discussed in this book in Chapters 4, 5, 6, and 7. The basis for macromolecular thermostability seems to reside in rather small differences in macromolecules which lead to increased intramolecular bonding. Subtle changes in hydrophobic interactions, hydrogen bonds, sulfur-sulfur bonds, and ionic bonds are sufficient to confer thermostability. In one case, a change in a single nucleic acid base in a gene has resulted in a change in a single amino acid in the corresponding protein which has led to an approximately 10°C increase in thermostability (for the enzyme kanamycin nucleotidyltransferase, see Chapter 7). In another study (see Chapter 6), a transfer RNA of a thermophile became more thermostable because of the change in a single atom in a critical base (the substitution of a sulfur atom for an oxygen atom at thymine-55 in a tRNA of *Thermus thermophilus*). Other changes in tRNA molecules which alter thermostability are reviewed by Oshima in Chapter 6. Although there are some factors such as polyamines which serve as stabilizing components in cells, it seems clear that the overall evolution of thermophily has resulted from numerous small mutational changes in proteins and nucleic acids.

This conclusion emphasizes the importance of studies on the genetics of thermophiles and thermophily. As described in Chapters 6, 7, and 8, considerable progress is being made in this area, but we still lack efficient and reproducible gene transfer systems for thermophiles. The advent of recombinant DNA technology promises, however, to permit circumvention of the painstaking development for each thermophile of a whole panoply of gene transfer techniques. Transformation alone (for which some progress is being made; see Chapters 6 and 7) should permit considerable progress in research on the genetics of thermophiles, when this technique is coupled with recombinant DNA research.

A critical structure in each cell is the cell membrane, and Langworthy and Pond (Chapter 5) describe the remarkable lipid chemistry of the membranes of thermophiles. Membrane and lipid chemistry has become of broader interest because of the strong relationships shown between the lipids of thermophilic archaebacteria and those of the nonthermophilic archaebacteria.

6.1. Evolution of Thermophily

The question of the origin of thermophilic microorganisms has intrigued scientists for over 100 years. Two major hypotheses have been advanced: 1) The first organisms arose in high-temperature environments, so that thermophilic organisms were primordial and subsequent organisms were derived from them by evolution. 2) The first organisms were not thermophiles, but were adapted to moderate temperature conditions, and thermophiles have had a secondary origin from more conventional types.

As we have indicated, *all* macromolecules of thermophiles seem to be stable at the high temperatures at which these organisms live. Therefore, if thermophiles arose from mesophilic organisms, many genetic changes would have to take place, more or less simultaneously. On the other hand, if the primordial organism were a thermophile, it is quite easy to imagine a temperature-sensitive mutant arising which was incapable of growth at the higher temperature. However, one difficulty with a primary origin for thermophily is that the thermophilic character is not restricted to a single phylogenetic line of bacteria, but is found in many groups of both archaebacteria and eubacteria (see Figure 2). Therefore, if thermophily were of primary origin, then the common ancestral organism of all life forms postulated in Figure 2, even the eucaryotes, would have to have been a thermophile. Without further evidence, speculation seems pointless. A single clear-cut experiment in which a thermophile was derived from a mesophilic organism by simple genetic means would do more to clarify this question than any amount of speculation.

7. BIOTECHNOLOGICAL APPLICATIONS OF THERMOPHILES

Thermophiles appear to offer some major advantages for biotechnology, and this is one theme that is explored extensively in this book. Weimer discusses, in Chapter 10, the possibilities for the production of chemicals and fuels using thermophiles, and Béguin and Millet, in Chapter 8, describe the possibilities for the genetic manipulation of these fermentative organisms. Ng and Kenealy, in Chapter 9, describe the industrial potential of thermophile enzymes. The most attractive attribute of thermophiles from a biotechnological point of view is that they produce enzymes capable of catalyzing biochemical reactions at temperatures markedly higher than those of conventional organisms. In addition, enzymes from thermophiles are more stable at conventional temperatures, thus prolonging the shelf life of commercial products.

Industrial enzyme processes have some distinct advantages if they can be carried out at high temperature. An increase in temperature results in an increase in the diffusion rate and in the solubility of most nongaseous compounds. An increase in temperature also reduces the viscosity and surface tension of water, which has some positive effects on enzyme-catalyzed reactions. Large-scale reactions are often limited by physical processes such as diffusion, and an industrial enzymatic process should therefore occur more rapidly if it can be carried out by a thermostable enzyme. In addition, microbial contamination of the process may occur less readily if it is carried out at high temperature, because most of the environmental contaminants that cause deterioration of biological processes are mesophiles.

Microbial fermentations carried out at high temperature also have some potential advantages (see Chapter 10). Metabolic activity results in heat production, and cooling of fermenters is a serious problem in large-scale processes. Since thermophilic fermentations need not be so extensively cooled, there is a consequent saving in energy. The industrial process of most potential interest is that for the manufacture of ethanol. There are three advantages of a thermophilic ethanol process: 1) the elevated temperature makes distillation of the ethanol more efficient; 2) the cooling requirement is reduced or eliminated; 3) some thermophilic bacteria can carry out a direct fermentation of polysaccharides to ethanol, whereas yeast (the conventional alcohol-producing organism) is incapable of hydrolyzing polysaccharide polymers. Weimer (Chapter 10) discusses the ethanol fermentation by *Clostridium thermocellum* and Béguin and Millet (Chapter 8) discuss the applied genetics of this organism.

The potential for waste treatment using thermophiles is discussed by Zinder in Chapter 11. Anaerobic waste-treatment processes using thermophiles appear to have the following advantages: 1) increased reaction rate and decreased retention time; 2) destruction of pathogenic microorganisms that might be present in the sewage; 3) lower viscosity so that less energy is required for mixing; 4) easier dewatering of the resulting sludge. Zinder also notes certain disadvantages of the thermophilic digestion process: 1) the high energy requirement for heating (insufficient heat is produced in the digestion process itself); 2) difficulty of maintaining a stable process for long periods of time, an especially important requirement for a waste-treatment process; 3) poor quality of the liquid effluent from a thermophlic digestion process in some instances, for unexplained reasons. Zinder describes several anaerobic treatment processes that have been successful, and notes the potential future applications.

Another potential practical use of thermophiles, for microbial leaching, is discussed by Brierley and Brierley in Chapter 12. Bacterial leaching of low-grade metal ores has been done for a long time using mesophiles, and it is only in recent years that it has become evident that in many commercial leaching operations, high temperatures develop in parts of the leach pile and that thermophilic microorganisms may thus be important. Brierley and Brierley indicate that leaching with thermophiles is much more rapid than with mesophiles, but that there are some serious problems in the use of these organisms because of the lack of detailed understanding of their physiology and ecology. Since the economic value of leaching is closely tied to world ore prices, it is especially important that the process be run in the most efficient manner.

8. CONCLUSION

The fascinating basic and applied research using thermophiles is admirably related by the authors of the various chapters of this book. We see that research is moving at an increasingly rapid rate, impelled first by perceived economic advantages of thermophiles, second by the recognition of archaebacteria as a distinct evolutionary group, and third by the remarkable discoveries of the deep-sea hydrothermal vents. The diversity of thermophiles is high, and new organisms continue to be discovered. Evolutionary, biochemical, and genetic studies have barely begun and should occupy scientists for many years.

Although biotechnological applications of thermophiles seem promising, the large-scale uses are to date rather modest. The industrial potential of thermophilic microorganisms has not yet been fully realized primarily because the requisite biotechnological research has barely begun. Hopefully, the material in this book will provide stimulus for further research, as well as a useful compendium of the status of current knowledge.

REFERENCES

1. T. D. Brock, *Thermophilic Microorganisms and Life at High Temperatures*, Springer-Verlag, New York, 1978.
2. R. W. Castenholz, in M. Shilo, Ed., *Strategies of Microbial Life in Extreme Environments*, Dahlem Konferenzen, Verlag Chemie, Weinheim, 1979.
3. M. R. Tansey and T. D. Brock, in D. J. Kushner, Ed., *Microbial Life in Extreme Environments*, Academic Press, New York, 1978, p. 159.
4. R. A. D. Williams, *Science Progress* **62**, 373 (1975).
5. T. D. Brock, *Science* **230**, 132 (1985).
6. G. E. Fox, E. Stackenbrandt, R. B. Hespell, J. Gibson, J. Maniloff, T. W. Dyer, R. S. Wolfe, W. E. Bälch, R. S. Tanner, L. J. Magrum, L. B. Zablen, R. Blakemore, R. Gupta, L. Bonen, B. J. Lewis, D. A. Stahl, K. R. Luehrsen, K. N. Chen, and C. R. Woese, *Science* **209**, 457 (1980).

7. C. R. Woese and R. S. Wolfe, Eds., *Archaebacteria*, Academic Press, New York, 1985.
8. W. R. Keefer, *The Geologic Story of Yellowstone National Park*, U. S. Geol. Survey Bull. 1347 (1972).
9. T. D. Brock, *Bull. Geol. Soc. Am.* **82**, 1393 (1971).
10. H. W. Jannasch and M. J. Mottl, *Science* **229**, 717 (1985).
11. T. D. Brock, M. L. Brock, T. L. Bott, and M. R. Edwards, *J. Bacteriol.* **107** 393 (1971).
12. K. O. Stetter, *Nature* **300**, 258 (1982).
13. W. J. Jones, J. A. Leigh, F. Mayer, C. R. Woese, and R. S. Wolfe, *Arch. Microbiol.* **136**, 254 (1983).
14. T. L. Bott and T. D. Brock, *Science* **164**, 1411 (1969).
15. T. D. Brock and J. L. Mosser, *Bull. Geol. Soc. Am.* **86**, 194 (1975).
16. J. L. Mosser, B. B. Bohlool, and T. D. Brock, *J. Bacteriol.* **118**, 1075 (1974).

2

METHODS FOR ISOLATION AND STUDY OF THERMOPHILES

JUERGEN WIEGEL

Department of Microbiology and Center for Biological Resource Recovery, University of Georgia, Athens, Georgia 30602

1. INTRODUCTION

Except for some basic considerations, most methods used for mesophilic bacteria can be directly used for thermophiles and extreme thermophiles without major changes. Examples of basic points which may have to be taken into account when working at elevated temperatures are:

1. Increased evaporation of medium in incompletely sealed culture vessels, especially when using aeration. Moreover, increased humidity in sealed containers causes internal condensation and can lead to smears on surfaces of solid medium.
2. Lower solubility of gases with increasing temperatures.
3. Melting of most plastic Petri dishes above 65 to 70°C.
4. Instability of agar at temperatures above 65 to 70°C.
5. Faster caramelization of sugars, especially under aerobic conditions and in the presence of high phosphate concentrations
6. Contamination of media components by heat resistant spores of thermophilic sporeformers which are normally not encountered at mesophilic growth temperatures

The following paragraphs will focus on some precautions and changes that are necessary when working with thermophiles or extreme thermophiles. In cases where brand names of equipment are used, they should be regarded as examples with which the author is familiar and should serve only as a starting point for an individual search for the most suitable equipment.

Where necessary, there will be a differentiation in the methods between those for thermophiles growing at the lower temperature range (thermophiles, thermotolerants) and those at the higher temperature range (extreme thermophiles, barothermophiles or pyrophiles); for definitions see Brock, Chapter 1, this volume, and Wiegel and Ljungdahl (1).

The present chapter deals with the growth of thermophiles in laboratory-scale equipment. Later chapters in this book deal with industrial applications in more depth and present the numerous advantages of thermophiles in industrial applications. For further discussions see also Wiegel and Ljungdahl (1). In general, the use of elevated temperatures does not create real problems for an industrial process, particularly when the temperature does not exceed 70°C. For processes using extreme thermophiles and temperatures above 70°C, the methods developed in the chemical industry can be easily adapted for a biotechnological application. These methods become even more important when significant amounts of aggressive substances like H_2S or acetic acid are present at the elevated temperatures.

2. INCUBATION EQUIPMENT

The variety of equipment to grow organisms at elevated temperatures varies from simple waterbaths to computer-controlled environmental chambers and pressurized titanium vessels.

In the author's laboratory, hot air incubators, modified for higher temperatures, are primarily used. Lunaire Environmental, Inc. (Williamsport, PA) modified their economical standard incubators by doubling the heating element and adding a door switch for shutting off the blower for the time the door is open. With these modifications, the incubator can be used at temperatures up to 100°C. After the door has been open for 5 minutes, the recovery time of a 34 cft (1 m³) incubator is around 5 minutes at 90°C (and less at lower temperatures). The use of less expensive drying ovens is not recommended since the temperature usually fluctuates too much. Air-heated incubators have the advantage that tubes can be immediately used for optical density measurements without elaborate glass cleaning. For small vials the electrically heated aluminum blocks (e.g., Thermoline, Sybron, Dubuque, IA) can be used. They have the advantage that the tops of the culture vials are easily accessible for sampling, pH control, or other additions to the culture without removing the vials from the heat block and thus without significantly altering the temperature in the vials.

For culturing thermophilic anaerobes, anaerobic chambers can be equipped with small incubators for use at temperatures up to 90°C (Model 77, Coy Laboratory Products, Inc., Ann Arbor, MI).

Many anaerobic chambers and jars contain palladium catalysts for removing oxygen. These and other catalysts do not function when wet. At the high humidity which frequently occurs at elevated temperature, the catalyst has to be exchanged very frequently. Furthermore, H_2S will poison most oxygen-removing catalysts. A lead acetate paper strip placed on top of the catalyst is a convenient indicator for the presence of sulfide and thus can prevent working with a poisoned catalyst. Layers of cotton impregnated with lead acetate can be used to remove small amounts of H_2S from the chamber atmosphere.

2.1. Temperature Range 40 to 75°C

For culturing organisms or performing experiments with cell-free extracts at temperatures up to 75°C, waterbaths are satisfactory. Evaporation is minimized by covering the bath with a (stainless steel) cover or having the water covered with air-filled plastic balls or (as a cheap substitute) Styropore packing material. Automatic water level controls in the form of overflow tubes are relatively easy to build and to add if the waterbath has a drainage line. The use of a dripping waterline to replace water lost by evaporation is simple but less reliable, and thus is not recommended.

For aerobic or gas-consuming cultures, several rotary and reciprocal shakers are available for temperatures up to 65°C (e.g., New Brunswick, NJ, USA; Braun, Melsungen, FRG) either as a waterbath or hot air incubator. At elevated temperatures, the solubility of gases is less than at mesophilic temperatures. Therefore, it is important to use culture vessels yielding a high ratio of surface to liquid volume. This can be obtained by using a shallow layer of culture fluid, or more effectively, a rotary shaker and baffled culture flasks. Baffles are easily added to Pyrex Erlenmeyer flasks or bottles so that maximal aeration can be obtained.

2.2. Temperature Range 70 to 100°C

At temperatures above 80°C, water should be replaced in baths by other liquids, such as glycerol, with higher boiling points. Oil and liquid paraffin are messy if the culture tubes have to be frequently removed to determine growth. Furthermore, with time, they can release toxic vapors.

For temperatures around 100°C, electric sandbaths with sufficiently accurate temperature regulations (± 1°C) are convenient.

If shaking is required, shaking waterbaths for temperatures up to 100°C are available. Presently, most commercially available hot-air shaking incubators have an upper limit of 60 to 80°C. One of the exceptions is the RF-1 model from Bioengineering Associates Inc. (Newton, MA, 02164) which has an upper limit of 100°C.

Since the solubility of gases at these higher temperatures is very low, gas overpressures of 1 to 3 atm at room temperature are recommended, in addition to the precautions for a high surface area discussed above, to avoid the possibility that the gas supply becomes the rate-limiting step. High agitation rates are normally required.

2.3. Temperatures Above 100°C

In aqueous solutions, temperatures above 100°C can be maintained for extended time periods only if the system can maintain sufficient pressure to keep the water in the liquid phase. Table 1 gives some examples of vapor pressures of pure water at temperatures above 100°C. Media components will influence these values, that is, they will lower the vapor pressure somewhat.

For temperatures around 125°C, where overpressures below 3 atm (45 psi) are required, the aluminum crimped Balch tubes (Bellco Glass Inc., NY; no. 2048–18150) or serum bottles can be used. If bottles larger than 100 ml are used with above 0.7 atm overpressure (10 psi), it is recommended that the culture vessels be kept in small wire baskets or plastic containers (e.g., used plastic bottles) to minimize the possibilities of injuries due to breakage. Furthermore, one has to keep in mind that the gas pressure that is applied at room temperature will be significantly increased at the elevated

TABLE 1. **Vapor pressures of water above 100°C**

Temperature (°C)	Vapor pressure	
	mmHg	psi
100	760	14.7
110	1,075	20.8
115	1,268	24.5
120	1,489	28.8
125	1,741	33.7
130	2,026	39.2
140	2,711	52.5
150	3,570	69.0
175	6,694	129.4
200	11,659	225.5

If an inert gas is used, apply about 5 psi (at the incubation temperature) above the vapor pressure. When $H_2/CO/CO_2$ gases are used as substrates, apply about 10–20 psi above the vapor pressure. To convert from mmHg to g/m^{-2} multiply the values by 1.3595 and for the conversion to atmospheres (atm) by 0.001316.
From CRC-Handbook (2).

incubation temperature. This is especially true for the strongly nonideal gas, carbon dioxide. Furthermore, at temperatures above 50°C, bicarbonate is unstable and is converted to carbonate with the release of carbon dioxide.

For enrichment studies and experiments at temperatures above 100°C, the commercially available Parr reactor with a variable motor drive (Parr Instruments, Moline, IL), usually used in chemistry for hydrogenation reactions, is well suited after slight modifications. Using a high-pressure liquid chromatography (HPLC) multichannel valve, samples can be taken without significantly disturbing the temperature and pressure of the cultures. At temperatures above 120°C and in the presence of hydrogen gas, the version made of 1.0 Cr/0.5 Mo–steel or 3.0 Cr/0.05 Mo–steel is preferable over the normal stainless steel model (3).

For determination of temperature ranges (T_{min} and T_{max}) and optimal temperatures for growth or temperature-dependent reactions, the use of a temperature gradient incubation device is very handy. Scientific Industries, Inc. (Bohemia, NY; Model 675TGI) has a variable temperature gradient device with a range from -5 to 110°C which is well suited for investigations of extreme thermophiles. Unfortunately, temperature variations of up to 2°C can occur over a 2-hour period. The following modifications can be made in an instrument shop: 1) replacement of the on/off switch with a proportional controller for the heater; 2) modification of the cooling system so that it runs continuously; 3) exchange of the thermistor with an ultra thermilinear composite thermistor (YSI 44018, Yellow Springs Instrument Co., Yellow Springs, OH). With these changes, the temperature stability of single slots in the device, depending on the temperature range, is less than $\pm 0.3°C$ over several days. Furthermore, the single temperature probe can be exchanged with a switchboard containing 12 temperature probes which have an accuracy of $\pm 0.75°C$ at temperatures between -30 and $100°C$ (Thermilinear YSI, Model 703; Yellow Springs Instrument Co., Yellow Springs, OH). Because in the author's laboratory anaerobes are primarily under study, the probes are placed in culture tubes (Hungate type) filled with 10 ml water (18 mm Bellco tubes do not fit into the slots). Since the temperature block has 30 slots on each side, some of the slots can be used to monitor the temperature gradient concomitantly during the experiment. For qualitative measure of growth responses and for studies with solidified media, glass tubing and long metal trays with glass covers, covering the whole gradient, are supplied by Scientific Industries, Inc. and can be placed in grooves on top of the aluminum block. For anaerobes and facultative anaerobes, the glass tubing is filled with inoculated medium containing an appropriate solidifying agent (see below). For aerobes, the trays are used like long Petri dishes. This setup is also convenient for the isolation of single colonies of temperature mutants at their maximum temperature.

3. ENRICHMENT AND ISOLATION METHODS

3.1. Collection of Samples

Thermophilic bacteria have been isolated since the end of the last century. Often, the samples used for enrichments had been transported at or close to the temperatures of their natural environment. The author has not experienced cases where organisms could not be isolated due to a temporary decrease in temperature. In several cases, non-spore-forming thermophiles have been isolated from samples kept for extended periods at 0 to 4°C. For example, *Methanobacterium thermoautotrophicum*, with a $T_{opt} = 68°C$ and $T_{max} = 75$ to $78°C$, has been isolated from lake sediment samples stored for 5 years at 0 to 6°C in a cold room. Enumerations by the most-probable-number (MPN) technique indicated little change in the number of thermophilic methanogens in these samples over 3 years (Wiegel, unpublished data). Castenholz (4) noted that in contrast to the mesophilic cyanobacteria which deteriorate quite rapidly in bottles and vials, samples of thermophilic cyanobacteria from hot springs stored in glass or other rigid vessels remained in good condition at room temperature for many days and even weeks, allowing isolation of thermophilic cyanobacteria. Castenholz has isolated thermophilic cyanobacteria from samples stored at 8 to 12°C for more than a year, whereas storing the samples below 5°C caused total lysis, sometimes in less than 24 hours. He recommended collecting the samples in large volumes of spring water in tightly closed vials, so that less than 5% of the volume is occupied with the cyanobacterial material. However, in other instances, especially when enumerations have to be done, temporary decreases of temperature should be avoided. Severe effects of temporary temperature decreases of about 10 hours on methanogenesis rates in samples from hot springs of Yellowstone National Park were observed by Ward (5). He concluded that a major reduction in the population density occurred as a result of cooling. Enumeration data on the effect of temporary temperature decreases (either to ambient temperature or below 10°C) on the survival rate of thermophilic bacteria in respect to time, temperature, and nature of sample material are rare, so that no generalization can be given. However, it is obvious that experiments on production rates and enumeration studies should be performed with as little as possible disturbance to the samples; optimally, they should be carried out in situ. Enrichment of new organisms should be similarly done with fresh samples whenever possible, but stored samples can be used in many instances.

Since most true thermophiles do not grow at ambient or room temperatures, storage of purified or enriched thermophilic cultures at temperatures around 20°C is sufficient to prevent growth and deterioration. To avoid complications caused by the growth of mesophiles in freshly collected samples, causing unfavorable pH shifts or formation of bacteriocides by mesophilic fungi, it is a good practice to keep the samples around 10°C during extended transport times. If airtight closed sample flasks are used, one should keep in mind that cooling of the sample will create an underpressure, and when the container is opened it may become contaminated. Furthermore, underpressure will change the carbon dioxide concentration in the sample, especially when the container is only partially filled. To avoid pressure and gas changes, the sample container should be completely filled with at most only a small gas pocket. The sample flask should be pressurized with sterile air, nitrogen, or argon gas before a sample is withdrawn with a syringe or a flask is opened. However, if samples are taken from hot spring areas, solfatara pools, geysers, or similar extreme habitats, the safety of the experimenter and the preservation of the location may need more attention than the samples.

Since cooling can occur relatively fast in small samples, causing changes in parameters such as pCO_2, pO_2, and pH, the temperature, pH, and concentrations of carbon dioxide, oxygen, etc. should be measured directly in the natural environments—and not in the sample—using easily readable portable pH meters and suitable electrodes and temperature probes mounted on a rod. If this is not possible, large samples should be taken and the sample container completely filled with the sample before measurements are made. Because most plastic materials are permeable to gases such as hydrogen and oxygen, misleading data may be obtained with samples transported in plastic containers.

To isolate thermophilic anaerobes, samples are best collected in glass containers with black butyl rubber stoppers. To monitor whether the samples in the tubes are under anaerobic conditions, the sample flask should contain a drop of sterile, 1% resazurin. Samples which get oxidized during the collection and stay oxidized for longer than 10 minutes (indicated by the pink color of resazurin) can be rapidly reduced by adding a drop of a sterile dithionite solution (1% w/v), indicated by a resazurin change from pink to colorless. However, resazurin cannot be used as indicator for acidic samples at pH values below 4.5, since it reacts under these conditions as a pH indicator instead of redox indicator.

3.2. Enrichment studies

As with mesophilic bacteria, there are many ways of enriching and isolating thermophiles. Besides the temperature, the selective conditions to be used in an enrichment will primarily depend on the physiological and nutritional type of the organism to be isolated. Most of the presently known thermophiles (1; Chapter 3, this volume) have been isolated via enrichment cultures. For aerobic organisms, at incubation temperatures above 50°C, the use of cotton plugs should be avoided. Metal-capped tubes are convenient for the temperature range up to about 60°C. Above this temperature, closed flasks are needed to prevent the evaporation of the liquid. To replenish the oxygen supply during growth, the flasks are flushed through sterile cotton filters with air or O_2-N_2 mixtures, the interval of flushing depending on growth and incubation temperature. In addition, pressurization up to 1 atm overpressure is recommended. Another method is to use large pressurized gas reservoirs connected to the flasks via small tubing and hypodermic needles. This is similar to the method described by Aragno and Schlegel (6) for growing hydrogen-oxidizing aerobes. When aerobic cultures or gas-utilizing cultures are continuously flushed with gas at temperatures above 50°C (in case of long incubation times even at 45°C) the incoming gas should be humidified by first passing it through a sterile cotton filter and then through a sterile water reservoir. The water reservoir should preferably be at the same temperature as the culture to obtain the proper humidification. When small fermentors are used, the water can be refluxed by means of water-cooled condensers placed directly in the exhaust line on top of the fermentor.

When anaerobic thermophiles are enriched from mesophilic sludge and mud samples only a low number of the thermophile is present, but the inoculum contains a considerable amount of fermentable substrates. The sample can be incubated in the enrichment medium without addition of a carbon and energy source until gas production has considerably decreased, usually occurring after 10 to 40 hours. Then, after adjustment of the pH, the selective substrate is added.

As discussed in Chapter 3, this volume, thermophiles and extreme thermophiles are widespread and many of them, especially the aerobic and anaerobic sporeformers, can be easily isolated from nearly any soil. Spore-forming thermophiles have even been isolated from Arctic ice. However, with decreasing environmental temperature, the number of thermophiles usually decreases, too. Thus, for the investigation of mesophilic environments, larger samples (e.g., 2–20 g of soil in 50–500 ml medium) have to be used. For example, in Georgia soil the number of organisms growing on 0.5% glucose-yeast extract medium under anaerobic conditions at 60° and at 72°C was around 1,000 to 10,000 and 1 to 10 per g soil sample, respectively, whereas in soil from around hot springs of Yellowstone National Park the numbers were 100- to 1000-fold higher (7). If water samples are used, a ratio of sample to buffered medium of 5:1 or at least of 1:1 is desirable. For organisms sensitive to changes in mineral content, etc., enrichment can also be carried out in undiluted samples supplemented with a nitrogen source and the selective carbon and energy sources. However, it seems that samples at least slightly buffered with medium give better results. To my knowledge no extensive comparison has been published. From New Zealand all organisms isolated directly from undiluted samples have also been isolated in inoculated artificial medium (pers. commun. H. Morgan). Nevertheless, it seems worthwhile to use both undiluted sample and dilutions in medium if an unknown sample is analyzed.

Enrichment cultures select for the organisms able to grow fastest under the culture conditions employed. Thus, the chance of isolating cultures of similar nutritional requirements but growing slower is slim. For the isolation of slow-growing organisms, as well as for organisms which initially do not grow well under laboratory conditions in liquid media, the isolation method of choice is direct isolation without enrichment culture, by using filtration as a concentration step (e.g., *Bacillus schlegelii* from 1 kg of melted Antarctic ice). The membrane filters (0.2 or 0.4 μm, depending on the size of the organism) are placed on and incubated on soldified media (6 and literature cited therein) using glass Petri dishes or special plastic Petri dishes withstanding 78–80° (Sarstedt, Inc., Princeton, NJ, No. 8211··series). The plates are incubated in an anaerobic chamber, an Atmosbag (Pierce Chemical Co.), a Gaspak, or other suitable container, depending on whether aerobes or anaerobes are isolated.

When transferring from enrichments, one should inoculate into prewarmed media to minimize stress and reduce long lag phases.

In cultures growing heterotrophically, or when bicarbonate is used as buffer, carbon dioxide can be produced in high concentrations and pressures. The use of a release check valve (e.g., Nupro no. 4CPA2–3 or –50 psi, Nupro Co., Willoughby, OH) set at the proper pressure, can function as a simple pH control, minimizing acidification by automatically releasing the carbon dioxide produced and preventing explosion of the culture vessels. For example, *C. thermosaccharolyticum* can produce pressures over 3 atm (45 psi) of CO_2 and H_2 from glucose-bicarbonate media at 60°C, thus causing explosions of tightly closed culture vessels.

Listed below are some illustrative examples of enrichments for some thermophilic organisms. For more in depth information on these and all other described thermophiles, the interested reader is referred to the original literature of the organisms as cited in Chapter 3 of this book, and in the review by Wiegel and Ljungdahl (1, Tables 3–6). A detailed description of the various enrichment and isolation procedures successfully used for thermophilic cyanobacteria are given by Castenholz (4). Furthermore, the handbook, *The Prokaryotes* (8), contains detailed descriptions for many thermophiles and extreme thermophiles isolated prior to 1980.

3.3. Examples of Isolation Procedures

1. Isolation of *Pyrodictium*: T_{min} 80°C; T_{max} 110°C (8)

Pyrodictium was isolated from samples of a submarine solfataric field near Vulcano, Italy. Several strains were obtained, representing two species, *P. occultum* and *P. brockii*. The geothermally heated water contained sulfurous deposits and had a temperature between 93 and 103°C. The samples were kept anaerobically at 4°C. Portions of filter-sterilized seawater (50 ml in 100 ml serum bottles) were supplemented with 0.75 g flowers of sulfur and inoculated with 2.5 ml of the samples. The serum bottles, sealed with black butyl-rubber stoppers, were pressurized to 300 kPa with hydrogen and carbon dioxide gas (80:20, v/v) and incubated without shaking at 85°C. After 3 days, disc-shaped organisms appeared and large amounts of H_2S were produced. The cultures could then be transferred into synthetic media (9). Six of the enrichments grew at 100°C. The strains were isolated using the repeated serial dilution technique.

2. Isolation of *Methanobacterium thermoautotrophicum* (10)

This anaerobe is methanogenic, thermophilic, chemolithoautotrophic, and ubiquitous. It is relatively easy to isolate from nearly all anaerobic fresh water mud samples or from sludge of thermophilic sewage digesters (10). It is frequently encountered in various anaerobic enrichments at 50 to 70°C without adding methanogenic substrates, since most mud samples contain some microorganisms producing hydrogen and carbon dioxide, thereby permitting growth of this methanogen. If the starting material contains a high concentration of fermentable biomass, for example cellulose, the pH has to be adjusted since *M. thermoautotrophicum* usually does not grow below pH 6.0. However, for the specific isolation of *M. thermoautotrophicum*, dilutions of the samples would be incubated under an atmosphere of hydrogen and carbon dioxide. Various mineral media can be used for enrichment and isolation. For optimal growth with high cell yields, the media described by Taylor and Pirt (11) and Schonheit et al. (12) are recommended. For enrichment and enumeration studies, a medium containing 25 mM phosphates pH 7.2 to 7.6, trace elements, and 0.2% bicarbonate is used (Wiegel, unpubl.) and is reduced with sodium sulfide/cysteic acid (each 0.5 g/l medium). About 50 ml medium in a 150 ml serum bottle is inoculated with 0.25 to 1 g of sample and incubated at 68 to 70°C under an atmosphere of H_2 and CO_2 (80:20 v/v, 10–20 psi pressure). Methane will be detectable after 2 to 3 days.

After 4 to 7 days, the culture can be diluted employing the same growth conditions. The highest dilution showing methane production is again diluted into agar (3% w/v) and agar shake roll tubes are prepared. After picking single colonies and repeating the agar shake roll tube procedure two to three times, one should obtain a pure culture. At the later purification stages, the mineral medium should contain a trace mineral mixture containing Ni^{2+} which is required for a high hydrogenase activity. Sulfide is used as the reducing agent, and it is also the required sulfur source. If an organism is enriched and isolated by applying the continuous culture technique, the isolation of single cell colonies on solid media is still mandatory. If the organism does not grow in/on solidified media, repeated dilution series have to be made, using either semisolid media (e.g., 0.3–1% soft agar) or liquid media with 1:2 dilution steps. It has been shown that strains of *C. thermohydrosulfuricum* can grow concomitantly with the methanogen under chemolithoautotrophic growth conditions, although the *Clostridium* does not grow in the medium as a pure culture. Apparently, *C. thermohydrosulfuricum* proliferates on some of the excretion products of the methanogen (11 and Wiegel, unpubl. data). *M. thermoautotrophicum* cultures enriched and isolated with the dilution technique or via a continuous culture approach may also still contain the chemolithoautotrophic homoacetogen *Clostridium thermoautotrophicum*. The clostridial contaminations can be easily detected by incubating media containing 0.5% (w/v) glucose and 0.2% yeast extract under nitrogen gas at 60°C. Under these conditions, heavy growth of the *Clostridium* should occur, but the methanogen will not grow. This procedure is a good control for assessing the purity of the thermophilic methanogen. Similar controls can be used for all obligate chemolithoautotrophs.

3. Isolation of *Thermus* (13)

Thermus is an extreme thermophilic heterotrophic aerobe which is widespread in neutral to slightly alkaline hot springs (first isolation) and also in water heaters with temperatures between 50 and 85°C. A basic medium for isolation is medium D described by Castenholz (14) or a modification thereof which contains, in milligrams per liter, the following components: Nitrilotriacetic acid, 100; $CaSO_4 \cdot 2H_2O$, 60; $MgSO_4 \cdot 7H_2O$, 100; NaCl, 8; KNO_3, 103; $NaNO_3$, 689; Na_2HPO_4, 111; $FeCl_3$, 0.28; $MnSO_4 \cdot H_2O$, 2.2; $ZnSO_4 \cdot 7H_2O$, 0.5; H_3BO_3, 0.5; $CuSO_4$, 0.016; $Na_2MoO_4 \cdot 2H_2O$, 0.025. The pH is adjusted to 8.2 with NaOH and 0.1% (w/v) each of yeast extract and tryptone added as carbon and energy source. The enrichment of *Thermus* requires low concentrations of organic compounds (reflecting the normal habitats of the organism). *T. aquaticus* does not grow well if the yeast extract and tryptone concentration are above 0.3% (w/v). However, *T. thermophilus* isolated by Oshima and Imahori (15) grows up to 85°C in the presence of up to 1% w/v yeast extract.

Incubation temperatures between 70 and 75°C proved to be best. After 2 to 3 days of aerobic incubation in closed flasks, growth should appear. Final purification can be achieved by repeated streaking on the above medium containing 3% agar. The streaked plates are incubated at 70°C in sealed plastic bags to prevent evaporation. The development of the compact but spreading, yellow colonies usually requires 1 to 2 days. The strains isolated from hot water heaters are rarely pigmented. The concentration of *Thermus* in various water heaters has been found to be as high as 24/ml tap water. The predominant growth occurs not in the water column but attached to the walls of the heater.

3.4. Selection of Media

Obviously, there is no such thing as a medium expressly formulated for the isolation of thermophiles. The media used depend on the type of organisms to be studied. The medium for a mesophilic bacterium may work with a physiologically similar thermophile in the same way that it may work with another similar mesophile. There are two exceptions: the nature of the solidifying agent for solid media and the effect of temperature on bicarbonate. Bicarbonate is not stable above 50°C and releases CO_2, thus a shift to alkaline pH occurs. To prevent this, an appropriate CO_2 pressure has to be maintained. Furthermore, temperatures above 70°C do not allow the use of agar to solidify media (see below).

Sterilization of media, especially complex media containing yeast extract, peptone, corn steep liquor, and similar compounds, requires special attention. Heat-resistant spores of thermophilic bacteria can survive the short sterilization times commonly employed with culture media for mesophiles. The spores of thermophiles usually do not germinate and grow with a significant rate at temperatures below 37°C. Therefore, thermophile contaminants are rarely encountered when working with mesophiles or during storage of the media. The heat resistance of *Bacillus stearothermophilus* spores has been known for a long time and spores of this organism are frequently used as biological indicators for successful sterilization processes (16). However, some spores are significantly more heat resistant. Hyan et al. (17) published D_{10}-times (time of reduction to 10% survivors) of 11 min for *C. thermohydrosulfuricum* strain E39 at 121°C. The heat resistance of spores is strongly dependent on the medium components and pH (18,19).

In the author's laboratory, 10 to 100 ml volumes are usually autoclaved for 30 to 40 minutes. Due to the longer sterilization times, particularly when the sterilization is done under aerobic conditions, the mineral components, the complex media components like yeast extract or peptone, and the sugars are each autoclaved separately. Sterile filtration (0.2 µm pore size) is used when temperature-labile compounds are involved. Most substrates used for the determination of a substrate spectrum of a newly isolated strain are filter sterilized. However, at temperatures above 60°C and incubation times of more than one day, most sugars will undergo some degradation.

3.5. Preparation of Solid Media

Several methods exist to obtain solid or semisolid media. The method to be used will depend on the type of experiment, the incubation temperature, and the response of the organisms. The various solidifying agents and their use are described in general terms by Krieg and Gerhardt (20). Below are some comments for the use of solidifying agents when working with thermophiles.

Agar This has been the most widely used component to solidify media for the purposes of isolating single colonies and purifying microorganisms. Melted agar solidifies at temperatures between 35 and 45°C (depending on the grade of polymerization), but temperatures between 95 and 100°C are necessary for remelting. Thus, agar can be used efficiently for the isolation of temperature tolerant and moderate thermophiles up to temperatures around 65 to 70°C, especially if higher agar concentrations are used (2–4%, w/v). However, some thermophiles, such as *Clostridium thermoaceticum*, do not grow well at high agar concentrations. Most of these organisms can be grown in agar shake tubes in media containing less than 1.5% agar, yielding a soft agar at high temperature. Growth can also be obtained in glass Petri dishes using a soft agar overlay technique, as well as in Bellco, Balch, or Hungate tubes (Bellco Glass, Inc., Vineland, NJ) for anaerobes. Colonies are picked using elongated Pasteur pipettes and a dissecting microscope. Thermophiles do not necessarily require a special purified grade of agar and usually standard bacteriological grade is satisfactory. Occasionally, agar from one lot has a lower thermostability than normal and thus cannot be used for temperatures at 65°C or above. Some agars may also exhibit a narrower pH range. At acidic pH (below 5.0) and alkaline pH (above 9.0) agar is hydrolyzed at elevated temperatures.

Carrageenan As a cheap substitute for agar, carrageenan ("Irish moss", "vegetable gelatin") can be used for moderate thermophiles. Specifically, the potassium salt of kappa carrageenan (Sigma Chemical, St. Louis, MO) can be used, since it forms a rigid, clear gel which solidifies below 55°C. For incubations below 50°C and below 65°C, 2 and 2.5%, respectively, are recommended (21).

Starch The use of starch for solidifying media is not very widespread, but it has been shown to be a good method for growing some organisms, for instance, the extreme thermophilic and acidophilic archaebacteria *Sulfolobus* (22, and A. Martin and H. Morgan, pers. comm.), and *Thermoproteus tenax* (A. Martin, H. Morgan, and W. Zillig, pers. comm.). *T. tenax* and

Sulfolobus were grown at 88 and 80°C, respectively, on their usual media but containing 12% starch (Merck no. 1252; 25 ml per plate) and using 2 ml medium containing 10% starch for the overlay technique. The overlay was poured on warm plates and dried for a few minutes at 65°C. Martin and Morgan (pers. comm.) found that if the starch overlay was allowed to gel in advance (2 ml sterile stock solution) and then remelted before adding the inoculum, the setting of the overlay was faster. *Sulfolobus* was grown at pH 3 to 3.5 on plates incubated in the inverted position and in sealed plastic bags.

Silica Gel   For temperatures above 65°C, silica gel has been successfully used. The potassium silicate from J. T. Baker Chemical Co. (Phillipsburg, NJ; reagent grade) or Fisher Scientific Co. (Pittsburg, PA; certified grade 923) can be used. The silica gel must be dissolved in a strong base (7% wt/vol KOH), and it is solidified by adding a strong mineral acid (2% o-phosphoric acid). Therefore, the solidified silica gel has to be extensively equilibrated with the corresponding growth media. Stetter (pers. comm.) found that several new extreme thermophiles could not be grown on silica gel. The disadvantage of silica gel is that its use is very laborious and time consuming.

Gelrite   A new alternative is a low-acetyl, clarified gellan gum, called Gelrite, prepared from a *Pseudomonas*, and marketed by Kelco Division of Merck and Co. (San Diego, CA). The polymer is solidified with the aid of divalent cations such as magnesium and calcium. Lin and Casida (23) described the use of Gelrite for aerobic thermophiles at temperatures up to 70°C. They obtained good plating efficiencies. Recently, Deming and Baross (24 and pers. comm.) developed a medium for studying thermophiles above 100°C using Gelrite shake tubes to isolate single colonies. In roll tubes, this medium still caused some problems at 95°C due to dehydration of the gel during 3 days of incubation. In shake tubes the following acetate-Gelrite medium of Baross and Deming is stable up to 120°C. Add to 100 ml H_2O: NaCl, 2.5 g; $MgSO_4$, 0.2 g; KCl, 0.08 g; Na_2HPO_4, 0.05 g; $FeSO_4$, 0.001 g; trace elements (25); $(NH_4)_2SO_4$, 0.1 g; $Na_2S_2O_3$, 0.3 g; sodium acetate, 0.5 g. Adjust the pH to 5.5 with HCl and add 0.8 g Gelrite. Autoclave 30 minutes. Solidification starts when the temperature decreases to 80°C and below, thus inoculation has to be done at temperatures above 80°C. The concentration of divalent ions indicated is specific for the specific medium. For other media, the correct concentrations of Ca^{2+} or Mg^{2+} ions have to be determined for different conditions such as pH, temperature, buffer, and substrates, as well as for the requirements and tolerance of the organisms. However, once the conditions are established, the use of Gelrite seems to be superior to silica gel. However, Gelrite is an organic compound and thus may be metabolized by some organisms. For critical experiments, such as verification of autotrophic growth, inorganic silica gel is the solidifying agent of choice.

When closed systems are used, such as roll tubes for anaerobes or agar slants for gas-utilizing organisms, the appearance of water of syneresis cannot be totally avoided with any of the above- mentioned solidifying agents. To minimize smearing of the individual colonies by water, the tubes are placed in the incubator at a vertical angle (15–25°). If the tubes are incubated in the inverted position, the water can be withdrawn with a syringe before the tubes are further handled.

4. ENUMERATION STUDIES

Although in some cases the viable counts of thermophiles do not change significantly during storage of the sample, it is strongly recommended that enumeration studies be done immediately or as soon as possible after collecting the samples, and without letting the samples cool. Except for cases where strain-specific antiserum is available, thermophiles can only be distinguished from contaminating mesophiles by culturing at the prospective temperatures.

Spores of most thermophiles require heat activation for efficient germination. The activation process is time dependent and the required time and temperature are strain dependent. For example, maximal germination of the spores from *C. thermohydrosulfuricum* (T_{max} for growth, 78°C) was obtained after incubations at 100°C between 0.3 and 1 hour. Without heat activation, spores of the same organism had lag periods of more than 2 weeks at 60°C before germination occurred (Wiegel, unpubl. data). If spore-forming and non-spore-forming thermophiles are enumerated at the same time, two parallel dilution series (either as MPN or as agar-roll tubes) should be performed, one with the heat-treated sample (for instance, 100°C for 15 minutes) and one without heat treatment.

5. ANTIBIOTIC SUSCEPTIBILITY

Thermophilic eubacteria and archaebacteria respond like their mesophilic counterparts to antibiotics. However, some antibiotics are not stable at the elevated temperatures. Table 2 gives some examples of the temperature stability of antibiotics.

Schenk and Aragno (26) used a graphical method to determine the real minimum inhibitory concentration (MIC) of the slightly unstable antibiotic penicillin G for the aerobic thermophile *Bacillus schlegelii* at 85°C. The compound was only temporarily inhibitory. The growth rate after the initial lag phase in the presence of the inhibitor was the same as the growth rate determined in the absence of the antibiotic. The logarithm of the various

TABLE 2. Effect of temperature on the stability of antibiotics

Antibiotic, mode of action	MIC* (μg/ml)		
	at 37°C	after 24 h at 70°C	after 15 min at 121°C
Cell wall:			
Penicillin G (1)	12.5	25	50
Ampicillin (2)	1.6	25	25
Bacitracin (1)	50	100	100
Cycloserine (2)	100	100	100
Protein synthesis:			
Chloramphenicol (2)	12.5	25	25
Tetracycline (2)	1.56	12.5	>200
Neomycin (2)	50	100	100
Streptomycin (2)	50	100	>200
DNA:			
Novobiocin (1)	6.25	6.25	6.25
Nalidixic acid (2)	25	25	25
Cell membrane/lipopolysaccharide:			
Polymyxin B (2)	1.56	25	50

*Minimum inhibitory concentration (MIC) values were obtained with *S. aureus* (1) or *E. coli* (2) dilution series (1:1 dilutions) in liquid media after the incubation of the antibiotics in water.

Vancomycin is unstable even at 37°C. However, it has successfully been used by Brock and co-workers for the specific enrichment of Thermoplasma (6). Gentamycin, lasalocid, and monensin are all quickly decomposed at temperatures above 100°C, whereas nisin and cycloheximide are stable at 121°C under slightly acidic conditions.

From H. Morgan et al., personal communication, manuscript in preparation.

lag phases plotted against the antibiotic concentration yielded a straight line, reflecting the breakdown rate of penicillin G under the conditions employed. The MIC was obtained by extrapolation to zero lag phase. This method can be used with organisms which grow relatively fast compared to the degradation rate of the tested antibiotic.

6. MAINTENANCE

6.1. Short-Term Storage

Since most thermophiles do not grow at 20 to 30°C, freshly inoculated or fully grown cultures can be kept for short-term storage at room temperatures. This is especially convenient for cultures grown in a closed system, since no medium evaporation can occur. However, with anaerobes the pH of fully grown cultures may have to be adjusted to keep the cultures in a viable stage. Most thermophiles can also be kept in the refrigerator, but some are sensitive to cold when grown up to the stationary growth phase, for example, stationary cells of *Thermoanaerobacter ethanolicus* will totally lyse in a few days in the refrigerator.

Keeping cultures at their optimal growth temperatures frequently leads to nonviable cultures after 3 to 7 days. The anaerobe *Thermoanaerobium brockii* has to be transferred every 2 to 3 days to keep the culture transferable. Such an effect may be partly because at elevated temperatures organic substrates in most media form inhibitory compounds (caramelization products).

A general observation is that cultures grown in chemolithoautotrophic media can be stored longer at 0 to 25°C than cultures grown in complex media. Some strains of *C. thermoautotrophicum*, when grown on glucose, do not significantly form spores and the cultures are often not viable after several weeks. However, when grown on methanol or H_2/CO_2, the cultures stay viable at 0 to 20°C for over a year, regardless of whether the cultures were at the logarithmic growth phase containing maybe few spores or at the stationary growth phase containing large numbers of spores.

Sporeformers should be stored as spore suspensions. In our hands, sufficient spore formation is obtained by growing the cultures at slightly suboptimal temperatures as smears on agar roll tubes using media with low amounts of sugars (0.1–0.2%), using sugars which are not utilized well, or using chemolithoautotrophic growth conditions. With several thermophilic and extreme thermophilic *Clostridia* (e.g., *C. thermohydrosulfuricum, C. thermoaceticum*), good sporulation is obtained when the incubation temperature of the cultures is allowed to decrease slowly, for example, gradual cooling during several hours from above 60 to below 50°C for *C. thermohydrosulfuricum* (T_{max} 78; T_{opt} 69; T_{min} 35°C), and from about 55 to 35°C for *C. thermoaceticum* (T_{max} 68; T_{opt} 59; T_{min} 35°C).

In the author's laboratory, short as well as long term storage in 50 to 60% sterile glycerol at −20°C (and recently at −70°C) has been used. Some thermophiles have been maintained for more than 7 years, including the nonsporeformer *Methanobacterium thermoautotrophicum* and *Thermoanaerobacter ethanolicus*. Stock cultures are prepared as follows: A culture in the mid- to late logarithmic growth phase is centrifuged in its growth vial to increase the cell concentration. The pellet is resuspended in 1 ml of sterile, fresh medium and thoroughly mixed with 1 ml of sterile glycerol. The glycerol must be sterilized by autoclaving for 50 or more minutes at 121°C, since glycerol can to some extent protect microorganisms against heat damage and hence is difficult to sterilize. In the case of anaerobic organisms, the glycerol is prereduced with a drop of the reducing solution used for the media.

The advantage of this method is that aliquots can be withdrawn via a syringe at any time without a significant change in the temperature of the stock culture and can be used as inoculum without further treatments. In our experience, most cultures grow up with only a short lag period. As storage vials we use 1 to 25 ml serum bottles or the Bellco tubes in which the organisms were grown. To prevent accidental loss of an adapted culture, we frequently store aliquots at $-18°C$ after adding 1 ml of the culture directly into glycerol without increasing the cell concentration by centrifugation.

6.2. Long-Term Storage

Thermophiles can be stored over long periods in the same way as mesophiles. Several methods are available; they are described in detail elsewhere (27). Storage in liquid nitrogen in capillary tubes, described by Hippe (27,28) and used in the German Culture Collection, seems to be one of the most effective methods.

7. MASS CULTURE

Thermophiles can be grown in sufficient quantities in the same kinds of large-volume fermentors used for mesophiles, except that several of the points discussed above have to be considered. High stirring rates are required for the transfer of gases into the liquid but they do not cause overheating problems as often encountered when growing mesophiles. Special attention has to be given to the gas transfer rate, since it is often the rate-limiting step in culturing gas-utilizing thermophiles.

Sonnleitner et al. (29) reported growth inhibition of thermophilic aerobes in new stainless steel fermentors, a phenomenon they did not observe with mesophiles. These authors overcame the problem by treating the vessel with concentrated nitric acid. Metal inhibition is usually not observed with thermophilic anaerobes. However, corrosion of stainless steel fermentors by higher concentrations of H_2 and by seawater has been commonly observed, even at the moderate thermophilic temperature of 55°C. Corrosion and growth inhibition in new fermentors can also be prevented by coating the inside of the fermentor with one of a variety of plastic materials, as has been done by Braun-Melsungen (Melsungen, FRG). Stetter et al. (9) successfully used the plastic coating "Haldan" in their Braun-Melsungen fermentor for the growth of *Pyrodictium* at 95°C. The use of titanium vessels is very expensive and does not eliminate totally the problem of the corrosion by sulfide.

Recently, Meyer and Charles (30) used a deep-jet fermentor for culturing *Thermonospora*. Although this thermophile forms very fragile hyphae, it could be grown with good yields, since the friction forces are relatively low in this type of fermentor.

The mass culture of thermophiles requires sterile conditions very similar to mesophiles. Although the contamination risks are lower than with mesophiles, spore-forming aerobic and anaerobic thermophiles are ubiquitous and can cause airborne infections. As already stressed above, special precautions have to be taken in respect to the time of heat sterilization. Due to the presence of cold pockets in large fermentors and the greater heat stability of thermophilic spores, especially in complex organic substrates, contamination can sometimes be difficult to overcome.

Since in many cases a gas stream is used for keeping anaerobic conditions or supplying a gaseous nutrient, the gas exhaust line has to be equipped with enlarged or additional condensers directly above the culture vessel to avoid large losses of liquid. With fermentors where the temperature is maintained by external waterbaths, for temperatures above 85°C, ethanediol or similar compounds may be substituted for water as a heat transmission fluid.

According to H. Morgan (pers. comm.), for mass culturing of sulfur-utilizing thermophiles it is better to add the sulfur without autoclaving, since during the heat treatment the sulfur melts and becomes less available for the organisms. So far, the addition of unsterilized elemental sulfur has never led to a contamination. Flowers of sulfur are apparently sufficiently antibactericidal. Stetter et al. (9) observed a similar problem when growing the sulfur-utilizing *Pyrodictium* above 110°C. The observed upper temperature limit for growth of this organism might be caused by a restricted availability of elemental sulfur at this temperature. Sulfur (S_8 melts at 112.8, 119, and about 120°C for the α, β, and γ form, respectively.

8. CELL-FREE EXTRACT EXPERIMENTS

If possible, experiments with cell-free extracts should be done at the optimum growth temperature of the organism studied. Most spectrophotometers have a jacketed multicuvette holder, so that the assay mixtures can be brought to the desired temperature in advance. Special care has to be given to the proper controls (e.g., the complete assay without extract and the assay with extract but without the various substrates) to ensure that the reaction is not totally or in part due to an abiotic reaction or that a slowly occurring protein precipitation does not cause an increase in turbidity. Unfortunately, at the present time there are not many thermophilic enzymes commercially available. In cases where coupled enzyme tests are employed, the temperature stability and temperature range of the commercially available mesophilic enzymes have to be determined. First, stability must be determined by preincubation at the elevated temperatures and assaying the remaining activity at 25 (or 30)°C. Second, the temperature range has to be determined by assaying the reaction directly at various temperatures. Many enzymes from mesophilic organisms have a T_{max} around or even above 50°C and can be used in assays for thermophilic enzymes above 35°C.

Purification of proteins and enzymes of thermophiles and extreme thermophiles can often be done at room temperatures since the proteases of most thermophiles are not active below 30°C. This eliminates the need for a cold room or refrigerated columns. However, if room temperature is used, most column material has to be poisoned against bacterial growth. The use of 0.01% azide prevents the growth of most aerobic organisms.

Enzymes of thermophiles are not necessarily cold labile. However, cold lability can occur if hydrophobic interactions in the protein are primarily responsible for the stability of the protein. Cold lability has also been found with several enzymes from mesophilic organisms (31). Therefore, the thermal stability of the enzyme in cell-free extract in the presence and absence of various protease inhibitors should be investigated before enzyme purification is started, regardless of whether a thermophile or mesophile is the source of the enzyme.

ACKNOWLEDGMENTS

I thank M. Aragno, R. Garrison, H. Morgan, and W. B. Whitman for reading and discussing the manuscript. This work was supported by DOE grant DE-FG09-84ER13248 from the United States Department of Energy.

REFERENCES

1. J. Wiegel and L. G. Ljungdahl, *CRC Crit. Rev. Biotechnol.* **3**, 39 (1986).
2. *Handbook for Chemistry and Physics*, 65th ed., CRC Press, Boca Raton, FL, 1984, p. D-193.
3. G. A. Nelson, Trans. *Am. Soc. Mech. Engr.* **373**, 205 (1951).
4. R. W. Castenholz, in M. P. Starr, H. Stolp, H. G. Trüper, A. Balows, and H. G. Schlegel, Eds., *The Prokaryotes*, Springer-Verlag, New York, 1981, Ch. 11.
5. D. M. Ward, *Appl. Environ. Microbiol.* **35**, 1019 (1978).
6. M. Aragno and H. G. Schlegel, in M. P. Starr, H. Stolp, H. G. Trüper, A. Balows, and H. G. Schlegel, Eds., *The Prokaryotes*, Springer-Verlag, New York, 1981, Ch. 70.
7. J. Wiegel and L. G. Ljungdahl, Proc. Third Int. Symp. Microbial Ecology, Abstract J-2, p. 52 (1983).
8. M. P. Starr, H. Stolp, H. G. Trüper, A. Balows, and H. G. Schlegel, Eds., *The Prokaryotes*, Springer-Verlag, New York, 1981.
9. K. O. Stetter, H. König, and E. Stackebrandt, *Sys. Appl. Microbiol.* **4**, 535 (1983).
10. J. G. Zeikus and R. S. Wolfe, *J. Bacteriol.* **109**, 707 (1972).
11. G. T. Taylor and S. J. Pirt, *Arch. Microbiol.* **113**, 17 (1977).
12. P. Schönheit, J. Moll, and R. K. Thauer, *Arch. Microbiol* **127**, 59 (1980).
13. T. D. Brock, *Thermophilic Microorganisms and Life at High Temperatures*, Springer-Verlag, New York, 1978.
14. R. W. Castenholz, *Bacteriol. Rev.* **33**, 476 (1969).
15. T. Oshima and K. Imahori, *Int. J. Syst. Bacteriol.* **24**, 104 (1974).
16. C. H. Lee, T. J. Montville, and A. J. Sinskey, *Appl. Environ. Microbiol.* **37**, 1113 (1979).
17. H. H. Hyan, J. G. Zeikus, R. Longin, J. Millet, and A. Ryter, *J. Bacteriol.* **156**, 1332 (1983).

18. C. A. Gauthier, G. M. Smith, and I. J. Pflug, *Appl. Environ. Microbiol.* **36**, 457 (1978).
19. J. Wiegel, L. G. Ljungdahl, and J. R. Rawson, *J. Bacteriol.* **139**, 800 (1979).
20. N. R. Krieg and P. Gerhardt, in P. Gerhardt, Ed., *Manual of Methods for General Bacteriology*, Amer. Soc. Microbiol., Washington, 1981, p. 143.
21. A. D. Lines, *Appl. Environ. Microbiol.* **25**, 8 (1977).
22. S. Yeats, P. McWilliam, and W. Zillig, *EMBO J.* **1**, 1035 (1982).
23. C. C. Lin and L. E. Casida, Jr., *Appl. Environ. Microbiol.* **47**, 427 (1984).
24. J. W. Deming, Presentation at Annual Meeting, Am. Soc. Microbiol., Las Vegas, March 1985.
25. R. Rippka, J. Deruelles, J. B. Waterbury, M. Herdman, and R. Y. Stanier, *J. Gen. Microbiol.* **111**, 1 (1979).
26. A. Schenk and M. Aragno, *J: Gen. Microbiol.* **115**, 333 (1979).
27. B. E. Kirsop and J. J. S. Snell, Eds., *Maintenance of Microorganisms*, Academic Press, London, New York (1984).
28. L. G. Ljungdahl and J. Wiegel, in A. L. Demain and A. Salomon, Eds., *Manual for Industrial Microbiology and Biotechnology*, Am. Soc. Microbiol., Washington, D.C., (1986), p. 84.
29. B. Sonnleitner, S. Cometa, and A. Fiechter, *Biotechnol. Bioeng.*, **24**, 2597 (1982).
30. H. P. Meyer and M. Charles, *Biotechnol. Bioeng.* **24**, 1905 (1982).
31. J. Wiegel and H. G. Schlegel, *Arch. Microbiol.* **112**, 239 (1977).

DIVERSITY OF EXTREMELY THERMOPHILIC ARCHAEBACTERIA

KARL O. STETTER

Institute for Microbiology, University of Regensburg, 8400 Regensburg, Universitätsstrasse 31, Federal Republic of Germany

1. INTRODUCTION

For a long time thermophilic bacteria with temperature optima above 45°C have been recognized to be widely distributed in soils, self-heated hay, and geothermally heated areas. Most of these bacteria are members of various genera which also contain mesophiles, such as *Bacillus* or *Clostridium*. The upper temperature limit of growth of these thermophiles is usually between 60 to 80°C, depending on the species (1,2). Since these species resemble their closely related mesophilic counterparts in many features, even on the level of their molecular biological properties (3), they may have become secondarily thermophilic.

About 15 years ago, bacteria living in the boiling springs of Yellowstone National Park were observed (4) and the first extremely thermophilic organism with a temperature optimum above 80°C and an upper growth temperature at around 92°C was isolated (5). Since that time, various extremely thermophilic organisms have been isolated from continental and submarine volcanic areas, such as solfatara fields, geothermal power plants, or geothermally heated sea sediments and hydrothermal vents (6,7). In this chapter extreme thermophiles all having a temperature optimum of growth well above 80°C will be described. As a rule, these bacteria do not grow at 60°C or below. *Pyrodictium*, the most extreme thermophilic organism existing in pure culture, does not even grow below 82°C (8).

2. PHYLOGENETIC CONSIDERATIONS

Volcanic activity has occurred on earth since the early outgassing phase about 4 billion years ago. Although few volcanic areas still exist today, these hot regions, continental as well as submarine, have remained almost unchanged and, if liquid water is present (1), they represent one of the oldest habitats for cellular life. The extremely high temperatures in many of these habitats have been restrictive for the evolution of the organisms thriving in these habitats, which suggests that these organisms may possibly still be similar to their archaic ancestors.

Phylogenetically, all extremely thermophilic bacteria with temperature optima above 80°C isolated up to now belong to the archaebacteria, the third kingdom of life discovered recently by Carl Woese (9). The phylogenetic relationships of the archaebacteria have been determined by comparing partial sequences of the 16S rRNA (10). From this "cataloging" method, similarity coefficients (S_{AB}) were obtained (11), showing quantitatively phylogenetic distances which could be illustrated by dendrograms (Figure 1). The branching point of the archaebacteria from the eukaryotes and the eubacteria is very deep and was determined to be at an S_{AB} value of around 0.1 (12,13). The archaebacterial kingdom itself also shows deep branchings into different groups, indicating distant relationships among ar-

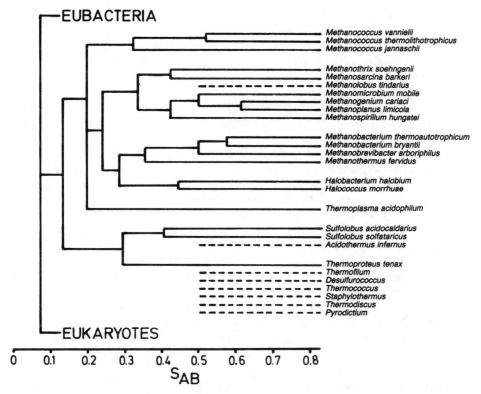

Figure 1. Phylogenetic tree of the archaebacteria. S_{AB}-values according to Woese et al. (9–15).

chaebacteria and large phylogenetic width. Two main branches are evident: a) the strictly anaerobic methanogenic bacteria comprising the orders *Methanococcales, Methanomicrobiales,* and *Methanobacteriales,* also including the aerobic extremely halophilic *Halobacteriales,* and b) the aerobic and anaerobic sulfur-metabolizing *Sulfolobales* and *Thermoproteales* (13–15). A further order, the *Thermoplasmales,* is up to now represented by only one species (16) and appears to be phylogenetically rather distant (13,17) and possibly an intermediate between the two main branches (6).

Due to their exceptional position as archaebacteria in evolution, it is important to compare the extreme thermophiles with archaebacterial mesophiles in order to discern the molecular principles critical to the thermophilic way of life. Within the methanogenic-halophilic branch, only *Methanococcus jannaschii* and *Methanothermus fervidus* are extremely thermophilic. In addition, *Methanococcus thermolithotrophicus, Methanobacterium thermoautotrophicum,* and *Thermoplasma acidophilum* are moderate thermophiles, while all others are mesophiles. In contrast, the branch of sulfur metabolizers contains up to now no mesophiles, but almost exclusively extreme thermophiles (see Figure 1). The only moderate thermophile within this group is *Sulfolobus brierleyi* growing at temperatures up to 75°C (18). The broken lines (see Figure 1) show taxonomic positions of several recently described new genera, determined qualitatively by DNA-RNA hybridization (19; and Fischer, Zillig, and Stetter, unpubl.), serology (König and Stetter, unpubl.) or by still incomplete cataloging (Stackebrandt and Woese, pers. comm.).

In addition to their low S_{AB} value in relation to the other two primary kingdoms, archaebacteria are unique because of the presence of additional features, for example, phytanyl ether lipids instead of fatty acid esters (see Langworthy, Chapter 5, this book), the replacement of ribothymidine by 1-methylpseudouridine or uridine in the "common arm" of their tRNAs (14), and a great variety of types of cell envelopes instead of murein (20). Due to the lack of murein, archaebacteria are resistant to cell wall antibiotics such as penicillin or vancomycin (21).

Although true prokaryotes, archaebacteria share many molecular features with eukaryotes, features which are not present in eubacteria: a) the replicating DNA polymerases are sensitive to aphidicolin (Zabel and Winter, pers. comm.; Prangishvilli and Zillig, pers. comm.; 22), b) the methionyl initiator tRNA is not formylated (14), c) the terminal base pair of the aminoacyl stem of initiator tRNA is AU (23), d) the translation elongation factor EFII is a target for ADP-ribosylation by diphteria toxin (24), e) archaebacterial "A-protein" sequences resemble that of yeast A-protein (25), f) translation is resistant to the antibiotics chloramphenicol and streptomycin, but sensitive to anisomycin (21,26), g) the DNA-dependent RNA polymerases are multisubunit enzymes insensitive to rifampicin and streptolydigin (27).

On the other hand, there are also typical eubacterial features within the archaebacteria, such as the existence of restriction enzymes (28,29; McWilliams, unpubl.), the occurrence of 70S ribosomes (30) which, however, differ from eubacteria in their three-dimensional structure (31), and the anti-Shine-Dalgarno sequences at the 3' end of the 16S rRNA of methanogens (32). Although unique, archaebacteria possess specific relationships to both of the two other kingdoms (33). However, too little information is yet available to decide whether the archaebacteria are more closely related at the molecular level to the eukaryotes or the eubacteria (14). Very recently, *Thermotoga maritima* was isolated, representing the deepest branch-off within the eubacterial kingdom and growing at temperatures up to 90°C (Huber et al., *Arch. Microbiol.*, in press, 1986).

3. EXTREMELY THERMOPHILIC ARCHAEBACTERIA OCCURRING IN CONTINENTAL VOLCANIC AREAS

3.1. The Genus *Methanothermus*

The first thermophilic methanogen, *Methanobacterium thermoautotrophicum*, was isolated from sewage sludge (34). Similar isolates can be obtained from thermal volcanic environments, mainly growing there on decomposing algal mats (35; Stetter, Thomm, and Huber, unpubl.). These rod-shaped methanogens belong to the order *Methanobacteriales* (36). They grow within a temperature range of 37 to 75°C with an optimum around 65°C and can therefore be considered moderate thermophiles. In contrast, members of the new genus *Methanothermus* thrive at temperatures up to 97°C in continental solfatara fields.

Ecology and Distribution   The genus *Methanothermus* was first isolated from a tiny hot spring with an original pH of 6.5 and a temperature of 85°C located in the Kerlingarfjöll, Iceland (37). No algal mats are present at these high temperatures and, therefore, *Methanothermus* most likely uses volcanic H_2 and CO_2 which are present in the gases emanating from the depths of the solfatara fields. The cooling overflow of the Kerlingarfjöll spring, in contrast to that of the surrounding springs, showed very rich algal growth, probably because of the high mineral content of the water. In 1984 in Iceland, another species of *Methanothermus, Methanothermus sociabilis* (Gebhard et al., in press), was isolated from a spring (10 cm diameter; 98°C; pH 6) again in the Kerlingarfjöll. In Grandalur close to Hveragerthi a further strain of *Methanothermus fervidus* was obtained from a depth of 30 cm within an acidic solfatara field. In contrast to the acidity of the surface water, the pH at depth was 6.0 and the temperature 100°C. Thus, *Methanothermus* seems to be present in the depth of solfatara fields which can be acidic on the surface (pH 1–3). The reduced environment is in agreement with the oxygen sensitivity of *Methanothermus*, which is extreme even for a methanogen (37). Concerning its distribution, *Methanothermus* could not be isolated (Stetter, unpubl.) from other solfataric areas such as Yellowstone National Park (Obsidian creek; Beryl spring; Firehole pool; West thumb; Lake Yellowstone), Italy (Solfatara; Larderello; Ischia; Vulcano), and the Azores (Caldeiras Barrentas; Furnas Caldeiras; Ribieira Grande). This possibly suggests that it may be restricted to Iceland.

Morphology and Chemistry   Cells of *Methanothermus* are immotile rods, about 1 to 4 μm long and 0.3 to 0.4 μm in width. Some isolates contain pilus-like appendages (Figure 2) up to about 10 μm in length. Cells of isolate KF-1/Fl stick together in liquid cultures even when shaken (Figure 3), forming flakes up to 1 mm in diameter. When growing on media solidified with polysilicate gel, slightly grayish colonies, 1 to 3 mm in diameter, are formed. Cells divide by the formation of septa. The bacteria are Gram-positive and contain a thick cell wall consisting of pseudomurein (37,38) which appears electron dense in thin sections (Figure 4). Pseudomurein is characteristic for all members of the *Methanobacteriales*. Outside the pseudomurein, an additional layer consisting of protein subunits is visible in *Methanothermus* (see Figure 4). This surface layer, or "S layer" (39), can be removed by sodium dodecylsulfate or pronase (37). On the ends of cells, channels are

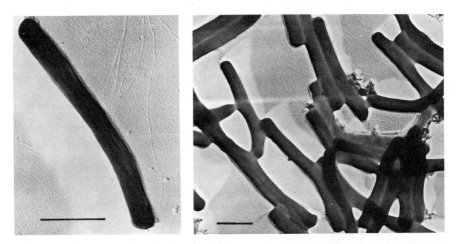

Figure 2 (left). *Methanothermus fervidus.* Electron micrograph, platinum shadowing. Bar, 1 μm.

Figure 3 (right). *Methanothermus sociabilis* KF-1/Fl. Electron micrograph, platinum shadowing. Bar, 1 μm.

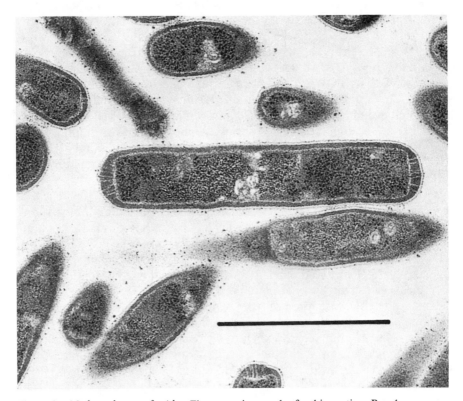

Figure 4. *Methanothermus fervidus.* Electron micrograph of a thin section. Bar, 1 μm.

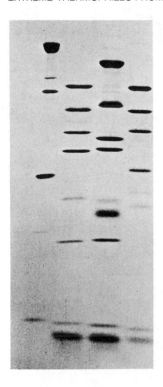

Figure 5. Protein patterns of DNA-dependent RNA polymerases after SDS polyacrylamide gel electrophoresis. From left to right, RNA polymerases from: *E. coli*; *Methanobacterium thermoautotrophicum*, strain Winter; *Methanobacterium thermoautotrophicum*, strain Marburg; *Methanothermus fervidus*. Coomassie staining.

visible within the pseudomurein layer, the function of which is presently unknown (Figure 4). The G-C content of its DNA, determined by its melting point and by direct analysis of its bases (18) is 33% (37).

Growth and Metabolism *Methanothermus fervidus* grows strictly anaerobically. It is most conveniently cultured by shaking in serum bottles (Type III glass) at pH 6.5 in medium 1 of Balch et al. (36), modified by omitting the organic components and pressurized to 200 kPa with a mixture of $H_2:CO_2$ (80:20). In serum tubes made of borosilicate glass, little or no growth occurs (37). *Methanothermus* grows within a temperature range from 65 to 97°C with an optimum around 83°C. Methane is formed from H_2 and CO_2. No growth occurs on formate, acetate, methanol, or methylamines. Organic material does not stimulate growth. Growth is inhibited by the antibiotic thiolutin (Neuner and Stetter, unpubl.).

Molecular Biology The DNA-dependent RNA polymerase of *Methanothermus fervidus* consists of eight subunits with different molecular weights, showing similar spacing in SDS polyacrylamide gels as does the same enzyme of *Methanobacterium thermoautotrophicum* (Figure 5) (Madon, Thomm, and Stetter, unpubl. data). In serological cross-reaction experiments employing antibodies against the denatured RNA polymerase subunits, the subunit pattern was determined to be AB'B"C, which is characteristic for members of the methanogenic-halophilic branch of archaebacteria (27). In spite of the high growth temperature of *Methanothermus*, its RNA polymerase shows in vitro a temperature optimum of transcription and stability at around 55°C (Thomm, pers. comm.). The stabilization in vivo is still unclear. This behavior is different from the homologous enzymes of the sulfur-dependent archaebacteria, which in vitro are stable at the (similarly high) growth temperatures (see below). It can be speculated that *Methanothermus* may have become secondarily thermophilic within the group of usually mesophilic rod-shaped methanogens, possibly by help of a weakly associated still unknown factor. The reason for the stability of the DNA double helix in vivo is still unknown. Due to its low G-C content, it melts in vitro at 82°C (0.1 × standard saline citrate) in sodium citrate buffer (0.0015 M) containing 0.015 M NaCl. There is some evidence that its structure may be stabilized by a high internal salt concentration in vivo (0.6 M NaCl; Schupfner, pers. comm.).

3.2. The Genus *Sulfolobus*

The genus *Sulfolobus*, the members of which live in the acidic hot springs of the USA, Italy, Dominica, and El Salvador, has been described by Brock and associates (5). A similar, but less thermophilic organism, was found in Yellowstone National Park (40), which was later described as *Sulfolobus brierleyi* (18). From the hot springs of Pisciarelli Solfatara (Italy), *Sulfolobus*-like organisms (the so-called "Caldariella-form habitat group") were isolated (41). One isolate from the same source is *Sulfolobus solfataricus* (18). Review articles describing *Sulfolobus* are elsewhere (1,6,42).

Ecology and Distribution   *Sulfolobus* has been found in many acidic solfatara fields, including Yellowstone National Park (5,40), Sulfur springs, New Mexico (Stetter, unpubl.), Solfatara and Pisciarelli Solfatara, Italy (5,18,42), Dominica, El Salvador, New Zealand (43), Iceland (Stetter and Zillig, unpubl), Japan (44), the Azores (Stetter and Stetter, unpubl), and Sumatra (Langururan; Albertshofer and Stetter, unpubl.). It therefore has a world-wide distribution in these hot acidic habitats. Different serotypes were found within the same spring in Yellowstone Park (45). On the other hand, several isolates from New Zealand were of the same serotypes as the Yellowstone strains (43). The same two species, *Sulfolobus acidocaldarius* and *Sulfolobus solfataricus* could be detected in USA (Yellowstone National Park) and Italy (Pisciarelli Solfatara) (6). *Sulfolobus* can be obtained usually from geothermally heated water and mud pots and from the surface of soils (6,46) with original temperatures from 60 to 95°C and a pH of 1 to 5. In mud pots, sometimes oily metal-like glimmering films can be seen which are especially rich in *Sulfolobus* (about 10^7 to 10^8 cells/ml). The ionic strength of these habitats are usually low (1; Stetter and Zillig, unpubl.).

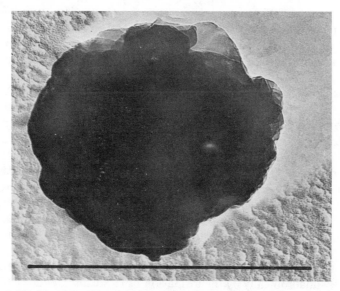

Figure 6. *Sulfolobus acidocaldarius.* Electron micrograph, platinum shadowing. Bar, 1 μm.

Morphology and Chemistry Under the microscope, *Sulfolobus* appears as irregular spheres, about 0.8 to 2.0 μm in diameter (Figure 6), which are often lobed (5,47). In thin sections, bodies of low electron density can often be seen which are composed of glycogen (48). In stationary cultures, translucent empty cells appear, containing dark granules about 0.2 μm in diameter. Pilus- and pseudopodium-like appendages were observed on cells attached to solid surfaces such as sulfur crystals (45) or sulfidic ores (47). *Sulfolobus* is Gram-negative and its cell envelope consists only of an S layer (39) which is composed of protein subunits in an hexagonal array (49–51). The dominating envelope protein of *Sulfolobus acidocaldarius* is a glycoprotein containing glucosamine (51).

The DNAs of *Sulfolobus acidocaldarius* and *Sulfolobus solfataricus* show GC contents of around 38 mol% (18), as determined by melting points or directly after hydrolysis and chromatography (18). *Sulfolobus brierleyi* shows a much lower GC content of 31 mol% (18) indicating that it is quite different.

Growth Requirements and Metabolism The growth temperature of *Sulfolobus* depends on the isolates. While *Sulfolobus solfataricus* grows maximally at 87°C (18), many other isolates show a maximum around 80°C or even lower (1,52). *Sulfolobus brierleyi* grows maximally at around 75° (18,40) and is therefore not extremely thermophilic. The optimal pH is around 3.

Most *Sulfolobus* isolates are able to gain their energy lithotrophically by oxidation of sulfur with the formation of sulfuric acid. Energy can also be obtained from the oxidation of sulfide to molecular sulfur (53). Many *Sul-*

folobus strains are able to oxidize ferrous iron (1,40). However, due to extensive precipitates of jarosite even at pH 2 and in uninoculated controls, no significant growth could be observed (1; Huber and Stetter, unpubl.). Under microaerophilic conditions, *Sulfolobus* is able to reduce ferric iron which therefore acts anaerobically as an electron acceptor (54). A few isolates are able to grow on sulfidic ores (55,56; Huber and Stetter, unpubl.). Usually these strains are moderate thermophiles with growth maxima around 70 to 75°C. However, *Sulfolobus*-like organisms growing on ores at temperatures up to at least 95°C have been isolated recently (Huber and Stetter, unpubl.).

Many *Sulfolobus* isolates can grow autotrophically. In the case of *Sulfolobus brierleyi*, CO_2 fixation most likely occurs via a reductive carboxylic acid pathway (57). Alternatively, most *Sulfolobus* isolates, including *Sulfolobus brierleyi*, are able to grow heterotrophically by aerobic respiration of organic material, including yeast extract, peptone, amino acids and sugars (1,6). Some isolates, like B 6/2 from Japan (W. Zillig, unpubl.) and NA 4 and Kra 23 from Italy and Iceland, respectively (Huber and Stetter, unpubl.) are obligate chemolithoautotrophs. *Sulfolobus brierleyi* (DSM 1651) can grow also strictly anaerobically by the formation of H_2S from H_2 and S_0 (58). This metabolic feature and its lower GC content (see below) suggest that *S. brierleyi* is possibly closely related to *Acidothermus* (see Section 3.3, below; 58).

Thermophilic Enzymes   A thermophilic β-galactosidase with a temperature optimum at 80°C was isolated from *Sulfolobus solfataricus* ("Caldariella acidophila"; 59). This enzyme shows a half time of 28 days at 70°C. *Sulfolobus acidocaldarius* and *Sulfolobus solfataricus* contain thermostable restriction endonucleases which are isoschizomers to Hae III (cleavage sequence; GG ↓ CC; McWilliam, unpubl.). The DNA-dependent RNA polymerases of *Sulfolobus acidocaldarius* and *Sulfolobus solfataricus* are stable in vitro at temperatures up to at least 80° (18). Both enzymes show a similar subunit pattern and are of the type "BAC" as indicated by immunochemical homologies (27). The RNA polymerases of *Sulfolobus*, like those of *Thermoplasma acidophilum* and the *Thermoproteales*, show a surprising resemblance to those of eukaryotes, especially to the rRNA transcribing RNA polymerase I situated within the nucleolus (60) and, similarly, are stimulated by the flavolignane silybin (61). *Sulfolobus acidocaldarius*, strain B12, contains a glucosyl transferase complex with a low substrate specificity, utilizing both ADP-glucose and UDP-glucose with high K_M values and low turnover numbers, possibly a primeval feature of an extremely thermophilic enzyme (48).

DNA and Viruses   The DNA of *Sulfolobus solfataricus* contains intervening sequences in the anticodon region of two tRNA genes (62). This was the first time that intervening sequences had been found in prokaryotes. *Sulfolobus acidocaldarius*, strain B12, contains a plasmid DNA of 13,000 base pairs able to integrate at a specific site of the cellular DNA (62). After ultraviolet induction, "lemon-shaped" viruslike particles containing closed circular DNA were liberated (63). Viruslike polyhedral particles with hexagonal dense packing seen as crystals within the cells have also been found in *Sulfolobus* isolate B6 (Zillig and Scholz, unpubl.).

Taxonomy As revealed by 16S ribosomal RNA cataloging (S_{AB} value below 0.2; see Figure 1; 15), *Sulfolobus* most likely represents an independent order rather than a single genus. Up to now, only *Sulfolobus acidocaldarius, Sulfolobus solfataricus,* and *Sulfolobus brierleyi* have been fully characterized by G-C content, serology, and temperature optima (18). By 16S rRNA cataloging, *Sulfolobus solfataricus* and *Sulfolobus acidocaldarius* are phylogenetically very distant ($S_{AB} = 0.41$; 15), indicating that both isolates may represent different genera, which, however, cannot be separated due to the absence of further distinguishing features. *Sulfolobus brierleyi*, in which the S_{AB} value is still unknown, may represent a third new genus due to its much lower G-C content, and its ability to grow anaerobically by reduction of sulfur with hydrogen (58; see Section 3.3).

3.3. *Acidothermus*, a New Genus Belonging to the *Sulfolobaceae*

Recently, facultatively aerobic organisms occurring in the same habitat as *Sulfolobus* have been isolated. These organisms are most likely related to *Sulfolobus* (58), but probably represent a new genus which will be called *Acidothermus*.

Ecology and Distribution Up to now *Acidothermus* have been isolated from solfataric springs and mud pots with pH values between 1.5 and 4.5 and temperatures between 85 and 93°C in Solfatara and Pisciarelli Solfatara (both in Italy), in Iceland (Kraffla, Hengill, Kerlingarfjöll, Hveragerthi areas), and in the sulfur springs within Yellowstone National Park, USA (58). One possibly similar organism has been found in Leirhnukur, Iceland (65). *Acidothermus*-like organisms have also been isolated from the Azores (Furnas Caldeiras and Ribieira Grande) (Segerer and Stetter, unpubl.) which, however, can grow only anaerobically.

Growth and Metabolism *Acidothermus* grows at 85°C anaerobically in Allen's medium (65) supplemented with sulfur and 0.02% yeast extract (not essential, but stimulating growth), gas phase H_2/CO_2 (80:20) at 300 kPa pressure in closed culture vessels (36). The medium is reduced by sodium sulfide or sodium dithionite (67). During growth, large amounts of H_2S are formed (Figure 7a). Growth is strictly dependent on sulfur and hydrogen, indicating that the organisms are thriving by hydrogen-sulfur autotrophy similar to *Thermoproteus* and *Pyrodictium* (68). No growth occurs at a redox potential of around 0 mV (determined by the indicator resazurin). However, in the same medium, but fully aerobically (without hydrogen), the organisms

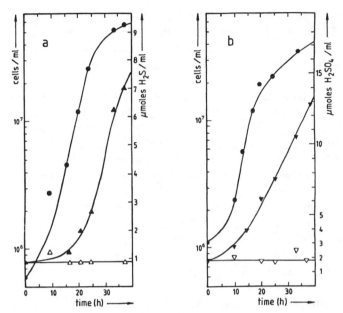

Figure 7. Aerobic and anaerobic growth of *Acidothermus infernus* So4a. (a) anaerobic growth on H_2 and S^0; (b) aerobic growth on S^0 and O_2, (●) growth curve; (▲) H_2S; (▼) H_2SO_4; (▽), (△) uninoculated controls.

can grow by the formation of H_2SO_4 from S_0 (Figure 7b). In contrast to *Sulfolobus*, no heterotrophic growth without sulfur is seen. *Acidothermus* has been purified extensively by serial dilutions and by plating. This is the first evidence that a highly anaerobic organism that requires the same low redox potential as methanogens can also grow fully aerobically after changing its metabolism to another type of chemolithotrophy using an energy pathway with an entirely different direction.

Morphology and Biochemistry   The new isolates (e.g., strain So4a from Solfatara, Italy) are irregular cocci, very similar to *Sulfolobus*, and are surrounded by an envelope composed of protein subunits (Figure 8). Anaerobically grown cells are smaller than aerobically grown cells (0.5–1 μm diameter for anaerobes compared to 2 μm for aerobes). Packed cells of the aerobes had an ochre color while the anaerobes are greenish black. DNA preparations from both aerobically and anaerobically grown cells had the identical G-C content of 31 mol% (69). In one isolate, a plasmid was detected which was amplified fivefold during anaerobic sulfur-reducing growth (65) and which turned out to be a prophage (W. Zillig, pers. comm.).

Taxonomy   DNA homology (Table 1) was established by comparison of isolate So4a (*Acidothermus infernus*) with *Sulfolobus acidocaldarius* and

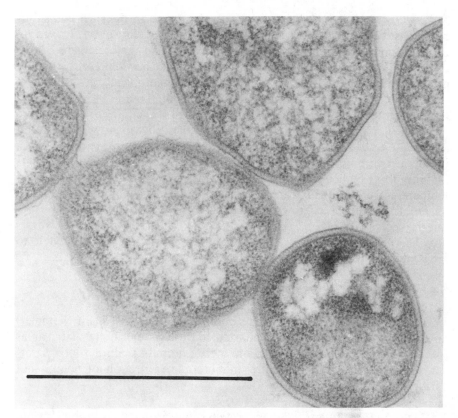

Figure 8. *Acidothermus infernus.* Electron micrograph of a thin section. Bar, 1 μm.

TABLE 1. DNA homology between *Acidothermus infernus* (So4a), *Sulfolobus brierleyi* (DSM 1651), and *Sulfolobus acidocaldarius* (DSM 639)

^{32}P-labelled DNA from	Filter-bound DNA from			
	A. infernus			
	H$_2$	O$_2$	*S. brierleyi*	*S. acidocaldarius*
A. infernus				
O$_2$	94	(100)	9	12
H$_2$	(100)	100	13	14
S. brierleyi	9	24	(100)	9
S. acidocaldarius	9	23	13	(100)

The amount of c.p.m. bound in the homologous filters was set as 100% homology (in parentheses) and compared with the activity of the heterologous filters (58). H$_2$ = grown anaerobically on H$_2$ and S^0; O$_2$ = grown aerobically on S^0

Sulfolobus brierleyi, which are similar in shape and in their aerobic sulfur metabolism. The hybridization between their DNA, however, showed an insignificant degree of homology (70), indicating that the organisms belong to different taxa above the species level, that is, they represent different genera. On the other hand, the aerobically and anaerobically grown cells of *Acidothermus infernus* show a DNA homology of 100 and 94%, indicating genetic identity in both genomes within the range of accuracy of this method.

The new genus *Acidothermus* is defined by its ability to grow by oxidation and reduction of sulfur, its high optimal growth temperature at around 90°C, its low pH optimum around 2, its low G-C content of 31 mol%, and its inability to grow heterotrophically. *Sulfolobus brierleyi* has a much lower temperature optimum, is capable of aerobic heterotrophic growth and of anaerobic lithotrophic growth. The anaerobic archaebacterial sulfur reducers, such as *Desulfurococcus, Thermoproteus, Thermofilum, Thermococcus, Thermodiscus, "Staphylothermus",* and *Pyrodictium* are all unable to grow aerobically and are therefore also unrelated.

3.4. The Genus *Thermoproteus*

The genus *Thermoproteus* contains rod-shaped (about 0.5 μm diameter) anaerobic bacteria which have been isolated from geothermally heated areas (68,71; Segerer, Huber, Gebhard, and Stetter, unpubl.). Similar-looking organisms were observed earlier by microscopic inspection of neutral hot springs by Brock and associates (4).

Ecology and Distribution   *Thermoproteus* can be isolated anaerobically (36) from various hot springs, mut pots and solfataric soils with pH values between 1.5 and 7 and temperatures of 70 to 100°C (6). Recently, similar organisms were obtained from a bore hole with slightly alkaline water (pH 8.5; 100°C) at the Kraffla geothermal power plant in Iceland (Gebhard, Huber, Kristjansson, and Stetter, unpubl.).

Within hot springs, *Thermoproteus* can exceed 10^6 cells/ml (6). In Iceland, about 25% of 267 samples taken anaerobically within a pH range from 1 to 8.3 and temperatures from 70 to 100°C yielded cultures of *Thermoproteus* (6).

Thermoproteus is widely distributed: It was isolated from Iceland (71; Gebhard, Huber, Kristjansson, and Stetter, unpubl.), Italy (Segerer and Stetter, unpubl.), the Azores (Stetter and Segerer, unpubl.), and USA (Zillig and Weber, unpubl.; Segerer and Stetter, unpubl.). Furthermore, similar-looking organisms have been observed in New Zealand (1).

Morphology   By phase-contrast microscopy, cells of *Thermoproteus* appear as stiff rods about 0.4 μm in diameter, which are highly variable in length, ranging from about 1 μm up to filaments of 80 μm. Normally, a length of about 3 to 5 μm predominates in cultures. As seen by electron microscopy,

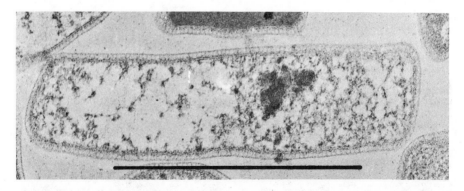

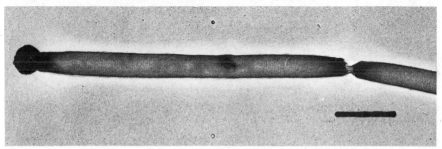

Figure 9 (top). *Thermoproteus neutrophilus.* Electron micrograph of a thin section. Bar, 1 μm.

Figure 10 (bottom). *Thermoproteus tenax.* Electron micrograph, rotation-shadowed (platinum). Bar, 1 μm.

the ends of *Thermoproteus* cells are angular with slightly rounded edges (Figure 9). During the exponential growth phase, spheres (0.3—0.5 μm diameter) are visible protruding terminally and (occasionally) laterally from the rods (Figure 10). Sometimes the rods show true branching and sharp bends with preferential angles of about 45, 90, and 135° (6).

A cell surface consisting of subunits about 16 nm thick and 20 nm wide in hexagonal array can be seen with the electron microscope in shadowed cells (Figure 11). This S layer (39) is situated directly outside the membrane (see Figure 9). Similar to most other archaebacteria, a rigid cell wall is missing (72). In contrast to *Sulfolobus*, the S layer of *Thermoproteus* cannot be disintegrated even by boiling in 2% SDS at alkaline pH but still maintains the original shape of the cells (König, pers. comm.). During exponential growth, formation of diaphragm-like septa is not observed, even within the branching and budding areas. Usually, *Thermoproteus* is immotile but shows pili protruding terminally and laterally. However, *Thermoproteus neutrophilus* possesses one terminal flagellum (6) and a new isolate (*Thermoproteus* "Geo 3") shows vigorous motility (Huber and Stetter, unpubl.).

Growth Requirements and Metabolism *Thermoproteus* must be grown strictly anaerobically by the application of the culture technique for meth-

Figure 11. *Thermoproteus tenax*, cell envelope. Electron micrograph, platinum shadowed. Bar, 1 μm.

anogens (36), since even traces of oxygen prevent growth. Good growth is obtained at temperatures between 80 and 92°C with an optimum around 88°C. The upper temperature limit is around 96°C, at which temperatures growth proceeds only slowly (71). Some isolates, such as H 10 and Geo 1 from Iceland, can grow at 102°C (Huber and Stetter, unpubl.). The pH for growth of *Thermoproteus tenax* ranges from 2.5 to 6 with an optimum around 5.5. However, there are also more neutrophilic isolates, such as *Thermoproteus neutrophilus* which grow within a pH range from 5.5 to 7.5 (Fischer and Stetter, unpubl.). *Thermoproteus* grows in Allen's medium (66) supplemented with sulfur and reduced by sodium sulfide (36). Most *Thermoproteus* isolates (e.g., *Thermoproteus tenax* DSM 2078, *Thermoproteus neutrophilus* DSM 2338) are able to grow chemolithoautotrophically on H_2, CO_2, and S^0 by H_2S formation (68). Under these conditions, a trace of yeast extract (e.g., 0.02%) stimulates growth but is not essential. However, there are also some *Thermoproteus* isolates unable to grow on molecular hydrogen (Segerer and Stetter, unpubl.). Alternatively, most *Thermoproteus* strains grow heterotrophically by sulfur respiration on organic material forming CO_2 and H_2S (exception: *Thermoproteus neutrophilus* is a strict autotroph). In the case of *Thermoproteus tenax* DSM 2078 (71), energy and carbon sources used include glucose, starch, amylose, amylopectin, glycogen, and, less efficiently, casamino acids. In the absence of sulfur, malate or cystine can serve as terminal electron acceptors (6). Some isolates do not form significant amounts of H_2S during heterotrophic growth, even in the presence of sulfur, while, as expected, there is always a strict connection between H_2S formation and autotrophic growth (Segerer and Stetter, unpubl.).

During growth, *Thermoproteus* is able to emulsify sulfur in the culture medium. The residual sulfur of autotrophically grown *Thermoproteus tenax* cultures contained compounds of the composition $(CH_2)_2S_3$ and $(CH_2)_2S_5$ which had adsorbed to it (Schäfer, unpubl.). Growth is not inhibited by vancomycin, chloramphenicol, streptomycin, rifampicin (71) or by thiolutin (Neuner and Stetter, unpubl.). Sulfate, thiosulfate, and sulfite are unable to replace sulfur as electron acceptors for growth. During chemolithoautotrophic growth on sulfur, H_2, and CO_2, about 1.5 moles H_2S are formed per gram of cells (dry weight). By heterotrophic sulfur respiration, 0.4 moles of H_2S are formed per gram dry weight (6). During incubation in the presence of H_2 and S^0 at temperatures above 85°C, low but significant amounts of H_2S are formed spontaneously, which must be considered when quantifying H_2S formation (Stetter, unpubl.).

Molecular Properties   The DNA of *Thermoproteus tenax* shows a G-C content of 56 mol%. The structural genes for the ribosomal RNAs are arranged on *Thermoproteus tenax* DNA in the order 16S, 23S, 5S with a large linker of 11 kilobase pairs between the 23S and 5S rRNA genes (73). The DNA-dependent RNA polymerase of *Thermoproteus tenax* is composed of 10 components and is of the "BAC" type (27). Like that of the other archaebacteria, it is resistant to rifampicin and streptolydigin. Like the RNA polymerase of *Methanobacterium thermoautotrophicum*, it is highly oxygen sensitive. It is completely stable at 95°C (71). *Thermoproteus tenax* DSM 2078 contains three unusual different rod-shaped temperate viruses in the same cell, TTV 1, TTV 2, and TTV 3, which contain linear double-stranded DNA. Virus production and cell lysis are induced by sulfur depletion of the culture medium (74).

Taxonomy and Phylogenetic Relationships   The type species of the genus *Thermoproteus* is *Thermoproteus tenax* (DSM 2078) (71). A second species, *Thermoproteus neutrophilus* (DSM 2338) has not been fully described (68). Both species show the same G-C content in their DNA and both are able to grow by hydrogen-sulfur autotrophy (68).

By rRNA-DNA hybridization (19) and by 16S rRNA cataloging (15), *Thermoproteus* was found to represent a new order, the *Thermoproteales* within the archaebacteria, which is clearly related to the *Sulfolobus* group by an S_{AB} value (11) of 0.30 (see Figure 1; 15).

3.5. The Genus *Thermofilum*

Hot acidic to neutral solfataric springs contain organisms of the genus *Thermofilum*. These filamentous bacteria are so thin that they can easily be overlooked. They grow anaerobically by heterotrophic nutrition (75). They are morphologically similar to organisms observed earlier in Boulder Spring, Yellowstone National Park (4).

Ecology and Distribution   *Thermofilum* can be isolated from various solfataric hot springs in Iceland, Italy, the Azores, and the USA with environmental temperatures between 55 and 100°C and pH values between 3 and 7. *Thermofilum* shares the habitat with *Sulfolobus, Thermoproteus, Desulfurococcus* (see Section 3.6) and, at pH values above 5.5, with *Methanothermus* (6; Fischer and Stetter, unpubl.). Similar bacteria have also been observed in ponds filled with geothermally heated seawater (85°C) containing decomposing algae in Ischia, Italy (Stetter, unpubl.). *Thermofilum* grows heterotrophically as a consumer of organic material and about 90% of the isolates are dependent on a cell component of other archaebacteria (see below; 75).

Morphology   *Thermofilum* is a filamentous rod (Figure 12), about 0.17 to 0.35 μm in diameter, depending on the isolate. Cells are 1 to more than 100 μm long with a median size between 5 and 10 μm (75). Vigorous shaking of cultures results in the formation of shorter cells. Like *Thermoproteus*, spherical bodies protrude from both ends. Very rarely, some branching has been observed. No diaphragmlike septa can be seen. Sometimes, piluslike structures are visible at both ends (75). The cell envelope consists of protein subunits arranged hexagonally and covering the membrane (Figure 13).

Growth Requirements   *Thermofilum* is a strictly anaerobic organism requiring similar precautions for cultivation as methanogens (36). *Thermo-*

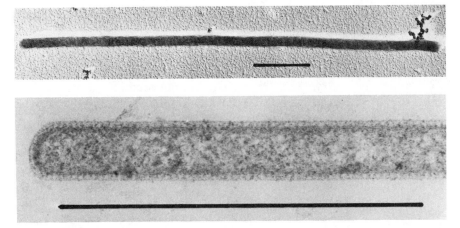

Figure 12 (top).   *Thermofilum librum.* Electron micrograph, platinum shadowed. Bar, 1 μm.

Figure 13 (bottom).   *Thermofilum librum.* Electron micrograph of a thin section. Bar, 1 μm.

filum pendens grows heterotrophically as a mixed culture together with *Thermoproteus tenax* in a medium (66) containing yeast extract (0.2 g/l), S^0 (5 g/l) and traces of H_2S. *Thermofilum* has unusual growth factor requirements, as shown by the following: It can be grown as pure culture by the addition of cell debris or a polar lipid fraction of *Thermoproteus tenax* to the culture medium (75). This fraction cannot be substituted by a similar one obtained from *Thermoplasma acidophilum* (75), but can be by those of methanogenic archaebacteria, including *Methanothermus fervidus, Methanobacterium thermoautotrophicum, Methanococcus thermolithotrophicus,* and *Methanosarcina barkeri* (Segerer and Stetter, unpubl.). There is no support of growth by eubacterial homogenates, such as from *Escherichia coli* or *Lactobacillus bavaricus*. About 10% of the hot spring sources having a pH 5–7 contain *Thermofilum librum*-like species which are independent of the lipid factor (Stetter, Fischer and Segerer, unpubl.). Due to its small diameter, *Thermofilum* can be easily enriched by filtration through membrane filters (Fischer and Stetter, unpubl.). As with typical archaebacteria, *Thermofilum* is not inhibited by the antibiotics vancomycin, chloramphenicol, rifampicin, and streptolydigin (75).

Molecular Biology *Thermofilum pendens* possesses an extremely heat-stable RNA polymerase resembling the "BAC" type characteristic of sulfur-dependent archaebacteria (27). In immunodiffusion assays (76), it does not cross-react with the RNA polymerase of *Thermoproteus tenax*. The G-C content of the DNA of *Thermofilum* is around 57 mol%. Cells do not contain muramic acid, which is in agreement with the lack of murein. Within the lipids, phytanol and C_{40} polyisoprenoid alcohols can be detected (71).

Taxonomy and Phylogeny By its resistance against antibiotics, lack of murein, RNA polymerase structure, and presence of phytanyl ether lipids, *Thermofilum* belongs to the archaebacteria. In spite of the very similar G-C content with *Thermoproteus tenax, Thermofilum pendens* shows only weak DNA-RNA hybridization with *Thermofilum tenax*, indicating, in addition to its different morphology, that this is a separate genus (19). The type species for the genus *Thermofilum* is *Thermofilum pendens* DSM 2475. Characterization of *Thermofilum librum*, a second species, is in progress and will be fully described elsewhere (Stetter et al., in preparation).

3.6. The Genus *Desulfurococcus*

Besides lobe-shaped *Sulfolobus* and *Acidothermus*, there are truly spherical organisms which grow strictly anaerobically and heterotrophically within hot solfatara fields, although these organisms are less extremely acidophilic (77). One group of such organisms has been described as the genus *Desulfurococcus*.

Ecology   *Desulfurococcus* has been isolated from hot solfataric springs in Iceland (77) and USA (Mt. Lassen National Park; Stetter, unpubl.) and from an Italian geothermal power plant (Stetter and Gebhard, unpubl.). More than half of the Icelandic springs yielding *Desulfurococcus* isolates had original temperatures above 90°C and pH values between 5 and 6.5 (78).

Growth Requirements   *Desulfurococcus* can be grown strictly anaerobically (36) at 85°C in Allen's medium (66) supplemented with sulfur and organic material such as yeast extract, tryptone, or casein (each 1 to 10 g/l). Very rarely (e.g., in the case of *Desulfurococcus saccharovorans*), isolates can be obtained which grow on glucose (0.5%) in the presence of 0.02% yeast extract (Stetter and Fischer, unpubl.). In a growth experiment, *Desulfurococcus mucosus* formed 8.8 μmoles H_2S and 6.6 μmoles CO_2 per ml of culture (0.2 mg dry weight of cells), indicating that it is able to grow by sulfur respiration (77). In the absence of sulfur, however, there is significant, although much slower growth, possibly due to another type of anaerobic respiration or fermentation. Growth occurs at pH values between 4.5 and 7 with an optimum around 6, whereas below 4.5 only cultures that have been adapted by stepwise lowering of the pH will grow (77). Streptomycin and vancomycin do not inhibit growth, while chloramphenicol and rifampicin cause inhibition (all at 100 μg/ml).

Morphology and Chemistry   Cells of *Desulfurococcus* are regular spheres, normally 0.5 to 1 μm in diameter (77). In the presence of 1% yeast extract *Desulfurococcus mucosus* forms giant cells up to 10 μm in diameter (6). Cells are surrounded by a protein subunit envelope (Figure 14) soluble in Triton-X100 (77). In *Desulfurococcus mucosus*, an additional mucoid layer is present covering the envelope and consisting of neutral sugars and a small fraction of amino sugars (6). *Desulfurococcus mobilis* is monopolar polytrichous flagellated with flagella of only 7 nm in diameter. Since no diaphragmlike septa formation could be observed during growth, cell division may occur by constriction or by budding (6). The lipids of *Desulfurococcus* contain phytanol and C_{40}polyisoprenoid dialcohols.

Molecular Biology   The DNA-dependent RNA polymerases of *Desulfurococcus mucosus* and *Desulfurococcus mobilis* show perfect serological cross-reaction in immunodiffusion, while there is no cross-reaction with polymerases of *Sulfolobus, Thermoproteus,* and *Thermofilum* (6). The enzymes of *Desulfurococcus* are of the "BAC" type (27) and are completely resistant to rifampicin, indicating that the target for rifampicin in vivo is different from the RNA polymerase, possibly similar to *Halobacterium* (78).

Taxonomy and Phylogeny   The type species of *Desulfurococcus* is *Desulfurococcus mucosus* (DSM 2162). Two species have been fully described, *Desulfurococcus mucosus* (DSM 2162) and *Desulfurococcus mobilis* (DSM

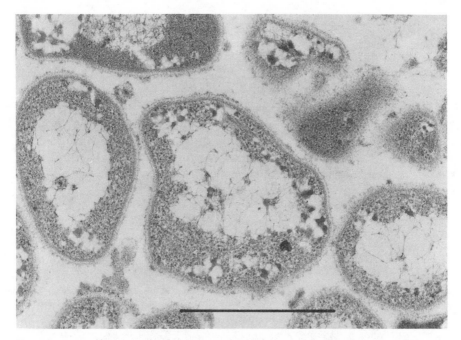

Figure 14. *Desulfurococcus saccharovorans.* Electron micrograph of a thin section. Bar, 1 μm.

2161), both showing a G-C content of their DNA around 51 mol%. A third species, *Desulfurococcus saccharovorans* (Figure 14) has a G-C content of 50 mol% and is different by its ability to use sugar (Stetter and Fischer, in prep.). The genus *Desulfurococcus* represents the family of *Desulfurococcaceae* within the *Thermoproteales* as indicated by the very low "fractional stability" of the hybrids (0.3) as measured by DNA-RNA hybridization with *Thermoproteus* (19).

4. EXTREMELY THERMOPHILIC ARCHAEBACTERIA OCCURRING IN SUBMARINE VOLCANIC AREAS

4.1. The Species *Methanococcus jannaschii*

The genus *Methanococcus* contains extremely anaerobic coccoid organisms occurring in sediments of natural waters, mainly in the mesophilic temperature range (36). A thermophilic species, *Methanococcus thermolithotrophicus*, grows at temperatures up to 70°C and was isolated from geothermally heated sediments at Stufe di Nerone, Italy (79). An extremely thermophilic species, *Methanococcus jannaschii*, obtained recently from deep sea hydrothermal vents along the East Pacific Rise by Jones and associates (80), will be described here in detail.

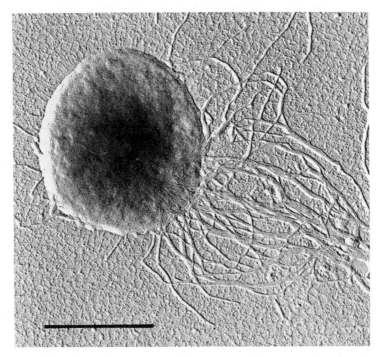

Figure 15. *Methanococcus jannaschii.* Electron micrograph, platinum shadowed. Bar, 1 μm.

Ecology *Methanococcus jannaschii* was isolated once from a sample taken from a "white smoker" chimney in the depth of 2,600 meters at 21°N (109°W) on the East Pacific Rise (80) but not from shallow submarine hydrothermal areas (e.g., Vulcano, Ischia, Naples, Ribieira Quente (Azores), and Reykjanes (Iceland)).

Growth and Metabolism *Methanococcus jannaschii* grows strictly anaerobically (37) in a mineral medium at pH 6 in the presence of H_2/CO_2 (80:20) within a temperature range between 50 and 86°C with an optimum around 85°C (80). No growth could be detected at pH 5 or 7 or at 37 or 95°C. No growth occurs on acetate, methanol, or methylamines, and, in contrast to the other methanococci, on formate (37,79,80). Growth is not stimulated by organic material, but by selenium ions. Sulfide is required at a concentration of around 1 to 3 mM. The optimal salt concentration is near 0.5 M Na^+.

Morphology and Ultrastructure Cells of *Methanococcus jannaschii* are irregular cocci (Figure 15) with a width of up to 1.5 μm (80). The cell envelope consists of a hexagonal array composed of protein subunits (80) which are not glycosylated (Nusser, König and Stetter, unpubl.). Occasionally, an additional layer at the interior of the cytoplasmic membrane with an unknown function is visible (80). The cells are motile due to two bundles of flagella (Figure 15) inserted close to the same cellular pole. The length of the flagella is about 5 μm (80).

Phylogeny By the G-C content (31 mol%) of its DNA (36), its shape, its metabolism, and its 16S rRNA oligonucleotide catalog (see Figure 1), *Methanococcus jannaschii* clearly belongs to the genus *Methanococcus* (80), although it differs from all other members by its high growth temperature.

4.2. The Genus *Pyrodictium*

The genus *Pyrodictium* contains the most extremely thermophilic bacterium known to date, growing optimally above 100°C. Bacteria of this genus have been isolated from a shallow submarine solfatara field (67).

Ecology *Pyrodictium* occurs within a submarine solfatara field close to the bay of Porto di Levante, Vulcano, Italy. The sea floor consists of heated sandy sediments and sulfur-coated cracks and holes where hot seawater and volcanic gases are emanating. The organism has also been isolated from hot sediments at the sandy shore of Vulcano and from a boiling black mud pot containing seawater, situated about 20 meters away from the beach. The temperatures of the sea floor range up to 103°C (67). Within the Porto di Levante area, *Pyrodictium* is very common. However, it could not be isolated from the hot sea floors of either Stufe de Nerone, Baja (Naples) or close to the Maronti beach, Ischia, both of which are also in Italy and are only about 400 km away from Vulcano. Similarly, *Pyrodictium* could not be isolated from the hot sea floors of the Azores (Ribieira Quente), Iceland (Reykjanes), or from a black smoker at the East Pacific Rise (however, see Section 4.5). It is presently unclear whether *Pyrodictium* is really an endemic organism or what kinds of barriers are possibly responsible for its narrow distribution.

Growth and Metabolism *Pyrodictium* grows strictly anaerobically (36) in synthetic seawater (8) supplemented with sulfur, with a gas phase of H_2/CO_2 (80:20; 2 bar) and grows within a temperature range between 85 and 110°C with an optimum around 105°C (Figure 16; 8,67). The pH range for growth is between 5 and 7. *Pyrodictium* shows an unusually wide salt tolerance, from 0.2 to 12% NaCl with a growth optimum at around 1.5%. Therefore, the organisms could possibly also be present in springs of low salt concentrations, such as continental solfatara fields. During growth, large amounts of H_2S are formed from H_2 and elemental sulfur indicating that *Pyrodictium* can grow by hydrogen-sulfur autotrophy (68). Cultivation in fermenters leads to a solid coating of the fermenter surface with pyrite, most likely due to spontaneous reaction of the H_2S formed by the organisms with

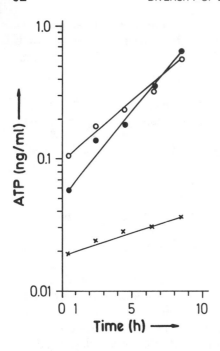

Figure 16. *Pyrodictium occultum*, growth curves. Growth was determined by adenosine triphosphate (ATP) measurements (67). (x)=85°C; (o)=100°C; (•)=105°C.

sulfur and Fe^{2+}, both present in the medium (8). In uninoculated controls, practically no H_2S is formed in this medium during incubation at the high temperatures. *Pyrodictium* is a strictly chemolithotrophic organism with an obligate dependence for growth on H_2 and S. Some isolates (e.g., *Pyrodictium brockii*) show stimulation of growth by yeast extract and peptone, but are unable to use these substrates as a single energy source without H_2. During growth, cells are extremely sensitive to oxygen and are inactivated by it within a few minutes. At 4°C, however, a part of the population is oxygen-resistant, at least for several months (8).

Morphology and Ultrastructure *Pyrodictium* grows as a mold-like layer upon the sulfur which can be removed by gentle shaking. The cells are irregularly disc and dish shaped, varying in diameter between 0.3 and 2.5 μm. Under dark field illumination, a huge network of fibers becomes visible within which the cells appear to be fixed (Figure 17). During the early exponential growth phase, only very few cells are visible, although the fibrous network is already present. By electron microscopy, the fibers have diameters of only 0.04 to 0.07 μm and are therefore invisible in the phase-contrast light microscope (8). They have lengths of up to 40 μm and are often associated in bundles (Figure 18). The fibers are hollow and, similar to flagella, are formed by subunits arranged in helical arrays with diameters of about 5 nm (Messner and Sleytr, pers. comm.). Granules of sulfur are seen frequently sticking to the fibers (Figure 18). In thin sections, the cells appear

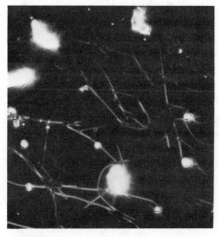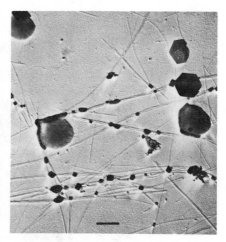

Figure 17 (left). *Pyrodictium occultum.* Darkfield micrograph.

Figure 18 (right). *Pyrodictium occultum.* Electron micrograph, platinum shadowed. Bar, 1 μm.

partially ultraflat and are surrounded by a protein subunit envelope about 30 nm in diameter (Figure 19). When stained by ruthenium red, an additional layer becomes visible between the membrane and the protein subunit envelope (Figure 20). The discs and the fibers show piluslike structures, 10 nm in diameter and up to 5 μm long.

Chemistry and Biochemistry The envelope of *Pyrodictium occultum* contains a dominating protein with an apparent molecular weight of 172,000 which stains periodate-Schiff positive and is therefore most likely a glycoprotein. The envelope can be dissolved in 1% SDS while the proteinaceous fibers are resistant even to boiling in 2% SDS at pH 12. The fibers are encrusted with zinc sulfide (Sprey, pers. comm.) which is formed (apparently selectively) from H_2S and the zinc ions present in the medium (8). Both pyrite and zinc sulfide formation may be important in the formation of ore deposits in hydrothermal systems, a process which is now occurring at Vulcano where *Pyrodictium* was isolated (81). *Pyrodictium* has a number of archaebacterial characteristics. The cell envelope does not contain muramic acid. The elongation factor EF2 from *Pyrodictium occultum* is ADP-ribosylated by diphteria toxin (Klink, pers. comm.). The lipids of *Pyrodictium occultum* contain about 45% biphytanyl tetraethers, 20% phytanyl diethers and some unknown minor compounds (Langworthy, pers. comm.).

Phylogenetic Relationships and Taxonomy *Pyrodictium occultum* was shown to be related to *Sulfolobus* when 16S rRNA cataloging was done (Stackebrandt and Woese, pers. comm.). However, the catalog is still in-

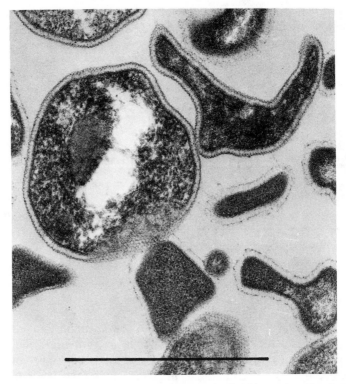

Figure 19. *Pyrodictium occultum.* Electron micrograph of a thin section. Bar, 1 μm.

complete due to a substantial number of posttranscriptionally modified bases. The exact taxonomic position can be determined after analysis of the modified oligonucleotides. The relationship of *Pyrodictium* to *Sulfolobus* is further substantiated by sequencing of a significant portion of the 5S RNA gene of *Pyrodictium occultum* (Kaine and Woese, pers. comm.). Up to now, 12 different *Pyrodictium* isolates have been obtained from Vulcano, all growing up to 110°C and showing the characteristic moldlike growth in unshaken cultures. Two species have been described, *Pyrodictium occultum* DSM 2709, which represents the type species, and *Pyrodictium brockii* DSM 2708. Both show a DNA base composition of 62 mol% GC. *Pyrodictium brockii* shows no glycoprotein in its envelope and growth is strongly stimulated by organic material.

4.3. The Genus *Thermodiscus*

Although similarly shaped, *Thermodiscus* differs from *Pyrodictium* in many metabolic, physiological, and biochemical properties.

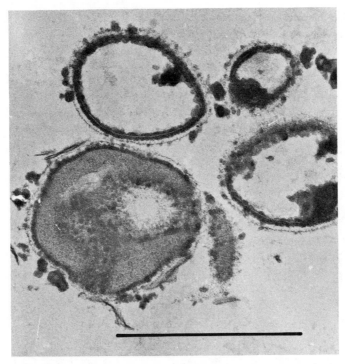

Figure 20. *Pyrodictium occultum.* Electron micrograph of a thin section, ruthenium red staining. Bar, 1 μm.

Distribution and Ecology   *Thermodiscus* has so far been isolated only from the submarine solfatara field close to Vulcano, where it occurs together with *Pyrodictium*. It is an anaerobic heterotroph most likely feeding on the cell components of lithotrophic primary producers such as *Pyrodictium*. In contrast to *Pyrodictium*, it does not occur at temperatures above 95°C. From its narrow salt tolerance at the ionic strength of seawater, *Thermodiscus* appears to be highly adapted to the marine environment.

Growth and Physiology   *Thermodiscus* grows strictly anaerobically within a temperature range between 75 and 98°C (optimum around 90°C) and between pH 5 and 7 (optimum around 5.5). Growth occurs by sulfur respiration in artificial seawater with NaCl concentrations between 1 and 4% (optimum around 2%) in the presence of sulfur and low amounts (optimum 0.02%) of yeast extract. In some isolates growth is stimulated by hydrogen, although no growth is obtained without yeast extract. All *Thermodiscus* isolates show good growth without sulfur and hydrogen on yeast extract. Sometimes, even at optimal growth, no significant amounts of H_2S are formed even in the presence of sulfur and hydrogen, suggesting a possibly unknown anaerobic respiration or fermentation.

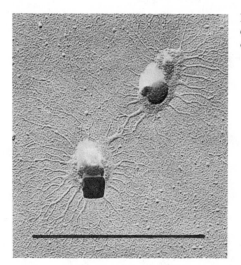

Figure 21. *Thermodiscus maritimus*, small cells. Electron micrograph, platinum shadowed. Bar, 1 μm.

Morphology   Cells of *Thermodiscus* are highly irregular dish to disc shaped, varying in diameter from about 0.3 to 3 μm, and are sometimes even smaller than 0.2 μm in diameter (Figure 21). The discs show a thickness of only about 0.1 and 0.2 μm (Figure 22). In contrast to *Pyrodictium*, no fibers are formed. Piluslike structures about 0.01 μm in diameter and up to 15 μm long are visible, sometimes connecting two individual cells. Flagella and motility are absent. In the stationary growth phase, empty cells containing dark granules reminiscent of *Sulfolobus* (40) can be seen. The Gram-negative cells are surrounded by an envelope composed of a hexagonal array of protein subunits (Figure 22), about 33 nm in diameter (König, pers. comm.).

Biochemical Properties   The DNA of *Thermodiscus* has a G-C content of 49 mol%. Crude extracts contain a protein, most likely EF2, which is ADP-ribosylated by diphtheria toxin (Klink, pers. comm.). The cell envelope contains large amounts of a periodate-Schiff positive protein with an apparent molecular weight of about 84,000 (König, pers. comm.). Neither muramic acid nor mesodiaminopimelic acid could be demonstrated. Isoprenoid ether-linked lipids are present (Langworthy, pers. comm.).

Phylogeny   Due to its biochemical features, *Thermodiscus* is an archaebacterium. By 16S rRNA cataloging, it is clear that *Thermodiscus* is specifically related to *Pyrodictium*, although the catalog is still incomplete (Stackebrandt and Woese, pers. comm.). *Thermodiscus* can be easily distinguished from *Pyrodictium* by its obligately heterotrophic growth in homogeneous cell suspensions, the lack of fibers, and the much lower G-C content. The type species is *Thermodiscus maritimus* (Stetter, manuscript in preparation).

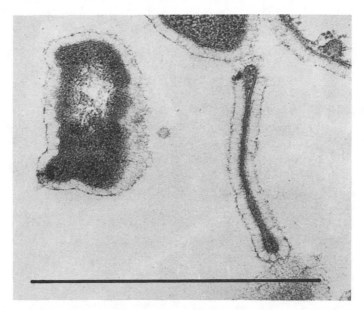

Figure 22. *Thermodiscus maritimus.* Electron micrograph of a thin section. Bar, 1 μm.

4.4. The Genus *Thermococcus*

Ecology and Distribution *Thermococcus* was first isolated from a hole containing hot marine water in Vulcano (Italy) (82). Organisms similar in shape and growth characteristics could be obtained from hot submarine sediments close to Vulcano, Ischia, and Stufe di Nerone, all situated in Italy, and at Ribiera Quente, Azores (Stetter, unpubl.). *Thermococcus celer* is an organism adapted to the marine environment, growing heterotrophically on peptides and protein and, therefore, is a consumer of organic material.

Growth Requirements and Metabolism Some isolates of *Thermococcus celer* are highly adapted to seawater, growing optimally within a range between 38 and 40 g NaCl/l. Below 35 g NaCl/l, cells lyse (82). However, other isolates have been obtained which do not lyse with even much less than 35 g NaCl/l (Stetter, unpubl.). *Thermococcus celer* utilizes tryptone or yeast extract and casein as carbon sources. Growth is stimulated by sucrose. *Thermococcus celer* grows optimally in the presence of sulfur, and about 1.5 moles of H_2S are formed per mole of CO_2 (82). As with *Desulfurococcus* and as with *Thermodiscus* (although less efficiently), *Thermococcus* can grow without sulfur with a tenfold lower rate of growth and yield. Under optimal conditions (88°C; pH 5.8; 40 g NaCl/l), the generation time is close to 50 minutes. Growth is inhibited by high concentrations of H_2S (82).

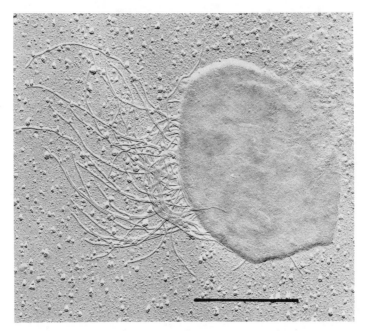

Figure 23. *Thermococcus celer.* Electron micrograph, platinum shadowed. Bar, 1 μm.

Morphology   Cells of *Thermococcus* are regular spheres showing diploid forms in growing cultures. Cell division seems to occur by constriction since diaphragmlike septa are absent. Sometimes, cells are connected by strings of cytoplasm surrounded by membrane and cell envelope (82). The envelope is composed of protein subunits. The cells are motile due to monopolar polytrichous flagellation (Figure 23).

Biochemistry and Molecular Biology   The G-C content of the DNA of *Thermococcus celer* is around 56 mol% (82). No muramic acid is present in the cells. The membrane contains phytanol and saturated polyisoprenoid C_{40} dialcohol (82). The DNA-dependent RNA polymerase is of the "BAC" type (27) and does not show serological cross-reaction in the Ouchterlony immunodiffusion test with antibodies prepared against the RNA polymerases of *Desulfurococcus, Sulfolobus, Thermoplasma* or *Thermoproteus* (76).

Taxonomy   *Thermococcus* is an archaebacterium which, due to the pattern of its RNA polymerase, belongs to the sulfur-dependent branch (see Figure 1). In DNA-RNA hybridization experiments, *Thermococcus* shows a fractional stability of 0.31 with *Sulfolobus* and of 0.41 with *Thermoproteus*, indicating only a distant relationship to both genera (19). The type species is *Thermococcus celer* DSM 2476. Very recently, thermophilic archaebacteria having a similar shape but with a G-C content of only 38 mol% and an upper growth temperature of 103°C were isolated, representing the new genus *Pyrococcus* (Fiala and Stetter, *Arch. Microbiol.*, in press, 1986).

4.5. The Genus *Staphylothermus*

Ecology *Staphylothermus* was isolated from a shallow submarine hydrothermal area at the base of the *Fossa* volcano, Vulcano, Italy and, later, from samples taken from a "black smoker" (10° 57′N, depth 2503 m; t = 338°C) (Jannasch, pers. comm.). It is therefore present in shallow, as well as in deep hydrothermal areas, living heterotrophically at the expense of organic material.

Growth Requirements and Metabolism *Staphylothermus* grows strictly anaerobically at temperatures between 65 and 98°C with an optimum around 92°C. At low temperatures (e.g., 4°C) in the presence of oxygen, cells survive for at least several weeks. *Staphylothermus* is an obligate heterotroph, growing optimally on a mixture of yeast extract (0.1%) and peptone (0.5%) in seawater (Fiala and Stetter, unpubl.). It shows an obligate dependence on elemental sulfur, although only relatively low amounts of H_2S are formed during growth (about 3.3 μmoles/ml). The type of metabolism is still unknown. As end products, CO_2, acetate, and isovalerate have been detected, suggesting a fermentative metabolism (83).

Morphology Under optimal conditions, *Staphylothermus* grows characteristically in aggregates up to 100 individuals (Figure 24), of slightly irregular cells, about 0.5 to 1 μm in diameter. In the presence of 0.2% yeast extract (without peptone), however, giant cells up to 15 μm in diameter are formed and can be transferred continuously (Figure 25). The cells are surrounded by an envelope.

Biochemistry The G-C content of the DNA of *Staphylothermus marinus* and of the "black smoker" isolate is 35 mol%, as determined by the melting point (69) and, after digestion, by high performance liquid chromatography (18) methods. The cell envelope contains four major protein components showing a periodate-Schiff positive staining capacity in polyacrylamide gels (Fiala and Stetter, unpubl.). The membrane lipids are composed of about 40% phytanyl tetraethers and about 20% diethers, with the remainder composed of an unknown component (Langworthy, pers. comm.). Crude extracts contain one protein which is ADP-ribosylated by diphtheria toxin (Klink, pers. comm.). No muramic acid can be detected (König, pers. comm.).

Taxonomy Although *Staphylothermus marinus* is an archaebacterium, its phylogenetic position is still unclear. It may possibly belong to the sulfur-dependent archaebacteria. Due to its very low G-C content and its char-

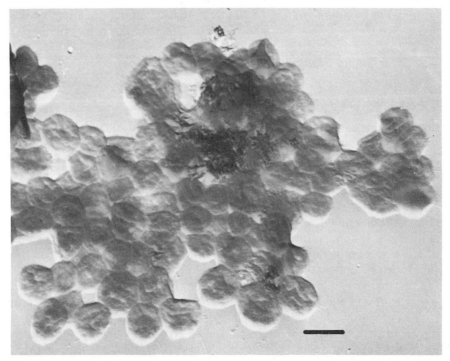

Figure 24. *Staphylothermus marinus.* Electron micrograph, platinum shadowed. Bar, 1 μm.

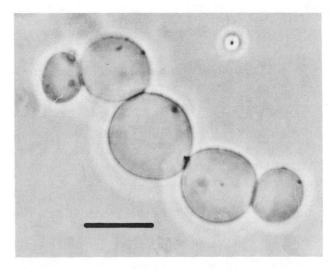

Figure 25. *Staphylothermus marinus*, giant cells. Phase-contrast micrograph. Bar, 10 μm.

acteristic growth in aggregates, *Staphylothermus* is different from the other coccoid marine archaebacteria, and therefore represents a new genus (83). The type species is *Staphylothermus marinus*, isolate F 1 (DSM 3639).

5. FACTORS LIMITING UPPER GROWTH TEMPERATURE OF EXTREME THERMOPHILES

Since some extremely thermophilic bacteria are growing even in superheated water, the question for a general upper temperature limit of life arises. On one hand, life depends on the existence of liquid water (1), which above 100°C is available only under elevated pressure, such as on the sea floor or in the depth of continental floors, and on the other hand, it is defined by the thermostability of cell structures (e.g., membranes and ribosomes), of macromolecules (e.g., nucleic acids, proteins, and lipids) and of small heat-labile organic molecules (e.g., NAD and ATP). At 110°C, where *Pyrodictium* is still able to grow, anhydrous bonds of phosphoric acid in ATP are rapidly hydrolyzed (half time below 30 minutes; Stetter, unpubl.), suggesting that as a strategy for survival such molecules might be continuously resynthesized by *Pyrodictium*. A report on the successful culture of "black smoker" bacteria from the East Pacific Rise in the laboratory at 250°C and 300 bar (84) is very doubtful, since uninoculated controls in similar experiments yielded the same organismlike particles which turned out to be precipitation artifacts (85). Up to now, no cultures of "black smoker" bacteria capable of growth at 250°C are available nor can growth be demonstrated or successfully reproduced by others. In contrast, the "black smoker" bacterium isolated in our laboratory (see Section 4.5, *Staphylothermus*) shows a maximal growth temperature of only 98°C and, therefore, was most likely growing at the cooler surface of the chimney and not in the 338°C hot water (Jannasch, pers. comm.). Under "black smoker" conditions (250°C; 260 bar = 26,000 kPa), biomolecules are hydrolyzed in an extremely short time and even the hydrolysis products (e.g., most amino acids) are unstable (86). For example, the half times for DNA are about 20 μseconds, for a protein with a molecular weight of 48,000 about 1 second, and for ATP less than 1 second. Therefore, a very fast resynthesis can be excluded as a strategy of survival under these conditions. The harmful effect of high temperature is further intensified by increased pressure which enchances hydrolysis of polymers such as DNA and proteins by shifting the equilibrium onto the side of the monomers due to the negative reaction volume (87). Even the "heat-stable" proteins of *Pyrodictium* are hydrolyzed very rapidly under "black smoker" conditions (88). Although the upper temperature limit of life is still unclear, it should be much lower than 250°C, possibly in a range between 110 and 150°C, where macromolecules are still sufficiently stable and small molecules can be successfully resynthesized. Possibly, fascinating organisms capable of growth at 150°C are waiting to be isolated by enthusiastic microbiologists.

ACKNOWLEDGMENTS

I wish to thank Dr. T. A. Langworthy, Dr. C. R. Woese, and Dr. E. Stackebrandt for providing unpublished data; Dr. H. König for the preparation of electron micrographs; and Dr. T. A. Langworthy for critically reading the manuscript.

REFERENCES

1. T. D. Brock, *Thermophilic Microorganisms and Life at High Temperatures*, Springer, New York, 1978.
2. M. Shilo, *Strategies of Microbial Life in Extreme Environments*, Dahlem Konferenzen, Verlag Chemie, Weinheim, 1979.
3. H. Zuber, in M. Shilo, Ed., *Strategies of Microbial Life in Extreme Environments*, Dahlem Konferenzen, Verlag Chemie, Weinheim, 1979, p. 393.
4. T. D. Brock, M. L. Brock, T. L. Bott, and M. R. Edwards, *J. Bacteriol.* **107**, 303 (1971).
5. T. D. Brock, K. M. Brock, R. T. Belly, and R. L. Weiss, *Arch. Microbiol.* **84**, 54 (1972).
6. K. O. Stetter and W. Zillig, in C. Woese and R. S. Wolfe, Eds, *The Bacteria*, Vol. 8, Academic Press, New York, 1985, Ch. 2.
7. K. O. Stetter, *Naturwissenschaften*, **72**, 291 (1985).
8. K. O. Stetter, H. König, and E. Stackebrandt, *Syst. Appl. Microbiol.* **4**, 535 (1983).
9. C. R. Woese, L. J. Magrum, and G. E. Fox, *J. Mol. Evol.* **11**, 245 (1978).
10. C. R. Woese, M. Sogin, D. Stahl, B. J. Lewis, and L. Bonen, *J. Mol. Evol.* **7**, 197 (1976).
11. G. E. Fox, K. R. Pechman, and C. R. Woese, *Int. J. Syst. Bacteriol.* **27**, 44 (1977).
12. W. E. Balch, L. J. Magrum, G. E. Fox, R. S. Wolfe, and C. R. Woese, *J. Mol. Evol.* **9**, 305 (1977).
13. G. E. Fox, E. Stackebrandt, R. B. Hespell, J. Gibson, J. Maniloff, T. W. Dyer, R. S. Wolfe, W. E. Balch, R. S. Tanner, L. J. Magrum, L. B. Zablen, R. Blakemore, R. Gupta, L. Bonen, B. J. Lewis, D. A. Stahl, K. R. Luehrsen, K. N. Chen, and C. R. Woese, *Science* **209**, 457 (1980).
14. C. R. Woese, in D. S. Bendall, Ed., *Evolution from Molecules to Men*, Cambridge University Press, Cambridge, 1983.
15. C. R. Woese, R. Gupta, C. M. Hahn, W. Zillig, and J. Tu, *Syst. Appl. Microbiol.* **5**, 97 (1984).
16. R. T. Belly, B. B. Bohlool, and T. D. Brock, *Ann. N. Y. Acad. Sci.* **225**, 94 (1973).
17. W. Zillig, R. Schnabel, J. Tu, and K. O. Stetter, *Naturwissenschaften* **69**, 197 (1982).
18. W. Zillig, K. O. Stetter, S. Wunderl, W. Schulz, H. Priess, and J. Scholz, *Arch. Microbiol.* **125**, 259 (1980).
19. J. Tu, D. Prangishvilli, H. Huber, G. Wildgruber, W. Zillig, and K. O. Stetter, *J. Mol. Evol.* **18**, 109 (1982).
20. O. Kandler, *Zbl. Bakt. Hyg., I, Abt. Orig. C* **3**, 149 (1982).
21. R. Hilpert, J. Winter, W. Hammes, and O. Kandler, *Zbl. Bakt. Hyg., I, Abt. Orig. C* **2**, 11 (1981).
22. P. Forterre, C. Elie, and M. Kohiyama, *J. Bacteriol.* **159**, 800 (1984).
23. Y. Kuchino, M. Hiara, Y. Yabusaki, and S. Nishimura, *Nature* **298**, 684 (1982).
24. M. Kessel and K. Klink, *Zbl. Bakt. Hyg., I, Abt. Orig. C* **3**, 140 (1982).

25. A. Matheson and M. Yaguchi, *Zbl. Bakt. Hyg., I, Abt. Orig. C* **3**, 192 (1982).
26. T. Pecher and A. Böck, *FEMS Letters* **10**, 295 (1981).
27. W. Zillig, K. O. Stetter, R. Schnabel, and M. Thomm, in C. R. Woese and R. S. Wolfe, Eds., *The Bacteria*, Vol. 8, Academic Press, New York, 1985, Ch. 11.
28. D. J. McConnell, D. G. Searcy, and J. G. Sutcliffe, *Nucleic Acids Res.* **5**, 1729 (1978).
29. K. Schmid, M. Thomm, A. Laminet, F. G. Laue, C. Kessler, K. O. Stetter, and R. Schmitt, *Nucleic Acids Res.* **8**, 2619 (1984).
30. C. R. Woese, *Zbl. Bakt. Hyg., I, Abt. Orig. C* **3**, 1 (1982).
31. J. A. Lake, *Cell* **33**, 318 (1983).
32. J. A. Steitz, *Nature* **273**, 10 (1978).
33. C. R. Woese and R. Gupta, *Nature* **289**, 95 (1981).
34. J. G. Zeikus and R. S. Wolfe, *J. Bacteriol.* **109**, 707 (1972).
35. J. G. Zeikus, A. Ben-Bassatt, and P. W. Hegge, *J. Bacteriol.* **143**, 432 (1980).
36. W. E. Balch, G. E. Fox, L. J. Magrum, C. R. Woese, and R. S. Wolfe, *Microbiol. Rev.* **43**, 260 (1979).
37. K. O. Stetter, M. Thomm, J. Winter, G. Wildgruber, H. Huber, W. Zillig, D. Janecovic, H. König, P. Palm, and S. Wunderl, *Zbl. Bakt. Hyg., I, Abt. Orig. C* **2**, 166 (1981).
38. O. Kandler and H. König, *Arch. Microbiol.* **118**, 141 (1978).
39. U. B. Sleytr and P. Messner, *Ann. Rev. Microbiol.* **37**, 311 (1983).
40. C. L. Brierley and J. A. Brierley, *Canad. J. Microbiol.* **19**, 183 (1973).
41. M. DeRosa, A. Gambacorta, G. Millonig, and J. D. Bu'Lock, *Experientia* **30**, 866 (1974).
42. T. D. Brock, in M. P. Starr, H. Stolp, H. G. Trüper, A. Balows, and H. G. Schlegel, Eds., *The Prokaryotes*, Vol. I, Springer-Verlag, New York, 1981, p. 978.
43. B. B. Bohlool, *Arch. Microbiol.* **106**, 171 (1975).
44. T. Furuya, T. Nagumo, T. Iloh, and H. Kaneko, *Agric. Biol. Chem.* **41**, 607 (1977).
45. B. B. Bohlool and T. D. Brock, *Arch. Microbiol.* **97**, 181 (1974).
46. C. B. Fliermans and T. D. Brock, *J. Bacteriol.* **111**, 343 (1972).
47. M. McClure and W. G. Wyckoff, *J. Gen. Microbiol.* **128**, 433 (1982).
48. H. König, R. Skorko, and W. Zillig, *Arch. Microbiol.* **132**, 297 (1982).
49. R. L. Weiss, *J. Bacteriol* **118**, 275 (1974).
50. H. Michel, D. C. Neugebauer, and D. Oesterhelt, in W. Baumeister and W. Vogell, Eds., *Electron Microscopy at Molecular Dimension*, Springer-Verlag, Berlin, 1980, p. 27.
51. K. A. Taylor, J. F. Deatherage, and L. A. Amos, *Nature* **299**, 840 (1982).
52. J. L. Mosser, A. G. Mosser, and T. D. Brock, *Arch. Microbiol.* **97**, 169 (1074).
53. T. D. Brock and K. O'Dea, *Appl. Environ. Microbiol* **33**, 254 (1977).
54. T. D. Brock and J. Gustafson, *Appl. Environ. Microbiol.* **32**, 567 (1976).
55. C. L. Brierley, *CRC Crit. Rev. Microbiol.* **6**, 207 (1978).
56. R. M. Marsh, P. R. Norris, and N. W. LeRoux, in G. Rossi and A. E. Torma, Eds, *Progress in Biohydrometallurgy*, Associazione Mineraria Sarda — 09016 Iglesias, 1983, p. 71.
57. O. Kandler and K. O. Stetter, *Zbl. Bakt. Hyg., I, Abt. Orig. C* **2**, 111 (1981).
58. A. Segerer, K. O. Stetter, and F. Klink, *Nature* **313**, 787 (1985).
59. V. Buonocore, O. Sgambati, M. DeRosa, E. Esposito, and A. Gambacorta, *J. Appl. Biochem.* **2**, 390 (1980).
60. J. Huet, R. Schnabel, A. Sentenac, and W. Zillig, *EMBO J.* **2**, 1291 (1983).
61. R. Schnabel, J. Sonnenbichler, and W. Zillig, *FEBS Lett.* **150**, 400 (1982).
62. B. P. Kaine, R. Gupta, and C. R. Woese, *Proc. Natl. Acad. Sci.* **80**, 3309 (1983).
63. S. Yeats, P. McWilliam, and W. Zillig, *EMBO J.* **1**, 1035 (1982).
64. A. Martin, S. Yeats, D. Janecovic, W. D. Reiter, W. Aicher, and W. Zillig, *EMBO J.* **3**, 2165 (1984).
65. W. Zillig, S. Yeats, I. Holz, A. Böck, F. Gropp, M. Rettenberger, and S. Lutz, *Nature* **313**, 789 (1985).
66. M. B. Allen, *Arch. Microbiol.* **32**, 270 (1959).
67. K. O. Stetter, *Nature* **300**, 258 (1982).
68. F. Fischer, W. Zillig, K. O. Stetter, and G. Schreiber, *Nature* **301**, 511 (1983).

69. J. Marmur and P. Doty, *J. Molec. Biol.* **5**, 109 (1962).
70. A. G. Steigerwaldt, G. R. Fanning, M. A. Sise-Ashbury, and B. J. Brenner, *Can. J. Microbiol.* **22**, 121 (1976).
71. W. Zillig, K. O. Stetter, W. Schäfer, D. Janecovic, S. Wunderl, I. Holz, and P. Palm, *Zbl. Bakt. Hyg., I, Abt. Orig. C* **2**, 205 (1981).
72. O. Kandler and H. König, in C. R. Woese and R. S. Wolfe, Eds., *The Bacteria*, Vol. 8, Academic Press, New York, 1985, Ch. 7.
73. H. Neumann, A. Gierl, J. Tu, J. Leibrock, D. Staiger, and W. Zillig, *Mol. Gen. Genet.* **192**, 66 (1983).
74. D. Janecovic, S. Wunderl, I. Holz, W. Zillig, A. Gierl, and H. Neumann, *Mol. Gen. Genet.* **192**, 39 (1983).
75. W. Zillig, A. Gierl, S. Schreiber, S. Wunderl, D. Janecovic, K. O. Stetter, and H. P. Klenk, *Syst. Appl. Microbiol.* **4**, 79 (1983).
76. Ö. Ouchterlony, in P. Kallos and B. H. Waksman, Eds., *Progress in Allergy VI*, Karger, Basel, 1962, p. 30.
77. W. Zillig, K. O. Stetter, D. Prangishvilli, H. Schäfer, S. Wunderl, D. Janecovic, I. Holz, and P. Palm, *Zbl. Bakt. Hyg., I, Abt. Orig. C* **3**, 304 (1982).
78. W. Zillig and K. O. Stetter, in S. Osawa, H. Ozeki, H. Uchida, and T. Yura, Eds., *Genetics and Evolution of RNA Polymerase, tRNA and Ribosomes*, University of Tokyo Press, 1980, p. 525.
79. H. Huber, M. Thomm, H. König, G. Thies, and K. O. Stetter, *Arch. Microbiol.* **132**, 47 (1982).
80. W. L. Jones, J. A. Leigh, P. Mayer, C. R. Woese, and R. S. Wolfe, *Arch. Microbiol.* **136**, 254 (1983).
81. A. Wauschkuhn and H. Gröpper, *Neues Jahrbuch Mineralischer Abhandlungen* **126**, 87 (1975).
82. W. Zillig, I. Holz, D. Janecovic, W. Schäfer, and W. D. Reiter, *Syst. Appl. Microbiol.* **4**, 88 (1983).
83. G. Fiala, K. O. Stetter, H. W. Jannasch, T. A. Langworthy, and J. Madon, *Syst. Appl. Microbiol.*, in press (1986).
84. J. A. Baross and J. W. Deming, *Nature* **303**, 423 (1983).
85. J. D. Trent, R. A. Chastain, and A. A. Yayanos, *Nature* **303**, 737 (1984).
86. R. H. White, *Nature* **310**, 430 (1984).
87. R. E. Marquis, *Advances in Microbial Physiology* **14**, 159 (1976).
88. G. Bernhardt, H. D. Lüdemann, R. Jaenicke, H. König, and K. O. Stetter, *Naturwissenschaften* **71**, 583 (1984).

PHYSIOLOGY AND GROWTH OF THERMOPHILIC BACTERIA

T. K. SUNDARAM

Department of Biochemistry and Applied Molecular Biology, University of Manchester Institute of Science and Technology, Manchester M60 1QD, United Kingdom

---

1. INTRODUCTION

Organisms which multiply optimally at high temperature have been known to exist for nearly a century (1). However, most of the significant studies on thermophily have been carried out during the last thirty years. The general definition of a thermophile and the diversity of thermophilic organisms have been discussed in Chapters 1 and 2. It is the purpose of the present chapter to present information on the general physiological characteristics of thermophilic bacteria, with emphasis on growth.

2. NUTRITION OF THERMOPHILIC BACTERIA

Thermophilic bacteria exhibit a wide range of nutritional capability: phototrophy and chemotrophy; autotrophy and heterotrophy; chemolithotrophy and chemoorganotrophy; aerobiosis and anaerobiosis. Examples of organisms capable of phototrophic and nonphototrophic (chemotrophic) growth are evident from Table 2 in Chapter 1. Bacteria that can grow autotrophically include many of the methanogens, *Sulfolobus* spp., *Bacillus schlegelii*, and *Clostridium thermoautotrophicum*. Some of these organisms, like the majority of thermophilic bacteria, will grow heterotrophically. A number of thermophiles, including most of the members of the genera *Bacillus, Clostridium, Thermus,* and *Thermomicrobium,* and the extremely thermophilic anaerobes, *Thermoanaerobacter, Thermoanaerobium* and *Thermobacteroides,* grow only heterotrophically (2). Heterotrophic thermophiles are capable of utilizing a variety of growth substrates, including sugars and sugar alcohols, polysaccharides such as cellulose, pectin, and starch, amino acids and peptides, organic acids and alcohols (2–4), aromatic compounds like phenols, cresols, and benzoate (5–9), hydrocarbons (10–12), carbon monoxide (13,14), and surfactants like Tween 80 (M. Kernick and T. K. Sundaram, unpubl. observations). There are also thermophiles which reduce nitrite (3), and Belay et al. (15) recently observed nitrogen fixation by *Methanococcus thermolithotrophicus* at over 60°C, a temperature higher than that at which nitrogen fixation occurs in thermophilic cyanobacteria. Not many thermophilic bacteria have been grown without a supplement of growth factors provided either as pure compounds such as amino acids and vitamins or as more complex mixtures such as yeast extract, tryptone, and similar preparations. However, some of them have proved capable of prototrophic or nearly prototrophic growth. For example, *Bacillus schlegelii*, an aerobic facultative chemolithoautotroph, and *Methanobacterium thermoautotrophicum*, an anaerobic autotroph, did not require any growth factors (5,16). A thermophilic *Bacillus* species, probably an amylase-negative variant of *B. stearothermophilus* (17), and a similar strain described by Daron (18) would grow in a defined salts medium with a single organic compound as carbon source; however, growth of the former strain was strongly stimulated by D-biotin (needed to form a functional pyruvate carboxylase) in medium containing L-lactate as carbon source (19). The extreme thermophile *B. caldotenax* could be grown in a defined salts medium with glucose as carbon source and a supplement of D-biotin and L-methionine (20). A growth requirement for methionine is quite commonly seen with thermophilic *Bacillus* spp. (2). The first strain of *Thermus aquaticus* isolated by Brock and Freeze (21) and another strain of the same bacterium (22) were reported to be prototrophic and able to grow in a synthetic medium; other *Thermus* strains, however, exhibit a requirement for several amino acids and vitamins (2,23). Some thermophiles, rather surprisingly, seemed to require an organic ingredient which was not readily replaceable. Thus the yeast extract in the *Thermoanaerobacter ethanolicus* growth medium could not be replaced by tryptone, casein hydrolysate or beef extract (24). The nature of this specific requirement remains to be clarified. *Thermoplasma acidophilum* apparently required for growth a polypeptide comprising 8 to 10 amino acids, which was present in yeast extract (25).

The growth of several thermophilic bacteria has been reported to be sensitive to organic nutrients such as yeast extract at concentrations that would normally not be regarded as unduly high. *Thermus* species are especially noted for such sensitivity (26,27). Other sensitive bacteria include *Thermomicrobium roseum* (28), *T. acidophilum* (29), and *Thermoproteus tenax* (2). There is no evidence at present suggesting that the sensitivity to organic nutrient is typical of thermophilic growth. Generally speaking, the organisms displaying such sensitivity belong to genera that do not form spores, and many of them were isolated from environments such as hot springs where the level of nutrient is relatively low and does not fluctuate. Adaptation to the hot spring environment probably explains the sensitivity of these thermophiles to organic material and also to inorganic salts: the growth of several thermophilic bacterial strains (*Thermus* is again prominent in this regard) is inhibited by sodium chloride and sometimes by phosphate (2).

From all the information available regarding the pH values at which thermophilic bacterial growth occurs (30–32), it appears that there are two distinct ranges of pH optimum: 1.5 to 4, where the thermoacidophiles grow optimally (*Thermoproteus tenax*, classed as an acidophile, has an optimum pH of 5), and 5.8 to 8.5, where the so-called neutrophiles grow optimally.

No thermophilic extreme alkalinophile is known as yet (2,3). Besides low pH, heavy metals which are usually present at relatively high concentration in acid hot waters may also constitute a deterrent to the widespread evolution of acidophiles at high temperature (33).

The thermoacidophiles known, which represent a relatively small number of genera, are rather heterogeneous in their growth habit. *B. acidocaldarius* is aerobic and heterotrophic, *Sulfolobus* can grow heterotrophically or autotrophically by oxidizing elemental sulfur or iron, *Thermoplasma* is an aerobic heterotroph, and *Thermoproteus* is anaerobic and sulfur respiring (34,35). Some of these strains are resistant to heavy metal ions at high concentration; for example, some *Sulfolobus* strains will cope with 0.3% Fe^{2+}, 0.25% Cu^{2+}, and 2,000 ppm Mo-VI (2). The use of such organisms for the extraction of metals from ores at high temperature has been suggested (see Chapter 12, this volume). The intracellular pH of acidophilic organisms has been estimated to be close to neutrality, implying the maintenance of a large gradient of 3.5 to 4.5 pH units across the cellular membrane/envelope (34,36). Low pH is essential for the cellular stability of *Thermoplasma* but not of *Sulfolobus* or *B. acidocaldarius* (34). The acidophilic nature of sulfur oxidizers like *Sulfolobus* may be understood as an evolutionary response to the generation of high acidity from the oxidation of sulfur compounds by these organisms, but this argument cannot be applied to acidophiles such as *Thermoplasma* and *B. acidocaldarius*, which do not oxidize sulfur compounds.

Oxygen is an important nutrient for the aerobic thermophiles. The solubility of oxygen in aqueous media is greatly diminished at the growth temperatures of thermophiles, and diffusion is generally the means of oxygen supply to the organism. It is therefore a difficult problem to obviate oxygen starvation or insufficiency in the growth media of thermophilic aerobes unless the cell density is low. Such a situation does prevail in hot springs, and thus it is not surprising that these environments have proved a happy hunting ground in the search for extremely thermophilic anaerobes in the recent past (24,37,38).

Metabolic aspects of thermophilic bacteria have been reviewed by Ljungdahl (4). In general, the nutritional and metabolic characteristics exhibited by thermophilic bacteria are similar to those associated with their mesophilic counterparts.

3. GROWTH CHARACTERISTICS OF THERMOPHILIC BACTERIA

3.1. Growth Rate

The data presented by Brock (39) on the maximal growth rates of several thermophilic bacteria with temperature optima in the range 55 to 70°C show that the generation times of these organisms in rich media at their respective

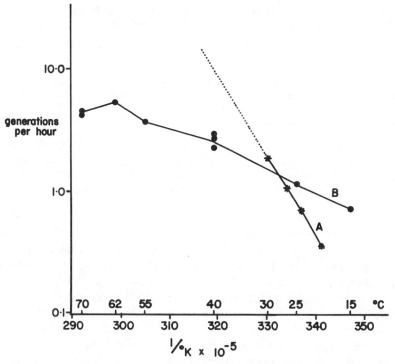

Figure 1. Arrhenius plots of growth rates of mesophilic and thermophilic bacteria. (A) Plot of the growth rates of *E. coli* at various temperatures. (B) Composite plot of the growth rates of various bacteria at their respective optimum temperatures; each point represents a separate bacterium. From Brock (39), copyright 1967 by the AAAS.

optimum temperatures fall between 11 and 16 minutes; the generation times given for the mesophilic *E. coli* and *Bacillus subtilis* are 21 and 26 minutes, respectively. Thermophiles, therefore, are capable of growing at rates that are similar to, and in many instances distinctly faster than, the growth rates of comparable mesophiles. This general conclusion is supported by the more recent and more comprehensive compilation of Sonnleitner (2). However, if the difference between the growth temperatures of the two types of bacteria is taken into account, it appears that thermophiles do not grow as efficiently as would be expected from theoretical considerations. This emerges clearly from the graphical presentation by Brock reproduced in Figure 1. The slope of the composite plot is considerably less steep than that of the plot for *E. coli*. Such Arrhenius plots for various individual mesophilic and thermophilic bacteria (below their T_{opt}) more or less resemble the plot for *E. coli*, suggesting that the temperature coefficients for the growth of all these bacteria are not dissimilar (39,40). Therefore it may be concluded that the effect of temperature on the growth rate of each individual bacterium broadly follows theoretical expectations but that, as a general rule, at a given tem-

TABLE 1. Activities of malate dehydrogenase (MDH) and isocitrate dehydrogenase (IDH) from mesophilic and thermophilic bacteria

	Specific enzyme activity	
Bacterium	MDH	IDH
Bacillus subtilis (Mesophile)	322	41
Bacillus stearothermophilus nondiastaticus (Moderate thermophile)	187	21
Bacillus caldotenax (Extreme thermophile)	83	4

MDH and IDH from the three bacteria were purified to homogeneity and their specific activities (μmol of product formed per min per mg of protein) were determined at 30°C.

From Smith et al. (43) and J. D. Edlin and T. K. Sundaram, to be published.

perature the thermophiles cannot sustain as fast a growth rate as the mesophiles. Accepting that the growth rate and the cardinal temperatures are genetically specified (39,41), one can infer that, in order to be able to grow at high temperature, thermophiles pay a price in terms of reduced growth efficiency. In a recent report, Ratkovsky et al. (42) observe that the Arrhenius-type plot (see Figure 1) for bacterial growth is not linear but curved and that a linear relationship is seen between the square root of the growth rate constant and the absolute temperature. This criticism, however, will not materially alter the conclusion drawn above regarding the efficiency of thermophilic growth.

Of the various explanations considered by Brock (39) for the slower (than predicted) growth rates of thermophiles, the lower catalytic efficiency of their enzymes is the most plausible and is best supported by experimental evidence (see also 3). In a number of instances, the activity of an enzyme from thermophilic sources has been found to be appreciably lower than that of its mesophilic counterparts; the data shown in Table 1 illustrate this point. Such reduced activity need not be an immutable property of all thermophile enzymes, but even a small number of key enzymes functioning less efficiently would account for the observed growth rates. The growth rates of thermophilic bacteria at their respective optimal temperatures cover a fairly wide range, and this includes some slow growers (2,3,39).

The accurate determination of the growth rates of thermophilic bacteria can present problems. In batch cultures, especially of extreme thermophiles, the exponential phase may be of short duration, as seen with *B. caldotenax* (20,44); one reason may be the inadequacy of the oxygen supply due to the high growth temperature and the increasing cell density. In such instances, it has been possible to obtain accurate results by the chemostat technique. Rather surprisingly, the value observed on the chemostat culture sometimes was severalfold higher than that derived from the batch culture for certain *Thermus* strains, even though the medium and culture conditions were the same for the two modes of growth (22,27). Another complication, seen mainly with *Thermus* species, was that growth in batch culture slowed down and stopped when there was no apparent substrate limitation (2,27). Incomplete utilization of substrate was reported in both batch and continuous cultures of these organisms (22,27). Our understanding of the growth characteristics of *Thermus* species is clearly unsatisfactory, as shown by the following examples: 1) these bacteria have been regarded as relatively slow growers (33) and, although the growth rates sometimes observed on batch cultures support this conclusion, those determined on continuous cultures of *T. aquaticus* and *T. thermophila* have been quite fast (doubling times of 12–15 minutes) (22,27); 2) conflicting reports have emerged from different laboratories on growth rates and growth requirements (22,27,45–47; see also 2); 3) the growth rate determined could depend on the experimental technique and the type of culture vessel used, and difficulties have been encountered in reproducing experimental results (22). Moreover, a number of bacterial strains reportedly belonging to the genus *Thermus* have been isolated, but the precise taxonomic relationships between them have not been elucidated (22). Cometta et al. (22) have also discussed the great strain variability they observed among a number of thermophilic aerobes belonging to the genera *Thermus* and *Bacillus* and suggested that it could be due to the genetic instability of extreme thermophiles. Other instances of instability of thermophilic bacterial strains are given by Sonnleitner (2).

The parameter K_s (substrate saturation constant) was estimated to be about 20 to 30 mg of glucose l^{-1} for *B. caldotenax* and a strain of *T. aquaticus* (2,22); this compares with about 5 mg of glucose l^{-1} for mesophilic bacteria. Certain other *Thermus* strains, such as *T. thermophilus*, may have much higher values (2,27).

As indicated above, the relationship of growth rate to temperature is commonly represented by an Arrhenius-type plot, and in many systems a linear plot has been presented over an appreciable range of temperature below T_{opt}. Deviations from linearity have been seen at lower temperatures (44), but some bacteria like *Thermoanaerobacter ethanolicus* and *Clostridium thermohydrosulfuricum* displayed a pronounced discontinuity and inflexion in the plot at about 60°C, only about 10°C below T_{opt} (24,48).

3.2. Growth Yield

The growth yields of mesophilic bacteria have been investigated extensively, and these studies have afforded useful insight into the physiology of these organisms (49,50). However, only a few reports have appeared on the quantitative determination of the growth yields of thermophilic bacteria related to substrate utilization.

Molar Growth Yield and Maintenance Coefficient Systematic studies of this kind have been mostly on aerobic, heterotrophic organisms, although some information has been presented about the growth yield of thermophilic anaerobes (2,38,51). Two approaches, viz. continuous culture and batch culture, have been used. A continuous culture has the advantage that it is grown under a defined, constant set of conditions and provides, by calculation, values for the true or maximum growth yield (it should be appreciated that this is a theoretical value which is not likely to be achieved in practice), for the maintenance requirement and for the death rate coefficient (52,53). However, this calculation is dependent on certain assumptions which are open to criticism, and therefore this method may produce values that are not valid (53,54). Batch cultures suffer from the disadvantage that their growth rates, which are usually the maximum possible for the conditions employed, cannot be altered at will and that the environment changes during growth as a result of metabolism; also, batch cultures will provide a value for the observed (actual) growth yield which is the resultant of the three parameters mentioned, but do not readily yield the values for those parameters. The results of some of the growth yield studies are summarized in Table 2.

The $Y_{glucose}$ value of 65 for the moderately thermophilic *B. stearothermophilus* grown in batch culture at its T_{opt} was comparable to, but somewhat lower than, the values reported for mesophilic bacteria at their T_{opt} under otherwise similar conditions; these values have been as high as 119 and as low as 48, with the majority in the range of 70 to 95 (49,57–59). The lower yield of the thermophile might be due, in part at least, to a higher maintenance requirement in comparison with mesophiles (see below). The $Y_{glucose}^{max}$ values for the four moderate and extreme thermophiles (these include the acidophile *B. acidocaldarius*) listed in Table 2, which were obtained from glucose-limited continuous cultures, were more or less similar to the values reported for mesophilic bacteria, with the value for *T. thermophilus* somewhat lower and that for *B. caldotenax* somewhat higher; for example, the $Y_{glucose}^{max}$ value for the mesophilic *Aerobacter cloacae* was observed to be 79 at 37°C by Pirt (60). There was, however, more divergence between the Y_{O2}^{max} values. The values for *T. thermophilus* and *B. acidocaldarius* were appreciably lower than those for *B. caldotenax*. For comparison, the reported values for some mesophiles were: 56 at 30°C and 28 at 35°C for *E. coli* B and 52 at 35°C for *Klebsiella aerogenes* (61,62).

The concept of maintenance energy requirement—energy postulated to be required for maintenance purposes such as turnover of cell material, cell mobility, and maintenance of solute gradients and of correct ionic composition and right pH inside the cell (53,60,63)—explains the decrease in the observed growth yields of bacteria that is found in chemostat cultures as the growth rate is reduced. Such a decrease in growth yield is seen with both mesophilic and thermophilic bacteria. It has been suggested that a high maintenance energy is an attribute of thermophiles as a consequence of their high-temperature growth habit (3,38). Both the $M_{glucose}$ and M_{O2} values for *B. caldotenax* (Table 2), which were nearly 10 times the values obtained for mesophiles (60–62), bear out this view. Curiously, however, the values for

TABLE 2. Growth parameters of thermophilic bacteria

Bacterium	T_{opt} (°C)	Culture technique	Temperature of parameter determination (°C)	$Y_{glucose}^{max}$	$Y_{glucose}$	$Y_{O_2}^{max}$	$M_{glucose}$	M_{O_2}	K_d	Reference
Bacillus stearothermophilus nondiastaticus	55–58	Batch	55	—	65	—	—	—	—	55
			45	—	86	—	—	—	—	
Bacillus sp.	62	Continuous	64.5	—	—	—	0.81	—	0.38	52
			60	76	—	—	0.57	—	0.075	
			52	74	—	—	0.23	—	0.03	
			45.5	78	—	—	0.18	—	0.01	
Bacillus caldotenax	65	Continuous	70	89	—	50	4.1(5.6)[a]	20(30)[a]	0.07	2,44
			65	89	—	50	3.8	20	0.03	
			60	79	—	47	2.2	15	—	
Thermus thermophilus	70	Continuous	78.5	54.9	—	21.7	0.38	0.95	0.013	56
			70	64	—	25.1	0.28	0.69	0.011	
			60	64.4	—	27.6	0.18	0.63	0.005	
Bacillus acidocaldarius	60 (pH 4.1)	Continuous	51 (pH 4.3)	83.5	—	28.1	0.53	2.3	0.012	54

$Y_{glucose}^{max}$, $Y_{O_2}^{max}$ = Growth yield (maximum) in g of cells · (mol of glucose utilized)$^{-1}$ or (mol of oxygen consumed)$^{-1}$

$Y_{glucose}$ = Observed growth yield in g of cells · (mol of glucose utilized)$^{-1}$

$M_{glucose}$, M_{O_2} = Maintenance energy requirement (coefficient) in mmol of glucose or oxygen · (g of cells)$^{-1}$ · h^{-1}

K_d = Specific death rate in h^{-1}. All the cultures were grown aerobically in a salts medium with glucose as the carbon source.

[a]Values within parentheses were those calculated for the viable cells in the culture.

T. thermophilus, an extreme thermophile, were extremely low—even lower than the values for mesophiles—and the $M_{glucose}$ value for the moderately thermophilic *Bacillus* sp. growing at 60°C was only about the same as for a mesophile. Other studies by Cometta et al. (22,27) yielded M_{O_2} values of 10 to 18 for *T. thermophilus* and *T. aquaticus* grown at their T_{opt} in a complex medium and led to the suggestion that high maintenance requirements are a basic characteristic of thermophiles independent of their ability to form spores. Sonnleitner (2) comments that maintenance requirements in a minimal salts medium, which McKay et al. (56) used in their experiments with *T. thermophilus*, would, if anything, be higher than in a complex medium.

Effect of Temperature on Growth Yield   The $Y_{glucose}$ value for *B. stearothermophilus* in batch culture progressively decreased with increasing growth temperature (55, and Table 2). At the higher temperature, it appeared that a greater proportion of the substrate was left incompletely utilized, mostly as acetate, and that therefore there was a less efficient coordination of the nonoxidative and oxidative phases of glucose metabolism. Acetate accumulation in the medium during growth on glucose was also seen with mesophilic bacteria (57,64); this acetate was subsequently oxidized but not used for cell growth. The $Y_{glucose}$ of the mesophiles *E. coli* and *Aerobacter aerogenes* in batch culture remained virtually constant from 22°C to the T_{opt} (58,59). With *A. aerogenes*, the growth yield dropped steeply above 37°C, and this was attributed to an energy uncoupling; with *E. coli*, a sharp decrease in the yield was observed below 18°C, and in this temperature range there was also a deviation of the growth rate from the Arrhenius relationship.

The observed growth yield ($Y_{glucose}$), represented by the steady-state concentration of bacterial cells in a substrate-limited continuous culture, decreased, at a given growth rate (dilution rate), with increasing growth temperature for the moderately thermophilic *Bacillus* sp. studied by Matsche and Andrews (52). This decrease in yield was presumably because of a higher maintenance requirement and possibly a higher cell death rate at the higher temperatures (Matsche, pers. comm.). It is worth noting that the effect of temperature on the observed yield in such continuous cultures cannot be equated with the temperature effect on yield in batch cultures reported by Coultate and Sundaram (55). In the continuous system, the growth (dilution) rate was the same at the various temperatures—and at each temperature the observed yield would increase with the growth (dilution) rate (see above)—whereas in the batch system the bacterial cells grew at different rates (the respective maximal rates) at the different temperatures. In the latter system, the higher maintenance requirement (and higher death rate if this was significant) at the higher temperature is likely to be counterbalanced wholly or partly by the faster growth rate at that temperature, as indicated by the equation (60), $1/Y = 1/Y^{max} + M/\mu$, where Y is the observed growth yield, Y^{max} the maximum growth yield, M the maintenance coefficient, and μ the specific growth rate. Moreover, it is unlikely that incomplete utilization of glucose resulting in acetate accumulation, which seemed to be a feature of growth in the batch culture, would be a significant factor in the substrate-limited continuous system, in view of the evidence (54,65) that there was little overflow metabolism during the growth of the mesophilic *K. aerogenes* and the thermoacidophilic *B. acidocaldarius* in glucose-limited continuous culture.

The effect of temperature on the maximum growth yield ($Y_{glucose}^{max}$ and Y_{O2}^{max}) has been studied with the following results. As outlined in Table 2, the yield related to glucose consumption remained more or less constant at temperatures between 45 and 60°C for the *Bacillus* sp. examined by Matsche and Andrews (52). For *B. caldotenax*, both $Y_{glucose}^{max}$ and Y_{O2}^{max} increased with temperature in the range 50 to 70°C, conforming to an Arrhenius-type relationship (44). For *T. thermophilus*, $Y_{glucose}^{max}$ at 60 and 70°C was essentially identical and at 78.5°C diminished appreciably; the maximum yield based on oxygen consumption, however, registered a slow, steady drop with increasing temperature. For *E. coli* B, Y_{O2}^{max} was reported to be constant between 17 and 32°C and then to decline sharply to half this value at 37°C (61). There is thus some contradiction between these conclusions. The maintenance coefficient and the specific death rate values went up with temperature for all three thermophiles (Table 2), but the values for *T. thermophilus* were appreciably lower than for the other two bacteria. It is not clear which of the maintenance functions are affected by temperature (see 44). It appears that the death rate is significant only at high temperatures, and for mesophiles the death factor during growth is usually ignored. Rather surprisingly, the death rate for the *Bacillus* sp. (T_{opt} 62°C) was extremely high (0.38) at 64.5°C.

Some of the growth parameters for *Thermothrix thiopara*, growing autotrophically in continuous culture under thiosulfate (source of energy and reducing power) limitation, are now available. Contrary to the trends observed with the other thermophiles just discussed, both Y^{max} and the specific maintenance rate (maintenance coefficient × Y^{max}) decreased with increasing temperature in the range 65 to 75°C (T_{opt} 72°C), and the growth yield was greater than the yields reported for mesophiles grown under similar nutritional conditions (66). Another thermophilic bacterium, a *Thermoactinomyces* sp., grown continuously in a salts-glucose medium containing yeast extract, was found to have a $Y_{glucose}^{max}$ of 76 and a maintenance coefficient of 0.13 at 55°C (67).

If a reduced observed growth yield and a high maintenance requirement turn out to be a feature of the growth of thermophiles at elevated temperature, then these organisms, especially the anaerobes, may accumulate, in larger quantity than their mesophilic counterparts do, catabolic products, some of which may be useful. This could invest thermophiles with a useful potential for industrial exploitation, as suggested by Zeikus (3).

3.3.   Minimum Growth Temperature (T_{min})

All bacteria have characteristic T_{min} and T_{max}, but for thermophiles these cardinal temperatures are considerably higher than for mesophiles. The question of what factors determine these temperatures is a general one applicable to all bacteria. The observation that heat-sensitive mutations can lower the T_{max} but not the T_{min} and cold-sensitive mutations can raise the T_{min} without altering the T_{max} adds support to the notion that these two limiting temperatures are specified by independent genetic determinants (41), and that changes in single proteins can alter T_{min} or T_{max}.

Cell Membrane and T_{min}   The structure of the plasma membrane of thermophilic bacteria, like that of biological membranes generally, is considered to conform to the model (68) of a lipid bilayer matrix, with the peripheral proteins loosely attached to the membrane surface and the integral proteins embedded in the membrane structure. Apparent exceptions are the membranes of the archaebacterial thermophiles (69,70), in which the lipid may be in the form of a covalently cross-linked bilayer structurally equivalent to an amphiphilic monolayer. A more detailed discussion of the membranes and lipids of thermophiles will be found in ref. 4 and Chapter 5, but some general features of membrane structure relevant to an examination of T_{min} will be given here. The cellular lipids in thermophiles tend to have a greater abundance of fatty acids with high melting points than do the lipids of mesophiles. Bacteria adjust the fatty acid composition of their lipids depending on the growth temperature, an increase in this temperature resulting in a greater proportion of fatty acids with higher melting points (71). The general relationship between the structure of a fatty acid and its melting point is briefly as follows (71). Saturated fatty acids with a straight chain structure have, as a class, the highest melting points, and an elongation of the carbon chain causes an elevation of the melting point. Isobranched fatty acids, which carry a methyl group on the carbon atom adjacent to the terminal methyl group, have melting points only slightly lower than their straight chain analogues. Anteisobranched fatty acids, in which the methyl group is more internal, have considerably lower melting points. Fatty acids with the lowest melting points are those with *cis*-double bonds near the center of the hydrocarbon chains.

   A significant property of the phospholipid bilayer of the bacterial membrane, which does not contain sterols, is the thermotropic, reversible transition it can undergo between an ordered, rigid gel or solid phase and a more fluid liquid-crystalline phase. The change from gel to liquid-crystalline phase involves the cooperative melting of the hydrocarbon chains in the interior of the bilayer. Therefore the temperature at which this transition takes place will be strongly determined by the nature of the fatty acids contained in the membrane lipid (71)—the nature of the polar group of the phospholipid also has an effect (72)—and it will be higher for membranes containing fatty acids which have higher melting points. Owing to the heterogeneity in the fatty acid and polar head group composition of biological membranes, the phase transition, rather than being sharp at a discrete temperature, occurs over a temperature range, in which domains of the gel and liquid-crystalline phases exist simultaneously; at temperatures below the lower boundary of the phase transition the membrane lipid is completely in the gel state and at temperatures above the upper boundary of the phase transition it is all in the liquid-crystalline state.

Although protein is a major component of the bacterial membrane (73), the evidence is conflicting as to whether protein has a bearing on the phase transition temperature. Working with a model lipid system, Papahadjopoulos et al. (74) found that proteins like cytochrome c bound to phospholipids, inducing a drastic decrease in the enthalpy and the temperature of the lipid phase transition. However, McElhaney and Souza (75) showed that the spheroplast membrane of B. *stearothermophilus* and an aqueous dispersion of the extracted total membrane lipid had very similar transition characteristics (upper and lower boundaries and midpoint of the phase transition and enthalpy of transition). They concluded that the bulk of the lipids in the membrane participated in the phase transition and that the membrane proteins did not significantly alter the thermal properties of the lipids. Similar results were obtained with the isolated membrane and the total membrane lipid from *Acholeplasma laidlawii* (71). Halverson et al. (see 4) observed, by freeze-fracture techniques, that membrane proteins in B. *stearothermophilus* spheroplasts were randomly dispersed at temperatures above the upper boundary of the phase transition but formed networklike patterns at lower temperatures; at temperatures below the lower boundary of the phase transition, large areas of the membrane were almost protein-free. This suggests a temperature-dependent interaction between lipid and protein in the membrane.

Besides an alteration in the fatty acid composition of the membrane (see above), changes in the membrane content of the cell and in the contents of lipid and of protein in the membrane have been reported with variations of the growth temperature of thermophiles (76). Thus, an increase in the membrane content of the cell, together with an increase in the protein content and a decrease in the lipid content of the membrane, was observed in the moderately thermophilic B. *stearothermophilus* as the growth temperature was raised (77,78). Rather different results were obtained with the extreme thermophile, T. *aquaticus*: an increase in growth temperature from 50 to 70°C was accompanied by a progressive increase in the contents of total lipid, phospholipid, glucolipid and carotenoid (79). The significance of the divergent findings in the two thermophilic systems is not clear.

The idea has gained support, based on work with the mesophilic prokaryotes, E. *coli* and the cell wall-less *Acholeplasma laidlawii*, that the liquid-crystalline state of the membrane lipid is an important condition for the occurrence of microbial growth and respiration and for the functioning of membrane transport and of membrane-associated enzymes (80–84). This can be construed to imply that the T_{min} and the T_{max} might be set by the phase transition behavior of the plasma membrane of the organism. This possibility was tested by McElhaney in the *A. laidlawii* system (71,81).

The situation in the *B. stearothermophilus* system is quite similar. This conclusion emerged from a detailed study of a wild type strain, YTG-2, of the thermophile and a temperature-sensitive mutant, TS-13, derived from it (75,85). T_{min}, T_{opt}, and T_{max} for strain YTG-2 were 37, 65, and 72°C, respectively, and for strain TS-13 were 37, 58, and 58°C, respectively. The mutant apparently was defective in lipid metabolism at high temperature. Consequently it was severely impaired in ability to adjust the fatty acid composition of its membrane lipid to the growth temperature in the normal way and so had a virtually constant membrane lipid phase transition temperature range of 20 to 40°C at all growth temperatures; it was also unable to maintain membrane integrity at high temperature and lysed at a temperature of 60°C or higher. For both strains, the T_{min} of 37°C was close to the upper boundary, and well above the lower boundary, of the lipid phase transition, strongly suggesting that this cardinal temperature and the lower end of the phase transition were unrelated. A further important observation was that the membrane lipid in strains YTG-2 and TS-13 was completely or predominantly in the liquid-crystalline state at all the growth temperatures (42–65°C for the wild type and 42–58°C for the mutant) used in the study.

Other Determinants of T_{min}   A number of factors which could determine the T_{min} value were considered by Babel et al. (86 and 4), and several of them, including accumulation of toxic metabolic products, enzyme repression, impairment of nutrient transport through the cell membrane, and defect in energy supply were ruled out by experimental evidence. Impaired membrane transport has also been suggested by other workers as the reason for cessation of growth below T_{min}. The evidence against this idea presented by Babel et al. was that respiration occurred at temperatures well below T_{min} in a strain of *B. stearothermophilus*. A similar result was obtained by Coultate and Sundaram (55) with another strain of this thermophile. Babel et al. favored the hypothesis that a change in the conformation of one or more essential enzymes, leading to loss of catalytic activity or alteration of an important regulatory characteristic, could arrest growth below T_{min}.

There is now a considerable body of evidence to corroborate the above hypothesis. As mentioned in Section 3.1, a number of enzymes from thermophiles are catalytically less efficient than analogous mesophile enzymes, and an inverse relationship between catalytic potential and the degree of thermophilicity of the source organism has been frequently observed (see for example Table 1). Ljungdahl and Sherod (87) have suggested that thermophile enzymes sacrificed part of their catalytic activity in order to acquire thermostability. Amelunxen and Murdock (40), however, prefer the theory that the lower activity may be the consequence of an adaptive mechanism to regulate metabolism and slow down the growth rate (see Section 3.1), in order to ensure the efficient utilization of available nutrient resources in the natural environment. Whatever the evolutionary implication here, allied to the low activity is the fact that a number of thermophile enzymes exhibit Arrhenius plots with two parts having different slopes. In most of these systems, this means a higher activation energy at temperatures below the break and hence a further drastic reduction in enzyme activity at these temperatures (87). A nonlinear Arrhenius plot is most commonly interpreted as indicating a change in the conformation of the enzyme molecule (88). Such plots are not peculiar to thermophile enzymes, but the significant feature in these enzyme systems is that the breaks in the plots occur at much higher temperatures than in the plots for mesophile enzymes. Moreover, enzyme reactions in vivo proceed mostly at subsaturating substrate concentrations, and therefore the effect of temperature on K_m may become crucial in determining the reaction rate in the growing cell at low temperatures. Although temperature effects on the K_m values do not follow a fixed trend, there are instances where K_m becomes less favorable as the reaction temperature is lowered (89). Such an effect would also reduce the activity of an enzyme under in vivo conditions at low temperatures. The study of some cold-sensitive mutants of mesophiles like *E. coli* has revealed that the mutant phenotype of an elevated T_{min} can be traced to an enormous enhancement in the sensitivity of an allosteric enzyme to feedback inhibition and a further accentuation of the sensitivity at lower temperatures (90). Conversely, the allosteric regulation of an enzyme could be abolished at lower temperatures. Several enzymes of this kind from thermophilic bacteria have been described (87), and the loss of their regulatory property may exert a deleterious effect on growth at the lower temperatures. All these observations constitute compelling reasons to believe that the relatively high T_{min} of a thermophile may be set by the loss of activity of one or more vital enzymes, resulting from an intrinsically low catalytic potential, which is further attenuated by a strongly unfavorable effect of low temperature, presumably caused by conformational changes in the enzymes, on this potential and perhaps also on kinetic and regulatory parameters.

A further possible determinant of the T_{min} of bacteria was identified by the study of another major class of cold-sensitive mutation in *E. coli* and *Salmonella typhimurium*, that raised the T_{min} (91,92). These mutations were in genes encoding ribosomal proteins and prevented ribosome synthesis at the lower temperatures. Ribosomal assembly in vitro requires incubation at a relatively high temperature in *E. coli* and an even higher temperature in *B. stearothermophilus* and has a large activation energy requirement (93,94). A large activation energy barrier is likely to be a feature of ribosome assembly in vivo as well. Thus, it is plausible that in the normal bacterial cell the formation of the ribosome from its components may become a growth-restricting step at temperatures below T_{min}.

To summarize, membrane lipid phase transition may be regarded as defining the ultimate lower temperature limit for growth in so far as a membrane with its lipid completely in the gel state will not function. But normally during the growth of bacteria, including thermophiles like *B. stearothermophilus, B. caldotenax*, and *T. aquaticus*, the membrane lipid is predominantly or wholly in the liquid-crystalline state (see below). This would rule out a direct relationship between T_{min} and membrane phase transition. If a homeoviscous adaptation mechanism operates, as it appears to do in *E. coli* and *B. stearothermophilus* (see Section 3.5), it will reinforce this conclusion. Current evidence strongly suggests that the low activity of key enzymes and the breakdown of processes like ribosome assembly will be growth restricting at sub-T_{min} temperatures. With so many potential determinants of T_{min}, convincing experimental evidence will be technically difficult to obtain.

3.4. Maximum Growth Temperature (T_{max})

Cell Membrane and T_{max} The observed increase in the melting point of membrane lipid in response to a rise in growth temperature (see Section 3.3) prompted Heilbrunn and Bělehrádek (see 71) to suggest that the melting of the lipid might be important in determining the T_{max}. However, the possibility that membrane lipid phase transition may directly define T_{max} can be discounted for the following reasons: a) the T_{max} of *A. laidlawii* remained constant at 44°C as the membrane fatty acid composition was manipulated with variation of the phase transition range in the region, 0–15°C to 25–53°C (71); b) as seen in Section 3.3, *B. stearothermophilus* strain TS-13 grew well at temperatures 42–58°C while the upper end of the phase transition remained unchanged at 40–41°C, and strain YTG-2 had its membrane lipid predominantly in the fluid state during growth at temperatures 42–65°C; c) another *B. stearothermophilus* strain maintained its membrane completely in a fluid state while growing at temperatures 45–65°C (95), and a similar inference was drawn for *B. caldotenax* by Kawada and Nosoh (96); d) the membrane lipid of *T. aquaticus* was wholly in the fluid state 40°C below the growth temperature (97).

Stability of the cellular membrane could be a determinant of the upper temperature limit for growth of organisms. Kuhn et al. (44) attributed the drop in the growth rate of *B. caldotenax* above the T_{opt} to an acceleration in cell death, a process which has been thought to be caused by damage to the cell membrane (39). The growth behavior of *B. stearothermophilus* mutant TS-13 is also consistent with a role for the cell membrane in fixing the T_{max}: this strain had its T_{opt} and T_{max} at 58°C, underwent lysis at 60°C, and had a membrane unstable at high temperature (see Section 3.3). Experimental evidence showing the greater stability of the thermophilic bacterial cell membrane as compared with the membrane of mesophilic bacteria was provided by Welker et al. (77,78) and Golovacheva et al. (98). Membrane stability was enhanced as the growth temperature of the thermophile was raised.

It is now established (see above) that *B. stearothermophilus* (moderate thermophile), *T. aquaticus* (extreme thermophile), and apparently also *B. caldotenax* (extreme thermophile) normally grow with their membrane lipids predominantly or wholly in the fluid form. This is true of other bacteria (95,99) including *E. coli*, which, although it can grow with about 50%, but not more, of its membrane lipid in the gel state, generally grows with the lipid completely in the liquid-crystalline form. There are some reports which are not compatible with this general conclusion regarding the physical state of the membrane at the temperatures of growth. Esser and Souza (100) proposed from a study of *B. stearothermophilus* strain YTG-2 and the mutant TS-13 (see Section 3.3) that the lipid phase transition boundaries set T_{min} and T_{max} and that for membrane function and assembly the simultaneous presence of solid and fluid lipid domains is essential. In a later investigation McElhaney and Souza (75) disposed of this discrepancy as a misinterpretation of data. More recently Janoff et al. (see 99) suggested that adaptive changes that maintain the lipids in a mixed gel-fluid state in the outer membrane, but not in the cytoplasmic membrane, of *E. coli* determine the growth temperature range of this bacterium. This conclusion, too, according to McElhaney (99), rested on an erroneous interpretation.

Accepting, then, the consensus view that bacteria normally have their membrane lipid predominantly in the liquid-crystalline state during growth, one can consider the question of whether there is an upper limit to the fluidity or disorder for the functioning of the membrane and cell growth. Several lines of evidence suggest that such a limit exists, although there is a fairly high degree of latitude with respect to the level of fluidity that is compatible with proper membrane function. Thus *E. coli* (80) and *A. laidlawii* (71) had their T_{opt} and T_{max} lowered when the membrane lipid transition temperature was drastically brought down through an incorporation of low-melting fatty acids, indicating that there is a limit to the gap which can exist between the transition temperature and these cardinal growth temperatures. Mutants of these same organisms which were unable to synthesize any fatty acid could not utilize for growth certain fatty acid supplements that presumably would have depressed the phase transition temperature too far and caused the membrane lipid to become too fluid at the growth temperature (99). The mutant TS-13 of *B. stearothermophilus* (see Section 3.3) had a nearly invariant phase transition temperature range and an upper boundary of 40 to 41°C throughout its growth temperature range, apparently because of its inability to adjust the fatty acid composition of the membrane lipid to the growth temperature. Above the T_{max} of 58°C, which was also the T_{opt}, it abruptly lost the ability to grow, and its membrane became unstable. This can be explained by the idea that at 58°C the upper limit to the degree of membrane fluidity was reached. By contrast, the wild type strain could raise the upper boundary of the phase transition to beyond

65°C at the higher growth temperatures, and its T_{max} of 72°C accordingly was considerably higher than that of the mutant. This upper limit to the membrane fluidity does not seem to have been defined in quantitative terms, however (99).

It is possible to envisage several ways in which excessive membrane fluidity could be detrimental to cell function and growth: a) the increased thermal motion of the fatty acyl chains may make the membrane leaky and unable to retain efficiently small molecules in the cell; b) the lipid bilayer structure may become unstable causing the breakdown of the permeability barrier and a consequent loss of macromolecules and small molecules; c) some of the membrane proteins which require lipid for function and/or stability may not operate properly if the membrane is too fluid (71). Thus, although the phase transition of the membrane lipid may not directly fix the T_{max}, the stability of the membrane, as stated earlier, could be a strong element in the determinant of the T_{max}, with the level of fluidity of the membrane perhaps being an indicator of the stability.

Other Determinants of T_{max}   The stability of macromolecules, especially of proteins including enzymes, is another very plausible factor in determining the T_{max}. A number of enzymes in thermophilic bacteria undergo rapid inactivation at temperatures just above the T_{max} (4). Cessation of growth above T_{max} through such inactivation of vital enzymes was envisaged by Babel et al. (86).

Friedman and Read (101, and Friedman, pers. comm.) recently described *B. stearothermophilus* mutants which carried double antibiotic-resistant markers and were retarded in growth at 65°C but not at 50°C. Preliminary analysis indicated that the mutations adversely affected the protein synthetic function, but not the stability, of the ribosome at the higher temperature. A possible interpretation is that a temperature-dependent regulation of ribosome function operated in the mutants. It might be suggested, on this basis, that in the wild type bacterium the ribosome function could be impaired at temperatures above T_{max}.

T_{max} of thermophiles, therefore, can be affected by several factors, such as membrane instability, protein instability, and possibly loss of ribosome function.

3.5. Thermal Regulation of Membrane Lipid Composition and Homeoviscous Adaptation

As already seen, many bacteria modify the fatty acid composition of their membrane lipid in response to changes in the environmental temperature, and this causes a shift in the phase transition temperature of the lipid. Another manifestation of this temperature-dependent regulation of the fatty acid composition is a phenomenon first described by Sinensky (102), who observed that in wild type *E. coli* the viscosity (fluidity) of the membrane lipid remained relatively constant irrespective of the growth temperature, as shown by the same amount of motional restriction experienced by spin labels at the various growth temperatures. Sinensky termed this phenomenon homeoviscous adaptation. This adaptation mechanism was also seen in B. stearothermophilus (75), and significantly the membrane viscosities in the two organisms were apparently very similar (103). It would be extremely interesting to establish how widely homeoviscous adaptation operates in bacteria belonging to the psychrophilic, mesophilic and thermophilic classes. The molecular mechanism of the thermal regulation of the fatty acid composition of the membrane lipid, which has been investigated in bacteria like E. coli, Bacillus, and Micrococcus cryophilus, is not fully understood but involves the temperature-dependent control of the activity, synthesis, and turnover of enzymes involved in the synthesis of fatty acids (95,104–106).

One can perceive several facets to this thermal regulation of the fatty acid composition of the microbial membrane lipid and consider their implications. The facility to shift the membrane lipid phase transition temperature will enable an organism to grow over a wide temperature range, in view of the notion discussed already that neither a completely solid nor an excessively fluid state of the lipid is conducive to efficient membrane function. The narrower growth temperature range of the B. stearothermophilus mutant TS-13 (in comparison with the wild-type strain) seen earlier supports this view. Beside this short-term advantage can be set another benefit, viz. the evolution, over the long term, of bacteria possessing adequately functional membranes for life and growth at temperatures ranging from the very cold to the very hot (71). The maintenance of a constant degree of fluidity (homeoviscous adaptation), is not, a priori, a vital requisite for growth, as is evident from the finding that, in organisms like A. laidlawii and E. coli, the membrane fluidity apparently could be varied by the forced incorporation of specific fatty acids without a significant effect on growth (81,95,99,107). It was obvious, at least in some of these systems, that the fluidity was not held constant at the level normally prevalent in the wild-type organisms. However, this adaptation mechanism, which operates during the normal growth of the wild-type organism (e.g., E. coli), can be regarded as a finer facet of the regulation of the fatty acid composition, which ensures that the membrane lipid is maintained between the two deleterious extremes of the wholly gel state and overfluidity. Another facet to consider is the level at which the fluidity is kept constant. This level is apparently similar in such diverse organisms as E. coli and B. stearothermophilus (103). It was estimated (72) that the normal degree of fluidity in E. coli is higher than required for the immediate purposes of growth. A possible explanation, suggested by Cronan and Gelman (72), is that the extra fluidity would enable the organism to adapt to sudden drops in the environmental temperature which might occur in nature and which, if the level of fluidity were lower, might convert the membrane lipid to the gel state.

3.6. Turnover of Macromolecules During Thermophilic Bacterial Growth

An early theory of thermophily (108) suggested that thermophilic micro-organisms are able to grow at high temperatures because they can replace by active metabolism cell material destroyed at those temperatures. On this basis, a high turnover of protein, one of the most heat-labile components in the cell, might be expected during thermophilic growth. The work of Bubela and Holdsworth (109) lent support to this theory by recording a fast turnover (30 to 60 times per hour) for protein in *B. stearothermophilus* at 63°C.

These results were contradicted by the investigation of Epstein and Grossowicz (110), who observed that there was little protein turnover during the growth of *B. stearothermophilus nondiastaticus* (T_{opt} 55–58°C) at 45, 55, and 65°C in a salts-glucose medium. Protein turnover during the growth of mesophilic bacteria is known to be very low (see 111). In this same moderately thermophilic *B. stearothermophilus nondiastaticus*, protein turnover was seen to be 2.5 to 4.5% per hour during growth in a salts medium with glucose or succinate as carbon source by Coultate et al. (111). In both these investigations, protein in growing cells of the thermophile was labelled with ^{14}C-amino acid. The loss of radioactivity from these cells was then monitored during their subsequent growth in medium containing ^{12}C-amino acid. In their experiments, Bubela and Holdsworth labelled their bacterial cells with ^{14}C-amino acids for 5 minutes under apparently nongrowing conditions and then followed the loss of radioactivity from the cells during subsequent incubation with excess ^{12}C-amino acids for 2 minutes. It is highly unlikely, therefore, that their results indicated protein turnover during growth (see 111). Coultate et al. also investigated the rate of denaturation of a single, specific enzyme during the growth of their thermophile. The experiments were done as follows: The enzyme, isocitrate lyase, was first induced by growth of the bacterium on acetate as carbon source, and the cells were then transferred to, and allowed to grow in, a medium containing glucose as carbon source. The synthesis of isocitrate lyase was repressed during growth on glucose, and it was therefore possible to examine the stability of the preinduced enzyme in growing cells under physiological conditions. During a period of active growth of almost 2 hours on glucose only a small amount of the enzyme was inactivated. More recently, Kenkel and Trela (112) showed that neither the bulk protein nor the enzyme, alkaline phosphatase, was degraded to a significant extent in growing cells of the extreme thermophile, *T. aquaticus*. It may be concluded, therefore, that, if there is protein turnover during thermophilic bacterial growth, it is moderate and not massive. The study by Coultate et al. further established that the turnover of RNA (not including messenger RNA) was low (about 1% per hour) in cells of *B. stearothermophilus nondiastaticus* growing in a salts-glucose medium. There was some indication from this work that protein and RNA turnover might be somewhat higher during more rapid growth in a nutrient broth medium.

A massive turnover of macromolecules such as protein and RNA would be reflected in a drastically reduced growth yield. As seen in Section 3.2, the molar growth yield of thermophilic bacteria is not very different from that of mesophiles. This provides additional evidence that high turnover of macromolecules is not a feature of thermophilic bacterial growth.

4. THERMOSTABILITY OF CELL COMPONENTS

The low protein turnover rate during thermophilic growth suggests that this macromolecular species, which is one of the most thermolabile components of cells, and other cell constituents must be stable at the high growth temperatures of thermophiles. This idea is indeed generally valid. Thermostability will only be briefly discussed here; further information on this topic, including the molecular mechanisms of thermostabilization, is presented in a number of publications (4,34,40,69,70,76,87,113–121).

4.1. Cell Membrane

One of the consequences of thermophily is the incorporation of fatty acids of higher melting points into lipid, a major membrane component. This and other aspects relating to membrane function in thermophilic bacteria have already been discussed (see Sections 3.3, 3.4, and 3.5). Moreover, it is worth noting that, unlike proteins, lipids do not undergo irreversible thermal inactivation. Membrane proteins, another important membrane component, are more stable, like other thermophile proteins, than their homologues from mesophiles, as exemplified by the ATPases from *B. stearothermophilus* and another thermophilic *Bacillus* (122–124). Some of these proteins could also be stabilized by association with the membrane (71,76). An exceptionally high carbohydrate content of lipids was observed by Pask-Hughes and Shaw (125) in several *Thermus* strains and was suggested to be a possible basis of thermophily (see also 116). In an earlier study, Ray et al. (79) reported an increase in the total lipid content, and most strikingly in the glycolipid content, as the growth temperature of *T. aquaticus* was raised.

Thermoacidophiles have to contend with not only high temperature but also the stress of high acidity. The membranes of these organisms, like the membranes of other thermophiles mentioned above, contain a high proportion of carbohydrate-containing lipids, and indeed it appears that, in thermophiles generally, both eubacterial and archaebacterial, these lipids form the major lipid class (115,116). The membranes and their components in thermoacidophiles and the mechanisms of thermoacidophily are discussed in several articles (4,34,69,115,116); see also Chapter 5, this volume, for a discussion of the membranes and lipids of thermophiles.

4.2. Nucleic Acids and Ribosome

DNA from even mesophilic sources is endowed with high thermostability, especially in an ionic milieu similar to that prevailing in vivo, and therefore its stabilization is unlikely to be a serious problem for thermophilic growth. It was reported by Stenesh (117) that the average G-C content of the DNA of mesophiles was 44.9 mol percent and that of the DNA of thermophiles 53.2 mol percent and that there was a correlation between this parameter and the T_{max} of the source organism. The DNA with the higher G-C content, as expected, had a higher melting temperature. This indicates that DNA in thermophiles may be further stabilized by extra G-C base pairs. Although a high G-C content of DNA may have been seen in some or many thermophiles, the sum total of the evidence currently available is that it is not an invariant characteristic of these organisms.

Transfer RNA (tRNA) is also normally likely to have adequate stability, but in some thermophiles it can be further stabilized by structural modifications. For example, *T. thermophilus* tRNA had a higher G-C content than *E. coli* tRNA, and additionally ribothymidylate, which is normally present in tRNAs, was replaced by a thiolated derivative, 5-methyl-2-thiouridylate, in the former tRNA. This substitution would enhance the stability of the tRNA by strengthening the base stacking interaction in the helical conformation of the RNA (126). Interestingly, the level of this substitution, as well as the melting temperature of the tRNA, increased with an elevation of the growth temperature of the thermophile.

Stenesh and Holazo (see 117) determined the average G-C contents of ribosomal RNA (rRNA) from mesophilic and thermophilic *Bacillus* strains to be 55.1 mol percent and 59.8 mol percent, respectively, and the greater G-C content was reflected in a higher melting temperature. The rRNA of *T. aquaticus* had a higher G-C content than the homologous RNA species from *E. coli*, and the 23S RNA, but not the 16S RNA, of the thermophile had a higher melting temperature than the analogous mesophile RNA (127). A study of the 5S RNA species from the thermophiles *B. stearothermophilus* and *T. aquaticus* and from some mesophilic bacteria led to the conclusion that eighteen nucleotide positions were unique to *B. stearothermophilus* as compared with the mesophiles and that ten were virtually unique to the two thermophiles (128,129). It was pointed out that these nucleotides might be involved in the adaptation of the thermophiles to growth at high temperature, possibly through an effect on the RNA structure or on its interaction with proteins in the ribosome. According to a paper by Douthwaite and Garrett (130), the presence of four bases, instead of three, in one of the loops in the secondary structure of 5S RNA is a common feature of thermophilic bacteria.

From the standpoint of thermophilic growth, the stability of the ribosome is more crucial than the stability of the free RNA species. Studies from several laboratories have provided evidence that thermophile ribosomes are more stable than mesophile ribosomes by the criterion of melting temperature and/or functional activity in protein synthesis (117,118,127,131). Especially with *B. stearothermophilus*, the ribosomes had a higher melting temperature than the free rRNA, suggesting that the RNA-protein interaction in the ribosomal particle made a significant contribution to stability. In line with this conclusion, it emerged from a study by Irwin et al. of psychrophilic, mesophilic, and thermophilic *Clostridium* species that, whereas the free RNA from the three bacterial classes had about the same thermal stability, the ribosomes from the thermophilic strains were more stable than the ribosomes from the psychrophilic and mesophilic strains (132). Nomura et al. (133) examined the thermostability of hybrid 30S ribosomal particles reconstituted from the RNA of *B. stearothermophilus* and the ribosomal proteins of *E. coli* and vice versa. Their results showed that the thermostability of the thermophile ribosomes was determined by both the RNA and the proteins; it was inferred that the proteins had special properties related to thermostability and that the RNA was required for thermostability by virtue of its special properties perhaps arising from its higher G-C content and/or of its specific interaction with the proteins.

There have been some indications that polyamines might stabilize the thermophile ribosome in vivo and thus facilitate its functioning in protein synthesis at high temperature. For instance, spermidine, which is widely distributed in living cells and is associated with the ribosomes from *B. stearothermophilus*, had a positive effect on the association of the ribosomal subunits from this thermophile in vitro, particularly at high temperature and low Mg^{2+} concentration (134). Thermine, a polyamine present in *Thermus* strains, appreciably enhanced protein synthesis at high temperature in an invitro system derived from *T. thermophilus* by promoting the formation of an active complex of messenger RNA, ribosome, and aminoacyl-tRNA (126) (see Chapter 6).

4.3. Proteins

Intrinsic Thermostability and Stabilization by Cell Components   A large number of proteins, many of them enzymes, have now been isolated in pure state from thermophilic sources, and generally they have turned out to be more thermostable than homologous proteins from mesophiles. This establishes that the functioning of proteins at the high growth temperatures of thermophiles is not merely dependent on the presence of transferable stabilizing factors in the thermophilic cell and that the intrinsic structural stability of the proteins is an important basis of thermophily. The observed inability of cell-free extracts of thermophiles to protect mesophile enzymes against thermal denaturation supports this conclusion (see also 40).

Inherent structural stability alone may not be sufficient to ensure the survival and functioning of proteins in the thermophilic cell. This view rests mainly on the relative heat-lability of several thermophile enzymes revealed by in vitro experiments (see 40). It must be borne in mind, however, that the stability of an enzyme observed in vitro need not necessarily be an accurate reflection of its stability in vivo because the experimental conditions in vitro may be very different from in vivo conditions, particularly with respect to the concentration of the protein in question and the nature and concentration of the other components in the milieu. Nevertheless, thermophile proteins display a rather broad spectrum of stability under more or less comparable conditions in vitro, suggesting that they are not all equally stable. Moreover, there are many instances of enzymes from thermophiles as well as from mesophiles being stabilized in vitro by supplements such as substrates, metal ions, allosteric effectors, and electrolytes, all of which will be present in the growing bacterial cell. The work on pyruvate carboxylase from *B. stearothermophilus nondiastaticus* (T_{opt} 55–58°C) illustrates some of these points. This carboxylase (molecular weight approximately 0.5 × 10⁶) is one of the few large enzymes from thermophiles to have been characterized in detail, and it is allosterically activated by acetyl coenzyme A (135,136). In a study of its thermostability in vitro, 71 percent enzyme inactivation occurred in 30 minutes at 55°C in phosphate buffer, pH 7 (136). Its thermostability in vivo was estimated as follows (111). The growth of the thermophile cells in exponential phase was arrested with chloramphenicol, and during continued incubation at 55°C, culture samples were withdrawn and the enzyme activity was determined in extracts of the harvested cells. The protein turnover of 1 percent per hour in cells after the chloramphenicol addition was similar to the turnover of 2.5 to 4.3 percent per hour determined in growing cells. The thermostability of pyruvate carboxylase in the chloramphenicol-treated cells was therefore assumed to be a fair representation of the enzyme stability in vivo during bacterial growth. The half-life of the enzyme in the antibiotic-treated cells was approximately 4 hours, which denotes a considerably greater stability in vivo than that observed in vitro. The stability in vitro was significantly enhanced by individual additions of the components of the enzyme reaction, viz. $KHCO_3$, sodium pyruvate, ATP, Mg^{2+}, and acetyl coenzyme A, and when a mixture of all these components was added, only 2 percent of the enzyme was inactivated in 30 minutes at 55°C. This strongly suggests that in vivo the enzyme would be stabilized by some or all of these components. The striking stabilization of glyceraldehyde-3-phosphate dehydrogenase from *B. coagulans* by electrolytes (137) is another example. The conclusion is warranted, therefore, that the intrinsic structural stability of thermophile enzymes, in some instances, may have to be augmented in vivo by one or more intracellular components, which may well include the other proteins. Stabilization of some proteins by the cell membrane has been mentioned earlier (see Section 4.1).

Molecular Basis of Stability Except for their greater structural stability, proteins from thermophiles are rather similar to their mesophilic counterparts in most respects, including size, subunit structure, gross higher order structural parameters such as levels of helicity and β-structure, indices of polarity and nonpolarity, and modulation of activity by metal ions and effectors (40,87). The current view is that structural stabilization of thermophile proteins is achieved through extra hydrogen bonds, ionic interactions and apolar (hydrophobic) interactions in certain crucial parts of the protein molecules (40,87). These types of interactions, of course, are normally involved in the maintenance of the native structures of all proteins, including those from mesophiles. The specific stabilizing interactions in thermophile proteins probably result from primary structural modifications in appropriate parts of the protein molecules. There is now clear evidence that the thermostability of a protein can be altered by a single amino acid change (138,139). It has been estimated that a free energy of stabilization of less than 13 kJ will account for the thermostability of most thermophile proteins (140,141). This relatively low energy demand can be met by a small number of the interactions referred to above. Thus subtle, rather than gross, structural alterations should be expected in thermophile proteins, and the precise modifications in each protein will have to be identified by a detailed study of that protein.

The study of the thermostability of thermophile proteins has been reviewed in several articles (4,40,69,87,113,114,119–121).

5. GENETICS OF THERMOPHILY

5.1. Origin of Thermophiles

The diversity of thermophilic bacteria, coupled with their strong similarity to mesophilic bacteria that is evident from this review, bespeaks a common origin for these two bacterial classes. Johnson (142) has argued that, considering the large number of components that are thermostable in the thermophilic cell, one must conclude that many genes will be involved in specifying the trait of thermophily. Although the question of whether thermophiles had a primary or secondary origin is still debatable, the present balance of opinion seems to favor a secondary origin (33,69). If this view is accepted, it follows that the generation of a thermophilic organism probably required many mutations.

It is noteworthy that thermophiles resemble mesophiles from their own taxa more than they do thermophiles from other taxa. This is well underlined by the finding that an enzyme such as malate dehydrogenase or isocitrate dehydrogenase from a thermophilic bacterium bears greater structural similarity to the homologous enzyme from a mesophilic bacterium belonging to the same genus than to the enzyme from another thermophile belonging to an unrelated genus (143; J. Edlin and T. K. Sundaram, results to be published). This supports the view that thermophily originated independently within each phyletic line (see also Chapter 1).

5.2. Genetic Determinants of Thermophily

About ten years ago, Lindsay and Creaser (144) reported that the mesophile *B. subtilis* yielded transformants that could grow at a high temperature (70°C) after treatment with DNA from the extreme thermophile *B. caldolyticus*. This would imply that thermophily is determined by a small number of genes and contradicts the hypothesis developed above. Friedman and Mojica-a (145) obtained similar thermophilic transformants after treating *B. subtilis* cells with DNA from *B. caldolyticus* or *B. stearothermophilus*. Johnson (142) has critically assessed these studies and concluded that the transformation to thermophily is a rare event, the state of the donor DNA in the recipient cell is unknown, the details are confusing, and the reproducibility of the transformation is equivocal. Furthermore, no follow-up work seems to have been published to provide a plausible explanation of the transformation. Other observations, which were interpreted to mean that mesophilic bacteria could be physiologically adapted to grow at high temperature or that thermophily is specified by plasmid-borne adapter genes (see 33), also have not been adequately explained or substantiated. It must be pointed out that all these studies, which subscribe to the intriguing and thought-provoking theory that a small number of genetic determinants are sufficient to encode thermophily, were carried out with *Bacillus* species. The possibility must not be ruled out, although there is little hard evidence for it at present, that there may be more than one genetic mechanism of thermophily operating in the various phyletic lines of thermophilic bacteria.

Experimental genetics of thermophily is still in its infancy and is an area where progress is greatly needed in order to provide answers to several basic questions. However, there have been some encouraging developments, as indicated by the following examples. A transfection system for *B. stearothermophilus* has been described by Welker (146). More recently Munster et al. (147 and personal communication) have reported successful transformation and transduction of *B. caldotenax* auxotrophs to prototrophy. A number of plasmids have been isolated from thermophilic *Bacillus* and *Thermus* species (148–150), protocols have been worked out for transforming *B. stearothermophilus* with some of these plasmids, and plasmid vectors have been developed and used for gene cloning in *B. stearothermophilus*.

6. SUMMARY AND CONCLUSIONS

Apart from their high temperature growth habit and the consequent modifications of some characteristics, thermophilic bacteria, by and large, resemble their mesophilic counterparts.

The diversity of thermophiles has expanded considerably during the past two decades. Some of the new isolates grow optimally at temperatures approaching or exceeding 100°C and are archaebacteria (see Chapter 3), which are characteristically distinct from eubacteria in their membrane structure (see Chapter 5). In view of the possible importance of the membrane in setting the upper temperature limits for the survival of living organisms, it may be speculated that the archaebacterial membrane is well suited to evolution towards extreme thermophily, especially in strongly acid conditions.

Although most thermophilic bacteria can grow rapidly at their high optimum temperatures, their growth rates, in general, fall short of what one would predict from the Arrhenius equation. An explanation for this is offered by the finding that many enzymes from thermophiles are catalytically less efficient than their mesophilic counterparts.

The growth of extreme thermophiles presents some anomalous features. Genetic instability, which may be a characteristic of thermophilic bacteria, and the consequent strain variability could explain some of them.

The growth yield of thermophiles is similar to that of mesophilic bacteria, if perhaps a bit lower. A higher maintenance requirement and a higher death rate appear to characterize thermophilic growth.

In comparison with mesophilic bacteria, thermophiles have higher-melting fatty acids in their membrane lipid and higher phase transition temperatures for the lipid; this is consistent with their higher growth temperatures. They normally maintain the lipid completely or predominantly in the liquid-crystalline state during growth, as other bacteria like _E. coli_ seem to do. It is therefore unlikely that their T_{min} will be determined by the phase transition of the membrane lipid. Probable factors restricting growth at sub-T_{min} temperatures include loss of activity of enzymes and of vital functions such as ribosome assembly. It appears that the T_{max} of thermophiles is not directly related to the phase transition of the membrane lipid. Instability of the cell membrane and inactivation of macromolecules such as enzymes will explain the cessation of growth at supra-T_{max} temperatures. Present evidence suggests that there is an upper limit to the degree of fluidity of the membrane, and overfluidity, which could cause the breakdown of membrane functions, may be regarded as a correlate of membrane instability.

Contrary to earlier ideas, no massive turnover of cell components like proteins occurs during thermophilic growth. On the other hand, thermophily is based primarily on the thermostability of these components. Besides the higher-melting fatty acids, a high glycolipid content may be an important factor stabilizing the cell membrane. The stability of the ribosome derives from both the RNA and the proteins and perhaps also from the polyamines associated with it. Thermophile proteins are invariably more thermostable than their homologues from mesophiles, and this intrinsic structural stability in some instances is complemented by the stabilizing effect of intracellular components and of the membrane. At the molecular level, it is thought that the intrinsic stability is achieved through extra hydrogen bonds and ionic and hydrophobic interactions in crucial parts of the protein molecule.

The question of whether thermophiles had a primary or secondary origin is still unsettled, although a secondary origin is currently favored. It is probable that thermophily is specified by a number of genes and many mutations occurred in the course of evolution to the high-temperature growth habit, and that this evolution took place independently in different phyletic lines.

Thermophile genetics is currently in a rudimentary state of development. This is an area where sustained effort is required and, if successful, can pay rich dividends in terms of answers to a number of fundamental questions concerning thermophily. A very promising recent development in this field is the application of recombinant DNA techniques.

ACKNOWLEDGMENTS

Part of the work from my laboratory, discussed in this review, was supported by grants from the Science and Engineering Research Council and the Royal Society, Great Britain. I am grateful to Miss Elizabeth Wright for her expert typing of the manuscript.

REFERENCES

1. L. L. Campbell and B. Pace, *J. Appl. Bacteriol.* **31**, 24 (1968).
2. B. Sonnleitner, *Adv. Biochem. Eng. Biotechnol.* **28**, 69 (1983).
3. J. G. Zeikus, *Enzyme Microb. Technol.* **1**, 243 (1979).
4. L. G. Ljungdahl, *Adv. Microb. Physiol.* **19**, 149 (1979).
5. A. Schenk and M. Aragno, *J. Gen. Microbiol.* **115**, 333 (1979).
6. J. A. Buswell and D. G. Twomey, *J. Gen. Microbiol.* **87**, 377 (1975).
7. J. A. Buswell and J. S. Clark, *J. Gen. Microbiol.* **96**, 209 (1976).
8. J. S. Clark and J. A. Buswell, *J. Gen. Microbiol.* **112**, 191 (1979).
9. R. S. Golovacheva and A. E. Oreshkin, *Mikrobiologiya* **44**, 470 (1975); through *Chem. Abstr.* **83**(No. 17), 242 (1975).
10. M. J. Klug and A. J. Markovetz, *Nature* **215**, 1082 (1967).
11. R. I. Mateles, J. N. Baruah, and S. R. Tannenbaum, *Science* **157**, 1322 (1967).
12. I. N. Pozmogova, S. D. Taptykova, and G. G. Sotnikov, *Mikrobiologiya* **41**, 299 (1972); through *Chem. Abstr.* **77** (No. 7), 225 (1972).
13. B. Krueger and O. Meyer, *Arch. Microbiol.* **139**, 402 (1984).
14. C. M. Lyons, P. Justin, J. Colby, and E. Williams, *J. Gen. Microbiol.* **130**, 1097 (1984).
15. N. Belay, R. Sparling, and L. Daniels, *Nature* **312**, 286 (1984).
16. J. G. Zeikus and R. S. Wolfe, *J. Bacteriol.* **109**, 707 (1972).
17. I. Epstein and N. Grossowicz, *J. Bacteriol.* **99**, 414 (1967).
18. H. H. Daron, *J. Bacteriol.* **93**, 703 (1967).
19. T. K. Sundaram, *J. Bacteriol.* **113**, 549 (1973).
20. H. Kuhn, U. Friederich, and A. Fiechter, *Eur. J. Appl. Microbiol. Biotechnol.* **6**, 341 (1979).
21. T. D. Brock and H. Freeze, *J. Bacteriol.* **98**, 289 (1969).

22. S. Cometta, B. Sonnleitner, and A. Fiechter, *Eur. J. Appl. Microbiol. Biotechnol.* **15**, 69 (1982).

23. T. Saiki, R. Kimura, and K. Arima, *Agric. Biol. Chem.* **36**, 2357 (1972); through ref. 2.

24. J. Wiegel and L. G. Ljungdahl, *Arch. Microbiol.* **128**, 343 (1981).

25. P. F. Smith, T. A. Langworthy, and M. R. Smith, *J. Bacteriol.* **124**, 884 (1975).

26. T. D. Brock, in M. P. Starr, H. Stolp. H. G. Trüper, A. Balows, and H. G. Schlegel, Eds., *The Prokaryotes*, Springer, Berlin, 1981, p. 978.

27. B. Sonnleitner, S. Cometta, and A. Fiechter, *Eur. J. Appl. Microbiol. Biotechnol.* **15**, 75 (1982).

28. T. J. Jackson, R. F. Ramaley, and W. G. Meinschein, *Int. J. Syst. Bacteriol.* **23**, 28 (1973); through ref. 2.

29. G. Darland, T. D. Brock, W. Samsonoff, and S. F. Conti, *Science* **170**, 1416 (1970).

30. T. D. Brock and G. K. Darland, *Science* **169**, 1316 (1970).

31. W. Zillig, K. O. Stetter, W. Schaefer, D. Janekovic, S. Wunderl, I. Holz, and P. Palm, *Zbl. Bakt. Hyg. I. Abt. Orig. C* **2**, 205 (1981).

32. M. De Rosa, A. Gambacorta, and J. D. Bu'Lock, *J. Gen. Microbiol.* **86**, 156 (1975).

33. R. W. Castenholz, in M. Shilo, Ed., *Strategies of Microbial Life in Extreme Environments*, Dahlem Konferenzen, Verlag Chemie, Weinheim, 1979, p. 373.

34. T. A. Langworthy, in M. Shilo, Ed., *Strategies of Microbial Life in Extreme Environments*, Dahlem Konferenzen, Verlag Chemie, Weinheim, 1979, p. 417.

35. W. Zillig, J. Tu, and I. Holz, *Nature* **293**, 85 (1981).

36. J. G. Cobley and J. C. Cox, *Microbiol. Rev.* **47**, 579 (1983).

37. J. G. Zeikus, P. W. Hegge, and M. A. Anderson, *Arch. Microbiol.* **122**, 41 (1979).

38. A. Ben-Bassat and J. G. Zeikus, *Arch. Microbiol.* **128**, 365 (1981).

39. T. D. Brock, *Science* **158**, 1012 (1967).

40. R. E. Amelunxen and A. L. Murdock, *CRC Crit. Rev. Microbiol.* **6, 343 (1978).**

41. J. L. Ingraham, in H. Precht, J. Christophersen, H. Hensel, and W. Larcher, Eds., *Temperature and Life*, Springer- Verlag, Heidelberg, 1973, p. 60.

42. D. A. Ratkowsky, J. Olley, T. A. McMeekin, and A. Ball, *J. Bacteriol.* **149**, 1 (1982).

43. K. Smith, T. K. Sundaram, M. Kernick, and A. E. Wilkinson, *Biochim. Biophys. Acta* **708**, 17 (1982).

44. H. J. Kuhn, S. Cometta, and A. Fiechter, *Eur. J. Appl. Microbiol. Biotechnol.* **10**, 303 (1980).

45. E. Degryse, N. Glansdorff, and A. Pierard, *Arch. Microbiol.* **117**, 189 (1978).

46. T. Oshima and K. Imahori, *J. Gen. Appl. Microbiol.* **17**, 513 (1971).

47. I. Lienert and D. Richter, *J. Gen. Microbiol.* **123**, 383 (1981).

48. J. Wiegel, L. G. Ljungdahl, and J. R. Rawson, *J. Bacteriol.* **139**, 800 (1979).

49. W. J. Payne, *Ann. Rev. Microbiol.* **24**, 17 (1970).

50. A. H. Stouthamer, *Methods in Microbiology* **1**, 629 (1969).

51. R. D. Schwartz and F. A. Keller, Jr., *Appl. Environ. Microbiol.* **43**, 1385 (1982).

52. N. F. Matsche and J. F. Andrews, *Biotechnol. Bioeng. Symp.* **4**, 77 (1973).

53. D. W. Tempest, *Trends Biochem. Sci.* **3**, 180 (1978).

54. S. G. Farrand, C. W. Jones, J. D. Linton, and R. J. Stephenson, *Arch. Microbiol.* **135**, 276 (1983).

55. T. P. Coultate and T. K. Sundaram, *J. Bacteriol.* **121**, 55 (1975).

56. A. McKay, J. Quilter, and C. W. Jones, *Arch. Microbiol.* **131**, 43 (1982).

57. L. P. Hadjipetrou, J. P. Gerrits, F. A. G. Teulings, and A. H. Stouthamer, *J. Gen. Microbiol.* **36**, 139 (1964).

58. J. C. Senez, *Bacteriol. Rev.* **26**, 95 (1962).

59. H. Ng., *J. Bacteriol.* **98**, 232 (1969).

60. S. J. Pirt, *Proc. Roy. Soc. Lond. B* **163**, 224 (1965).

61. S. E. Mainzer and W. P. Hempfling, *J. Bacteriol.* **126**, 251 (1976).

62. O. M Neijsell and D. W. Tempest, *Arch. Microbiol.* **107**, 215 (1976).

63. S. J. Pirt, *Principles of Microbe and Cell Cultivation*, Blackwell Scientific Publications, Oxford, 1975.

64. W. H. Holms and P. M. Bennett, *J. Gen. Microbiol.* **65**, 57 (1971).

65. O. M. Neijssel and D. W. Tempest, *Arch. Microbiol.* **106**, 251 (1975).

66. D. K. Brannan and D. E. Caldwell, *Appl. Environ. Microbiol.* **45**, 169 (1983).

67. S. E. Lee and A. E. Humphrey, *Biotechnol. Bioeng.* **21**, 1277 (1979).

68. S. J. Singer and G. L. Nicolson, *Science* **175**, 720 (1972).

69. T. A. Langworthy, T. D. Brock, R. W. Castenholtz, A. F. Esser, E. J. Johnson, T. Oshima, M. Tsuboi, J. G. Zeikus, and H. Zuber, in M. Shilo, Ed., *Strategies of Microbial Life in Extreme Environments*, Dahlem Konferenzen, Verlag Chemie, Weinheim, 1979, p. 489.

70. A. F. Esser, in M. Shilo, Ed., *Strategies of Microbial Life in Extreme Environments*, Dahlem Konferenzen, Verlag Chemie, Weinheim, 1979, p. 433.

71. R. N. McElhaney, in M. R. Heinrich, Ed., *Extreme Environments: Mechanisms of Microbial Adaptation*, Academic Press, New York, 1978, p. 255.

72. J. E. Cronan, Jr. and E. P. Gelman, *Bacteriol. Rev.* **39**, 232 (1975).

73. M. R. J. Salton, in L. A. Manson, Ed., *Biomembranes*, Plenum Press, New York, 1971, p. 1.

74. D. Papahadjopoulos, M. Moscarello, E. H. Eylar, and T. Isac, *Biochim. Biophys. Acta* **401**, 317 (1975).

75. R. N. McElhaney and K. A. Souza, *Biochim. Biophys. Acta* **443**, 348 (1976).

76. N. E. Welker, in M. R. Heinrich, Ed., *Extreme Environments: Mechanisms of Microbial Adaptation*, Academic Press, New York, 1978, p. 229.

77. H. Bodman and N. E. Welker, *J. Bacteriol.* **97**, 924 (1969).

78. C. Wisdom and N. E. Welker, *J. Bacteriol.* **114**, 1336 (1973).

79. P. H. Ray, D. C. White, and T. D. Brock, *J. Bacteriol.* **108**, 227 (1971).

80. P. Overath, H. U. Schairer, and W. Stoffel, *Proc. Natl. Acad. Sci. U.S.A.* **67**, 606 (1970).

81. R. N. McElhaney, *J. Mol. Biol.* **84**, 145 (1974).

82. N. A. Machtiger and C. F. Fox, *Ann. Rev. Biochem.* **42**, 575 (1973).

83. H. K. Kimelberg and D. Papahadjopoulos, *Biochim. Biophys. Acta* **282**, 277 (1972).

84. B. De Kruyff, P. W. M. Van Dijk, R. W. Goldbach, R. A. Demel, and L. L. M. Van Deenen, *Biochim. Biophys. Acta* **330**, 269 (1973).

85. K. A. Souza, L. L. Kostiw, and B. J. Tyson, *Arch. Microbiol.* **97**, 89 (1974).

86. W. Babel, H. A. Rosenthal, and S. Rapoport, *Acta Biol. Med. Ger.* **28**, 565 (1972).

87. L. G. Ljungdahl and D. Sherod, in M. R. Heinrich, Ed., *Extreme Environments: Mechanisms of Microbial Adaptation*, Academic Press, New York, 1978, p. 147.

88. C. R. Middaugh and R. D. MacElroy, in M. R. Heinrich, Ed., *Extreme Environments: Mechanisms of Microbial Adaptation*, Academic Press, New York, 1978, p. 201.

89. S. M. Libor, T. K. Sundaram, and M. C. Scrutton, *Biochem. J.* **169**, 543 (1978).

90. G. A. O'Donovan and J. L. Ingraham, *Proc. Natl. Acad. Sci. U.S.A.* **54**, 451 (1965).

91. C. Guthrie, H. Nashimoto, and M. Nomura, *Proc. Natl. Acad. Sci. U.S.A.* **63**, 384 (1969).

92. P. C. Tai, D. P. Kessler, and J. L. Ingraham, *J. Bacteriol.* **97**, 1298 (1969).

93. M. Nomura and V. Erdmann, *Nature* **228**, 744 (1970).

94. K. H. Nierhaus, *BioSystems* **12**, 273 (1980).

95. D. L. Melchior, *Curr. Top. Membr. Transp.* **17**, 263 (1982).

96. N. Kawada and Y. Nosoh, *FEBS Letters* **124**, 15 (1981).

97. D. L. Melchior and J. M. Steim, *Ann. Rev. Biophys. Bioeng.* **5**, 205 (1976).

98. R. A. Golovacheva, A. P. Kalyuzhnaya, V. I. Biryuzova, and G. N. Zaitseva, *Mikrobiologiya* **38**, 679 (1969); through ref. 4.

99. R. N. McElhaney, *Curr. Top. Membr. Transp.* **17**, 317 (1982).

100. A. F. Esser and K. A. Souza, *Proc. Natl. Acad. Sci. U.S.A.* **71**, 4111 (1974).

101. S. M. Friedman and H. Read, *Proc. 28th Wind River Conf. Genet. Exchange*, Saugerties, New York, 1984, Abstr. 48.

102. M. Sinensky, *Proc. Natl. Acad. Sci. U.S.A.* **71**, 522 (1974).

103. A. F. Esser and K. A. Souza, in M. R. Heinrich, Ed., *Extreme Environments: Mechanisms of Microbial Adaptation*, Academic Press, New York, 1978, p. 283.

104. C. O. Rock and J. E. Cronan, Jr., *Curr. Top. Membr. Transp.* **17**, 207 (1982).

105. D. de Mendoza and J. E. Cronan, Jr., *Trends Biochem. Sci.* **8**, 49 (1983).

106. N. J. Russell, *Trends Biochem. Sci.* **9**, 108 (1984).

107. M. B. Jackson and J. E. Cronan, Jr., *Biochim. Biophys. Acta* **512**, 472 (1978).

108. M. B. Allen, *Bacteriol. Rev.* **17**, 125 (1953).

109. B. Bubela and E. S. Holdsworth, *Biochim. Biophys. Acta* **123**, 364 (1966).

110. I. Epstein and N. Grossowicz, *J. Bacteriol.* **99**, 418 (1969).

111. T. P. Coultate, T. K. Sundaram, and J. J. Cazzulo, *J. Gen. Microbiol.* **91**, 383 (1975).

112. T. Kenkel and J. M. Trela, *J. Bacteriol.* **140**, 543 (1979).

113. R. E. Amelunxen and A. L. Murdock, in D. J. Kushner, Ed., *Microbial Life in Extreme Environments*, Academic Press, London, 1978, p. 217.

114. R. Jaenicke, *Ann. Rev. Biophys. Bioeng.* **10**, 1 (1981).

115. T. A. Langworthy, in S. M. Friedman, Ed., *Biochemistry of Thermophily*, Academic Press, New York, 1978, p. 11.

116. T. A. Langworthy, *Curr. Top. Membr. Transp.* **17**, 45 (1982).

117. J. Stenesh, in M. R. Heinrich, Ed., *Extreme Environments: Mechanisms of Microbial Adaptation*, Academic Press, New York, 1978, p. 85.

118. S. M. Friedman, in S. M. Friedman, Ed., *Biochemistry of Thermophily*, Academic Press, New York, 1978, p. 151.

119. R. Singleton, Jr. and R. E. Amelunxen, *Bacteriol. Rev.* **37**, 320 (1973).

120. H. Zuber, in M. Shilo, Ed., *Strategies of Microbial Life in Extreme Environments*, Dahlem Konferenzen, Verlag Chemie, Weinheim, 1979, p. 393.

121. H. Zuber, in H. Eggerer and R. Huber, Eds., *Structural and Functional Aspects of Enzyme Catalysis*, 32. Colloquium der Gesellschaft für Biologische Chemie, Springer-Verlag, Heidelberg, 1981, p. 114.

122. C. Marsh and W. Militzer, *Arch. Biochem. Biophys.* **60**, 433 (1956).

123. A. Hachimori, N. Muramatsu, and Y. Nosoh, *Biochim. Biophys. Acta* **206**, 426 (1970).

124. N. Sone, M. Yoshida, H. Hirata, and Y. Kagawa, *J. Biol. Chem.* **250**, 7917 (1975).

125. R. A. Pask-Hughes and N. Shaw, *J. Bacteriol.* **149**, 54 (1982).

126. T. Oshima, in M. Shilo, Ed., *Strategies of Microbial Life in Extreme Environments*, Dahlem Konferenzen, Verlag Chemie, Weinheim, 1979, p. 455.

127. J. G. Zeikus, M. W. Taylor, and T. D. Brock, *Biochim. Biophys. Acta* **204**, 512 (1970).

128. C. A. Marotta, F. Varrichio, I. Smith, and S. M. Weissman, *J. Biol. Chem.* **251**, 3122 (1976).

129. R. N. Nazar and A. T. Matheson, *J. Biol. Chem.* **252**, 4256 (1977).

130. S. Douthwaite and R. A. Garrett, *Biochemistry* **20**, 7301 (1981).

131. B. Pace and L. L. Campbell, *Proc. Natl. Acad. Sci. U.S.A.* **57**, 1110 (1967).

132. C. C. Irwin, J. M. Akagi, and R. H. Himes, *J. Bacteriol.* **113**, 252 (1973).

133. M. Nomura, P. Traub, and H. Bechmann, *Nature* **219**, 793 (1968).

134. M. García-Patrone, N. S. González, and I. D. Algranati, *Biochim. Biophys. Acta* **395**, 373 (1975).

135. S. Libor, T. K. Sundaram, R. Warwick, J. A. Chapman, and S. M. W. Grundy, *Biochemistry* **18**, 3647 (1979).

136. J. J. Cazzulo, T. K. Sundaram, and H. L. Kornberg, *Proc. Roy. Soc. Lond. B* **176**, 1 (1970).

137. J. W. Crabb, A. L. Murdock, and R. E. Amelunxen, *Biochemistry* **16**, 4840 (1977).

138. K. Yutani, K. Ogasahara, Y. Sugino, and A. Matsushiro, *Nature* **267**, 274 (1977).

139. M. G. Grulter, R. B. Hawkes, and B. W. Matthews, *Nature* **277**, 667 (1979).

140. M. F. Perutz and H. Raidt, *Nature* **255**, 256 (1975).

141. S. Wilkinson and J. R. Knowles, *Biochem. J.* **139**, 391 (1974).

142. E. J. Johnson, in M. Shilo, Ed., *Strategies of Microbial Life in Extreme Environments*, Dahlem Konferenzen, Verlag Chemie, Weinheim, 1979, p. 471.

143. K. Smith, T. K. Sundaram, and M. Kernick, *J. Bacteriol.* **157**, 684 (1984).
144. J. A. Lindsay and E. H. Creaser, *Nature* **255**, 650 (1975).
145. S. M. Friedman and T. Mojica-a, in S. M. Friedman, Ed., *Biochemistry of Thermophily*, Academic Press, New York, 1978, p. 117.
146. N. E. Welker, in S. M. Friedman, Ed., *Biochemistry of Thermophily*, Academic Press, New York, 1978, p. 127.
147. M. J. Munster, R. J. Sharp, A. Vivian, S. Ahmed, and T. Atkinson, *13th International Conference of Microbiology*, Boston, 1982, Abstract.
148. A. H. A. Bingham, C. J. Bruton, and T. Atkinson, *J. Gen. Microbiol.* **114**, 401 (1979).
149. T. Imanaka, *Trends Biotechnol.* **1**, 139 (1983).
150. B. S. Hartley and M. A. Payton, *Biochem. Soc. Symp.* **48**, 133 (1983).

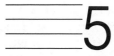

MEMBRANES AND LIPIDS OF THERMOPHILES

THOMAS A. LANGWORTHY AND JEAN L. POND

Department of Microbiology, School of Medicine, University of South Dakota, Vermillion, South Dakota 57069

1. INTRODUCTION

The cytoplasmic membrane is an essential element of all living cells. It occupies a unique position in cellular physiology because it must control the flow of substances between the internal and external cellular environment. It has a complex molecular organization comprised of proteins, ions, and lipids which interact in a geometrically and thermodynamically optimal manner (1). Of the major membrane components, the lipids make the main contribution to the molecular architecture. Lipids provide the hydrophobic matrix of the membrane and maintain the appropriate fluidity for the function of membrane proteins under a variety of changing external parameters, especially temperature. Because of the paramount effect of temperature on membrane stability, the recent increase in the number of new thermophiles has stimulated a general interest in thermophile lipids with the goal of discerning how these membranes are stabilized and function at temperatures ranging from about 55 to 105°C.

There is now a fairly wide range of thermophiles, including both aerobic and anaerobic eubacteria and archaebacteria. However, an idea of the nature and function of the lipids from all of these organisms is far from complete. We would like to present here a general description of the more unusual lipids which have been found to date. These demonstrate the diverse ways in which thermophilic bacteria maintain the integrity of their membranes at high temperatures.

2. THE ORGANISMS

A brief description follows of the major genera of thermophilic eubacteria and archaebacteria in which some aspects of the lipids have been investigated.

2.1. Eubacteria

Aerobic eubacteria studied include members of three different genera of heterotrophs: *Bacillus, Thermus,* and *Thermomicrobium* species. Among *Bacillus* species investigated, *B. stearothermophilus* has been the most extensively studied. It is a rather moderate thermophile growing optimally at 50 to 65°C but between 37 and 70°C (2). *B. caldolyticus, B. caldovelox* and *B. caldotenax* are more extreme thermophiles, growing well between 75 and 85°C (3). *B. acidocaldarius* is thermophilic, growing optimally at 60 to 65°C and within a range of 40 to 70°C. It is, however, also acidophilic with optimal growth at pH 3 and within the range of pH 2 to 6 (4). Members of the genus *Thermus* in which lipids have been examined include *T. aquaticus* (5), *T. flavus* (6) and *T. thermophilus* (7). These are Gram-negative, non-spore-forming rods. *T. aquaticus* and *T. flavus* grow best at 70°C and between 40 and 80°C, whereas *T. thermophilus* has a slightly higher optimum of 75°C. The genus *Thermomicrobium* includes *T. roseum* (8) which is unusual since it is pleomorphic, lacking a typical peptidoglycan (9), but is sensitive to penicillin. It has an optimal growth temperature of 70 to 75°C. Some possibly related species of thermophilic hydrocarbon utilizers have been looked at by Phillips and Perry (10).

Anaerobic eubacterial thermophiles studied include mainly *Clostridium* species. *C. tartarivorium* and *C. thermosaccharolyticum* are rather moderate thermophiles, growing optimally at 55 to 60°C (11). *C. thermocellum* (12) grows well between 5 and 60°C, but does have an upper maximum temperature of 70°C. *C. thermohydrosulfuricum* (13) grows over a temperature range of 40 to 78°C with the optimum at 68°C. *C. thermosulfurogens* is a Gram-negative organism that is fermentative, but produces elemental sulfur from thiosulfate. It grows between 35 to 75°C with a temperature optimum of about 60°C (14). An example of an aerobic thermophilic sulfate reducer is *Thermodesulfotobacterium commune* (15). It is a Gram-negative nonsporulating rod which grows optimally at 70°C and between 40 and 85°C. This organism can utilize H_2, lactate, or pyruvate as electron donors and sulfate, thiosulfate, or pyruvate as electron acceptors.

2.2. Archaebacteria

Archaebacteria differ from eubacteria in a variety of unique aspects, including cell wall structure, RNA polymerases, 16S rRNA structure, tRNA architecture, and lipids (16). The lipids are in fact so distinctive that they are a good taxonomic marker for distinguishing archaebacteria (17–19). The actual number of thermophilic archaebacteria in which aspects of the lipids have been studied is, however, quite small.

The most extensively studied archaebacteria are the aerobic, thermophilic, and acidophilic members of the genera *Thermoplasma* and *Sulfolobus*. *T. acidophilum* is unique among archaebacteria because it completely lacks a cell wall (20,21). It grows optimally at pH 2 and 59°C and within the limits of pH 1 to 4 and 40 to 62°C. Its cellular membrane is directly exposed to the hot acidic environment and it is acidophilic, undergoing lysis at neutrality (22). *Sulfolobus* species, such as *S. acidocaldarius, S. solfataricus*, and *S. brierleyi*, are facultative autotrophs capable of growth on sulfur and iron (23–25). They grow within the limits of 55 to 85°C and pH 2 to 5 depending upon strains. *S. acidocaldarius* and *S. solfataricus* grow optimally at 75 to 85°C while *S. brierleyi* grows best near 60°C. In contrast to *Thermoplasma*, these organisms possess a cell wall structure and are stable, but do not grow, at neutral pH.

The thermophilic, anaerobic archaebacteria investigated include several genera of methanogens. *Methanobacterium thermoautotrophicum* grows at 70°C and between 40 and 75°C (26). *Methanothermus fervidus* is a rod-shaped methanogen growing between 65 and 97°C with an optimum temperature around 83°C (27). *Methanococcus jannaschii* is a motile irregular coccus growing best at 85°C and between 50 and 85°C (28). Nonmethanogenic anaerobes include species of the genera *Thermoproteus, Desulfurococcus, Thermodiscus* and *Pyrodictium. Thermoproteus tenax* grows between 80 and 96°C and optimally at 88°C (29). It is a rod-shaped organism able to form H_2S and CO_2 from elemental sulfur and organic substrates. *Desulfurococcus mucosus* and *D. mobilis* grow optimally at 85°C and between 75 to 90°C (30). These cells have coccal morphologies with variable diameters and utilize organic carbon sources by sulfur respiration with the production of H_2S and CO_2, or grow by fermentation. *Thermodiscus maritimus* is a heterotroph that grows optimally at 85°C (31). It is a flat disc-shaped organism with a thick envelope and long protrusions. The final, and perhaps most intriguing thermophile, is *Pyrodictium occultum* (32). It is a disc- to dish-shaped organism with a filamentous network connecting the cells. This obligate hydrogen-sulfur autotroph grows between 80 to 110°C and currently holds the distinction of having the highest proven temperature optimum, 105°C, of any known organism.

3. MEMBRANE LIPIDS: THE APOLAR CHAINS

The apolar hydrocarbon chains, which are capable of hydrophobic interaction and provide the core of the membrane, have been the most extensively studied components of thermophile lipids. The majority of the hydrocarbon chains, usually as pairs, are bound to glycerol molecules through one of three classes of chemical bonds as acyl esters (fatty acids), vinyl ethers (fatty aldehydes), or alkyl ethers (alcohols). The derived glycerolipids in turn provide the typical bilayer assembly (33). When substituted with polar head groups, the glyerolipids also provide the hydrophobic region of the more complex glycolipid and phospholipid structures. In general terms, the hydrocarbon chains of the thermophilic bacteria follow the trends observed in their mesophilic counterparts which have been subjected to short-term shifts in temperature (34); that is, increasing temperature of cultivation increases average chain length, decreases percent unsaturated fatty acids, and/or increases the proportion of methyl-branched chains. Thermophiles tend to possess longer hydrocarbon chains and methyl-branched chains. Unsaturated hydrocarbon chains, which significantly lower melting points, are rare. What follows is a general overview of the structure of the hydrocarbon chains, how they are linked to glycerol, how these lipids might impart membrane stability at high temperatures, and the exceptions to general principles that a number of thermophiles have evolved to deal with high temperatures. For details regarding the chemical methodology, the reader is referred to the papers cited and to recent reviews (35,36).

3.1. Fatty Acids

The most common apolar chains in thermophilic eubacteria are fatty acids. They usually contain 14 to 20 carbons with the 15 to 18 carbon chains predominating. The characteristic types encountered are normal (n) straight chains, **1**, or monomethyl-branched chains in the *iso*, **2**, or *anteiso*, **3**, configuration.

$$CH_3 [CH_2]_n COOH$$

1

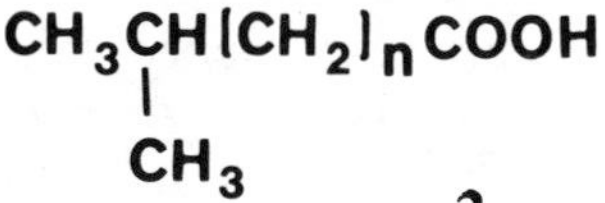

$$CH_3 CH [CH_2]_n COOH$$
$$| $$
$$CH_3$$

2

$$CH_3 CH_2 CH [CH_2]_n COOH$$
$$|$$
$$CH_3$$

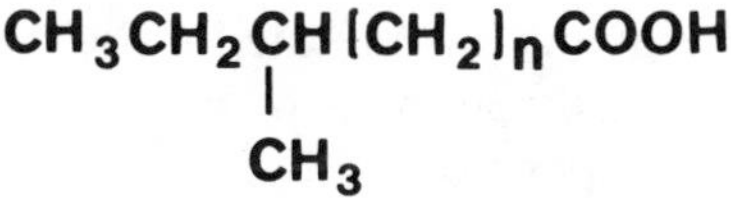

3

Iso- and *anteiso*-branched fatty acids are not restricted to thermophiles, but are also quite common in many mesophilic bacteria, particularly Gram-positive organisms (37). Like their mesophilic counterparts, the fatty acids of thermophilic eubacteria may be linked to polar head groups such as sugar residues or, most commonly, may be ester-linked to glycerol to form typical diacylglycerols, **4**.

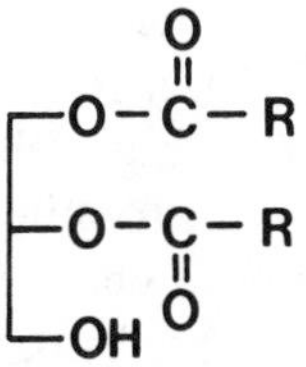

4

Since acylglycerols possess an asymmetric center at the central carbon of the glycerol molecule, as do all other types of bacterial glycerolipids, the acylglycerols are optically active. The eubacterial acylglycerols are levorotary and by convention (19) the fatty acids are attached to the number 1 and 2 carbons of glycerol (1,2-diacylglycerol). Thus, when polar head groups are present they are linked to the number 3 carbon of the glycerol molecule. The acyl ester-linkage is sensitive to both acid and alkaline hydrolysis.

As shown by numerous studies, *iso-* and *anteiso*-branched fatty acids predominate in the thermophilic eubacteria. Some examples among the aerobes include *B. stearothermophilus* which contains *iso*-C_{15} and *iso*-C_{17} acids as major components, accounting for about 34 to 64% of the total, but considerable amounts of *anteiso*-C_{16}, *anteiso*-C_{17}, and nC_{16} are also present (38–43). In the more extremely thermophilic *Bacillus* species, *B. caldolyticus, B. caldovelox* and *B. caldotenax, iso*-C_{15}, *iso*-C_{16}, and *iso*-C_{17} acids represent about 80% of the total fatty acids. Upon increasing growth temperatures from 45 to 80°C, a pronounced shift from *iso*-C_{15} to *iso*-C_{17} and *iso*-C_{16} to nC_{16} has been observed, suggesting chain extension at higher temperatures (44–46). In the more extremely thermophilic *Thermus* species, *iso*-C_{17} is the most abundant acid, accounting for 50 to 61%, followed by the *iso*-C_{15} acid. Together, this *iso*-branched pair represents about 85% of the total fatty acid content (43,47,48). In *T. thermophilus* grown at increasing temperatures from 48 to 82°C, Oshima and Miyagawa (43) observed an increase in the ratio of *iso*-C_{17} to *iso*-C_{15} fatty acids, again indicating chain elongation at higher temperatures. However, when *T. aquaticus* was grown between 50 and 75°C, Ray et al. (48) noted a slight decrease in the *iso*-C_{17} content and an increase in the amount of nC_{16} and *iso*-C_{16} fatty acids. Although the fatty acid composition of several hydrocarbon-utilizing thermophiles examined by Merkel and Perry (49) varied depending upon growth substrates, in these organisms *anteiso*-branched C_{15}, C_{16}, and C_{17} fatty acids were prevalent.

Like the aerobes, the thermophilic anaerobic eubacteria are also enriched in branched-chain fatty acids. In *Clostridium thermosulfurogenes, iso*-C_{15} acid represents 19% and the *iso*-C_{17} acid accounts for 75% of the total acids, whereas in *Clostridium thermohydrosulfuricum*, the *iso*-C_{15} acid accounts for 97% or nearly all of the fatty acids present (Langworthy, Holzer, Lovett, and Zeikus, in preparation). *Clostridium thermocellum* contains mainly *iso*-C_{16}, nC_{16}, and *anteiso*-C_{17} fatty acids. Examination of ethanol-induced changes in the lipid composition of this organism by Herrero et al. (50) has shown that ethanol causes an increase in straight chain and *anteiso*-branched acids and the unusual synthesis in a thermophile of unsaturated $C_{14:1}$, $C_{16:1}$, and $C_{18:1}$ fatty acids. In addition, branched-chain unsaturated hydrocarbons could be detected. The authors suggested that the ethanol-induced shift to fatty acids of lower melting points produces a more fluid membrane which in fact may be detrimental to the growth of *C. thermocellum*. An unsaturated fatty acid has also been detected in a moderately thermophilic clostridium examined by Chan et al. (11). Although nC_{14}, nC_{16}, and predominantly *iso*-C_{15} were the main acids encountered, about 10% of the total consisted of an unsaturated cyclopropane fatty acid, **5**, identified as 12,13-methylene-9-tetradecenoic acid which provided the first evidence of an unsaturated cyclopropane fatty acid in bacteria.

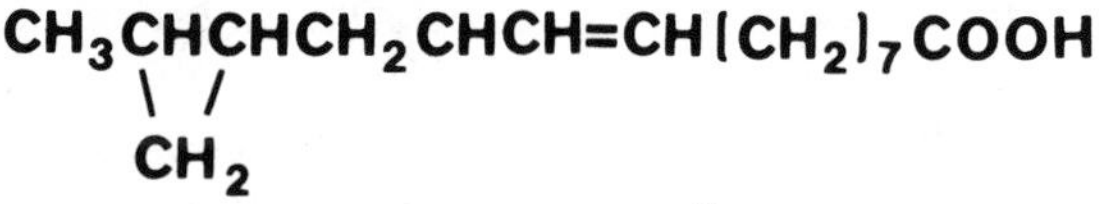

$$CH_3CHCHCH_2CHCH=CH[CH_2]_7COOH$$
$$\backslash \ /$$
$$CH_2$$

5

An exception to the general trend of *iso-* and *anteiso*-branched fatty acids is found in the aerobic thermophile, *Thermomicrobium roseum*. Although straight chain nC_{18}, nC_{19}, and nC_{20} acids are present, internally methyl-branched fatty acids predominate (51). The 19-carbon fatty acid, 12-methyl-C_{18} is the major component representing nearly 68% of the total acids, **6**.

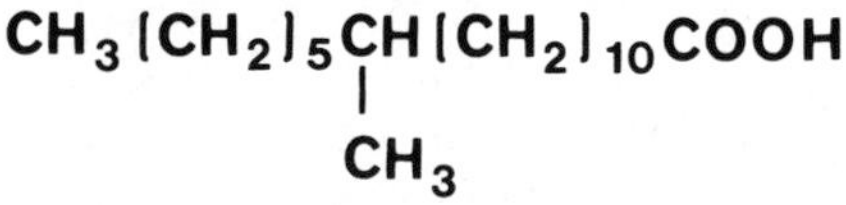

$$CH_3[CH_2]_5CH[CH_2]_{10}COOH$$
$$|$$
$$CH_3$$

6

Also present are lesser amounts of 10-methyl-C_{16}, 10 methyl-C_{17}, 12-methyl-C_{17}, 12-methyl-C_{19}, and 14-methyl-C_{20}, as well as a small amount of 7,11-dimethyl-C_{17} fatty acids. Why *T. roseum* synthesizes internally branched fatty acids is not clear but may be related to the fact that *T. roseum* does not possess glycerolipids. Rather, the fatty acids are either ester linked to long chain diols or are amide linked to amino sugars which constitute the major polar head groups in this organism (see Section 3.4).

Another exception is found in the thermoacidophile *Bacillus acidocaldarius*. This organism contains *iso-* and *anteiso*-branched fatty acids, but the major components, accounting for 50 to 90% of the total, are alicyclic ω-cyclohexyl fatty acids which possess a cyclohexyl ring on the terminal end of the hydrocarbon chain as represented by the general structure, **7**.

$$[C_6H_{11}]CH_2[CH_2]_nCOOH$$

7

There are two cyclohexyl C_{17} and C_{19} acids present, 11-cyclohexylundecanoic acid and 13-cyclohexyltridecanoic acid, respectively, which have been shown to be biosynthesized from glucose via the shikimate pathway (52–54). Additionally, the thermoacidophilic bacteriophage NS11 (55), for which *B. acidocaldarius* serves as the host, also contains cyclohexyl fatty acids (along with other host membrane lipids) in its envelope. The advantage of cyclohexyl fatty acids in the thermophily or acidophily of *B. acidocaldarius* is not clear. Oshima and Ariga (53) could find no significant changes in the fatty acid composition when cells were grown at different temperatures and pH values, but De Rosa et al. (56) noted changes which had a relationship between temperature, pH, and metabolism. At lower pH, increasing temperature raised the proportion of *iso-* and *anteiso*-branched fatty acids. At higher pH, increasing temperature raises the proportion of cyclohexyl fatty acids. Mutants of *B. subtilis* which are able to synthesize cyclohexyl fatty acids when supplied with appropriate precursors are unable to grow at either high temperatures or low pH (57). Furthermore, cyclohexyl fatty acids are not restricted to *B. acidocaldarius*. They also occur in certain hydrocarbon-utilizing mesophilic species of *Mycobacterium* and *Nocardia* grown on *n*-alkyl-substituted paraffins (58), as well as in the nonacidophilic mesophile *Cutobacterium pusillum* (59). Cyclohexyl fatty acids constitute about 81% of the total in *C. pusillum* and, when subjected to increasing temperatures, this mesophile does appear to increase its cyclohexyl fatty acid content. One functional possibility for the cyclohexyl fatty acids in the *B. acidocaldarius* membrane (57) is that the bulky cyclohexyl ring in the central interior of the membrane may provide optimal packing parameters for orientation of the numerous triterpene derivatives present in this organism (see Section 3.6). On the other hand, studies employing artificial membranes, consisting of phosphatidylcholine-containing ω-cyclohexyl fatty acids, show only a weak or no phase transition when measured with fluorescence probes (57,60). However, it has been shown that these artificial lipids possess a relatively high viscosity and that liposomes prepared from these lipids possess an exceptionally low permeability (61–63). This suggests that the cyclohexyl acyl chains do not necessarily serve as a means of membrane thermal stabilization, but rather provide a mechanism for maintaining barrier functions in response to the acidic environment.

In addition to the cyclohexyl fatty acids, two new alicyclic fatty acids have been identified in a thermoacidophilic *Bacillus* species isolated from garden soil (64). The major fatty acid is a naturally occurring cycloheptyl fatty acid, ω-cycloheptaneundecanoic acid, that contains a terminal seven-member ring (64). Another, representing about 10 to 20% of the total acids, has been identified (65) as an hydroxyl cycloheptyl fatty acid, ω-cycloheptyl-α-hydroxyundecanoic acid, **8**, which has the D-configuration.

$$[C_7H_{13}]\ CH_2\ [CH_2]_8\ \underset{\underset{\displaystyle OH}{|}}{C}HCOOH$$

8

In contrast to the monocarboxylic fatty acids of other eubacterial thermophiles, a new type of dicarboxylic fatty acid has been detected in two thermophilic clostridia, *C. thermohydrosulfuricum* and *C. thermosulfurogenes* (Langworthy, Holzer, Lovett, and Zeikus, in preparation). This dicarboxylic acid accounts for about 10 and 23%, respectively, of the total apolar chains of these organisms. It possesses a total of 30 carbons which is made up of two "head-to-head" condensed *iso*-C_{15} fatty acids resulting in a *bis*(isopentadecyl)-1,1'-dicarboxylic acid, **9**.

$$\underset{\substack{| \\ CH_3 \\ 9}}{\overset{\substack{CH_3 \\ |}}{HOOC[CH_2]_{11}CHCH_2CH_2CH[CH_2]_{11}COOH}}$$

How this acid is biosynthesized, or how it is linked to glycerol is unknown. It is very likely ester-linked to two glycerol molecules on each terminal end of the chain in the form of a diglycerol tetraester, a structure analogous to the diglycerol tetraethers of archaebacteria (see Section 3.5), although this has not been proved. If this assumption is correct, such tetraesters would aid in membrane stabilization at high temperatures, again in a manner analogous to the archaebacterial tetraethers. Similar types of C_{30} and C_{34} dicarboxylic acids have been detected in two as yet unnamed thermophilic, nonclostridial eubacteria (Langworthy, Holzer, Huber, and Stetter, in preparation). The only other known occurrence of similar dicarboxylic acids is in a mesophilic *Butyrivibrio*, species S2 (66–67). In this organism, dicarboxylic acids, called "diabolic acids", are not made up of "head-to-head" condensed chains, but rather the two internal methyl branches are vicinal (on two adjacent central carbons of the chain). In this organism, it has been shown that a single diacyl chain is ester-linked to the central carbon atom of a glycerol molecule on each end of the chain.

3.2. Fatty Aldehydes

Fatty aldehydes contain a more reduced carbonyl functional group, -CHO, at the terminal end of the hydrocarbon chain, replacing the -COOH group of the fatty acids. Fatty aldehydes are less common as apolar chains and so far appear to be restricted to anaerobic eubacteria. They are especially prevalent among the mesophilic clostridia (37). Normally, a single aldehydic chain is linked via an acid-sensitive, though alkaline-stable, vinyl ether bond to the number one carbon of glycerol, resulting in a 1-alk-enyl glycerol commonly termed a plasmalogen. In the natural form, a conventional fatty acid is usually ester-linked to carbon 2 of the glycerol molecule, **10**.

$$\left[\begin{array}{l} O - CH = CH - R \\ O - \underset{\underset{O}{\|}}{C} - R \\ O \end{array} \right.$$

10

Although the fatty acid compositions of a number of thermophilic clostridia have been reported (see Section 3.1), there is little information available on the nature of the fatty aldehydes in these organisms. Three representative thermophilic clostridia do possess fatty aldehydes as do their mesophilic counterparts (Langworthy, Holzer, Lovett, and Zeikus, in preparation). These include *C. thermocellum*, in which fatty aldehydes account for 33% of the total apolar chains, and *C. thermosulfurogenes* and *C. thermohydrosulfuricum* where aldehydes represent 19 and 10%, respectively. As with the corresponding fatty acids, *iso*- and *anteiso*-branched chains predominate. The major aldehydes of *C. thermocellum* consist of *iso*-C_{15} (21%), *anteiso*-C_{15} (17%), nC_{16} (17%), *anteiso*-C_{16} (17%), and *iso*-C_{17} (16%). *C. thermosulfurogenes* contains mainly *iso*-C_{15} (38%) and *iso*-C_{17} (59%) aldehydes as does *C. thermohydrosulfuricum*, *iso*-C_{15} (81%), and *iso*-C_{17} (11%). Whether plasmalogens impart any special advantage at moderately high temperatures is not clear since plasmalogens are also common constituents of mesophilic anaerobes.

3.3. Fatty Alcohols

Fatty alcohols provide the most reduced form of apolar chain in which an -OH group replaces the terminal -COOH of the fatty acids. The fatty alcohols are normally bound to glycerol via an ether, C-O-C, bond, which is resistant to both acid and alkaline hydrolysis. The usual form in eubacteria is a glycerol monoether (1-*O*-alkylglycerol) in which the fatty alcohol is ether-linked to the number 1 carbon of glycerol, **11**.

$$\begin{array}{l} \text{O} \!-\!\!-\!\!- \text{R} \\ \text{O} \!-\! \underset{\underset{\text{O}}{\|}}{\text{C}} \!-\! \text{R} \\ \text{OH} \end{array}$$

11

Like plasmalogens, the monoethers are usually substituted at carbon 2 of the glycerol molecule with an ester-linked fatty acid. Within the eubacteria, glycerol ether lipids are uncommon and appear to be restricted to those organisms that grow anaerobically (37). They are normally apolar constituents of mesophilic clostridia, although they are not present in substantial amounts. Like their mesophilic relatives, glycerol monoethers have been detected in at least three thermophilic clostridia (Langworthy, Holzer, Lovett, and Zeikus, in preparation). *C. thermocellum* contains about 8% monoethers possessing *iso*-C_{16} (35%), nC_{16} (47%), and *iso*-C_{17} (18%) *O*-alkyl chains. Glycerol monoethers account for about 15% of the apolar residues in *C. thermosulfurogenes* and 13% in *C. thermohydrosulfuricum*. In *C. thermosulfurogenes*, the *iso*-C_{15} (31%) and *iso*-C_{17} (66%) chain monoethers predominate and the *iso*-C_{15} chain monoether is the sole species present in *C. thermohydrosulfuricum*. Thus, branched-chain monoethers again predominate in these organisms, and have a distribution similar to the fatty acid and aldehyde composition. Glycerol monoethers have also been reported to represent about 11% of the apolar residues in the thermophilic, anaerobic sulfate reducer, *Thermodesulfotobacterium commune* (68). They are comprised mainly of the *anteiso*-C_{17} chain monoether with lesser amounts of the *iso*-C_{16}, nC_{16}, *iso*-C_{17}, *iso*-C_{18}, *anteiso*-C_{18}, and *iso*-C_{19} species. Whether glycerol monoethers endow any benefit for membrane stabilization at high temperatures is also unclear at this time.

The second type of eubacterial ether lipid is the glycerol diether (1,2-di-*O*-alkylglycerol) in which two variable chain length fatty alcohols are ether-linked at carbons 1 and 2 of the glycerol molecule, **12**.

$$
\begin{array}{l}
\rule[-0.3em]{0.05em}{2.2em}\!\!-O-R \\
\rule[-0.3em]{0.05em}{2.2em}\!\!-O-R \\
-OH
\end{array}
$$

12

Such a glycerolipid in eubacteria is rare, in fact unknown until recently (68). The only known organism to possess such diethers with the 1,2-stereo-configuration is *Thermodesulfotobacterium commune*. The diethers account for 82% of the apolar residues and, along with the glycerol monoethers (see above), indicate a eubacterial membrane comprised of glycerol ether lipids rather than the omnipresent eubacterial glycerides. Five principal diethers occur containing C_{16}:C_{16}, mixed C_{16}:C_{17}, C_{17}:C_{17}, mixed C_{17}:C_{18}, and C_{18}:C_{18} *O*-alkyl side chains. The C_{17}:C_{17} diether with two *anteiso*-branched chains is predominant. The alkyl chains of the other diether species include mainly *iso*-C_{16}, *iso*-C_{18}, and nC_{18} side chains. Why *Thermodesulfotobacterium* has evolved ether lipids for the assembly of its membrane architecture, or what special physical properties they may impart, is not known. Perhaps the ether bonds render the membrane more resistant to disruption by the huge quantities of H_2S produced by this organism at high temperatures. Other sulfate reducers, however, appear to contain normal glyceride residues (37).

3.4. Long-Chain Diols

Long-chain 1,2-diols are natural constituents of surface waxes of plants (69) and the sebum of animals, such as wool wax or lanolin (70). They have not been known to occur as constituents making up the hydrophobic domain of cellular membranes. However, such compounds have now been found to replace glycerolipids in the thermophilic aerobe, *Thermomicrobium roseum* (51). This is the first evidence of an organism in which glycerol-derived lipids are absent. *T. roseum* possesses instead a series of C_{18} to C_{23} straight-chain and internally methyl-branched 1,2-diols of general structure, **13**.

$$CH_3 [CH_2]_n \underset{\underset{OH}{|}}{CH_2} CH_2 OH$$

13

The major diol is the nC_{21} compound, 1,2-eicosanediol (49%), followed by the internally methyl-branched C_{20} diol, 13-methyl-1,2-nonadecanediol (21%), and the nC_{19}diol, 1,2-nonadecanediol (11%). The diols account for about 72% of the apolar residues, and fatty acids (see Section 3.1) account for about 28%. The fatty acids are either ester-linked to carbon 2 of the diols or amide-linked to amine radicals of the glycolipids of the organism, but not to glycerol. Polar head groups, when present, are linked to the terminal -OH of the diols at carbon 1. The replacement of glycerolipids by 1,2-diols is unique. They can, however, be considered to be structural analogues of glycerolipids if the stereoconfiguration of these molecules is considered to be turned at carbon 3 rather than thought of as a straight chain as represented by structure **14**.

$$
\begin{array}{l}
\rule{1cm}{0.4pt}\ R \\
\rule{0.4pt}{0.6cm}\!\!-O-\underset{\underset{O}{\|}}{C}-R \\
\rule{0.4pt}{0.6cm}\!\!-OH
\end{array}
$$

14

The diols are thus analogous to the most reduced glycerolipid counterpart, 1-*O*-alkylglycerol (see Section 3.3). The difference is that at carbon 3 of the diols, which corresponds to carbon 1 of eubacterial glycerolipids, no functional group is present. The diols thereby have a "glycerol-like" function already built into their structure. It is also noteworthy that in this configuration, the length of the extended hydrocarbon chains is reduced by 3 carbons so that the chain length of the major C_{21} diol, for example, becomes 18 carbons long which corresponds to the chain length of the major 12-methyl-C_{18} fatty acid of *T. roseum*. This is now within the nominal chain length range of most thermophilic membrane apolar chains. Whether the diols form a lipid bilayer is not known, but in the double-chain form this seems likely because of their close analogy to the double-chain glycerolipids. What physical properties the diols provide, their stereochemistry, and their biosynthesis, is unknown at this time.

3.5. Isopranyl Alcohols

Isopranoid alcohols comprise the apolar chains of both mesophilic and thermophilic archaebacteria. They are a hallmark of these organisms (17–19). A fully detailed review of archaebacterial lipids in general has recently appeared elsewhere (19).

Structure Mesophilic and thermophilic eubacteria can alter the specific apolar chain composition of their lipids in response to altered growth temperature, thus achieving a constant membrane viscosity. The apolar chains of archaebacteria are, however, almost invariably fixed at either 20 or 40 carbon atoms. The C_{20} isopranoid monoalcohol is a fully saturated phytanol molecule while the fully saturated C_{40} isopranoid chains occur as 1,1'-biphytanyl diols with two terminal -OH groups. The biphytanyl diols are the structural equivalent of two "head-to-head" condensed phytanol molecules. When ether-linked to glycerol, these isopranoid alcohols provide two basic types of glycerol ethers in archaebacteria, phytanyl glycerol diethers **15** and dibiphytanyl diglycerol tetraethers **16** (19,71–76).

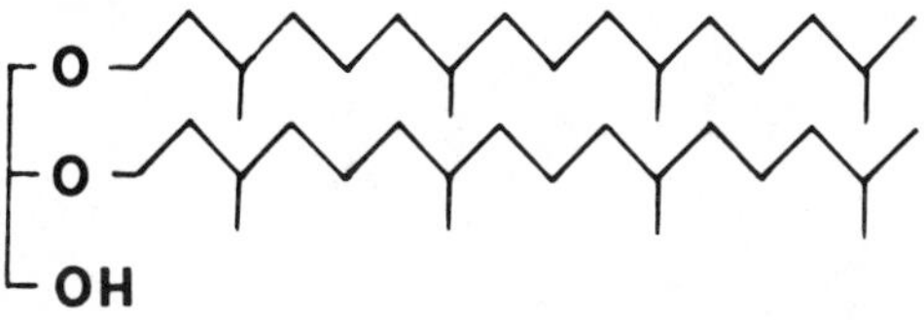

15

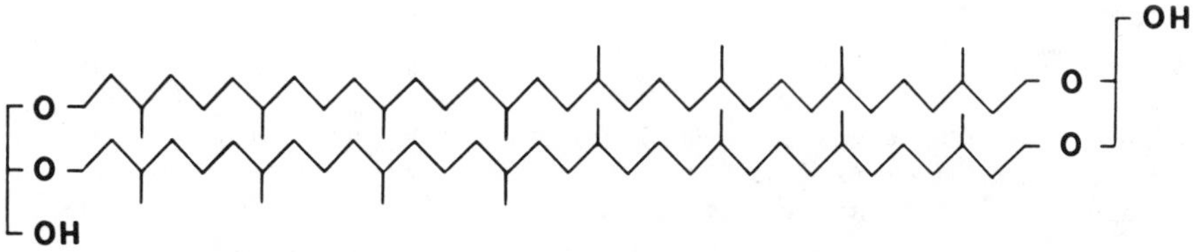

16

Although the diethers bear a structural resemblance to eubacterial glycerolipids, with the exception that two identical pairs of C_{20} phytanyl chains are present, the tetraethers can be considered to be the structural analogue of two phytanyl diether molecules that have been covalently linked through the terminal ends of the phytanyl side chains. The terminal -OH groups of the tetraethers occur in the *trans* configuration.

The isopranoid glycerol ethers of archaebacteria also possess an optically active center and are dextrorotary. They are the stereochemical mirror image of eubacterial glycerolipids. The isopranoid chains are thus ether-linked to the number 2 and 3 carbon atoms of glycerol and polar head groups are bound to carbon 1. The diethers therefore occur as 2,3-di-*O*-phytanylglycerol and the tetraethers as 2,3,2',3-tetra-*O*-dibiphytanyldiglycerol.

Diethers are the only glycerolipids present in the mesophilic halophilic archaebacteria, but both diethers and tetraethers occur in varying ratios in mesophilic as well as thermophilic archaebacteria (17,19). Within the thermoacidophilic archaebacteria, *Thermoplasma* and *Sulfolobus*, however, tetraethers comprise almost all (90–95%) of the glycerolipids and the remainder are diethers. The tetraethers of these two organisms are more

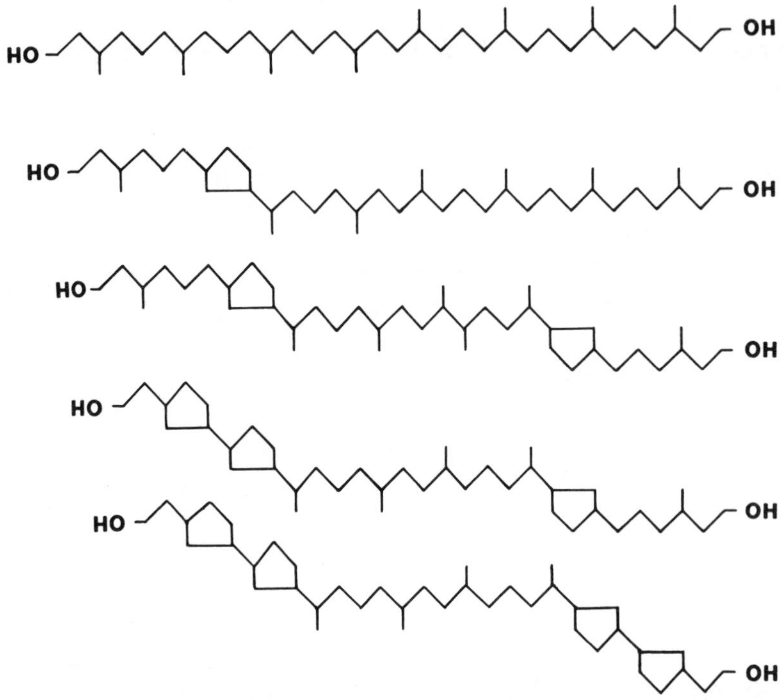

Figure 1. Structure of the C_{40} biphytanyl diols of thermoacidophilic archaebacteria. These constitute the apolar chains of the tetraethers of the organisms and may be either acyclic (top) or contain up to four pentacyclic rings (bottom).

specialized (i.e., variable in structure) than those of archaebacteria in general (19,72–75) because they can alter the acyclic structure of the C_{40} biphytanyl chains to include from one to four cyclopentane rings (Figure 1). Like the acyclic biphytanyl chains, the cyclic chains usually occur as an identical pair linked to glycerol in an antiparallel configuration (19,72–75). However, a limited amount of tetraether species in *Sulfolobus* have recently been shown to possess nonidentical pairs of biphytanyl chains (77).

A second variety of tetraether structure also occurs which is peculiar to *Sulfolobus* (78,79). De Rosa and associates (79) have established that the tetraether is a nonitol-glycerol tetraether, made up of a branched, 9-carbon alcohol containing nine hydroxyl groups which replaces one of the two glycerol molecules in the basic diglycerol tetraether structure, **17**.

It accounts for about 75% of the total tetraethers in *Sulfolobus* grown autotrophically on sulfur as the energy source, but only about 50% in heterotrophically grown cells (80). This lipid is thus far characteristic only of *Sulfolobus* species. It has not been detected in a large number of nonacidophilic thermophilic or methanogenic archaebacteria which have been examined (Langworthy, Menath, and Stetter, unpublished observations).

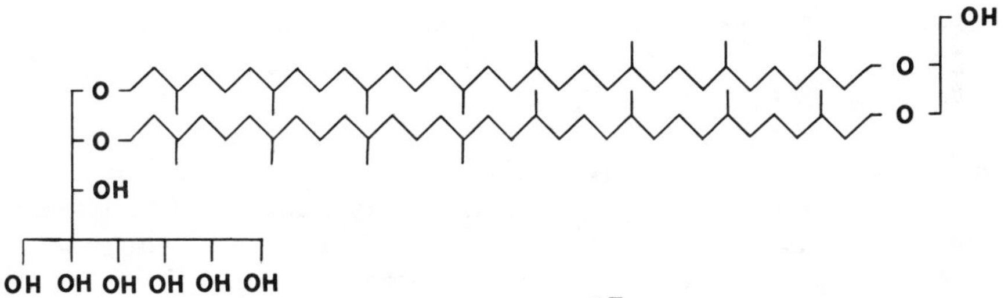

17

A third variety of tetraether has been described which is also known only in *Sulfolobus* (77). It is a minor component, representing only about 3% of the total ether lipids. It could be viewed as a "hybrid" between a diether and a tetraether. It is comprised of two glycerol molecules bridged by one C_{40} biphytanyl chain which is ether-linked to the respective carbon 3 of the two glycerols, but it possesses two "noncondensed" C_{20} phytanyl chains that are ether-linked to carbon 2 of the two glycerols, **18**.

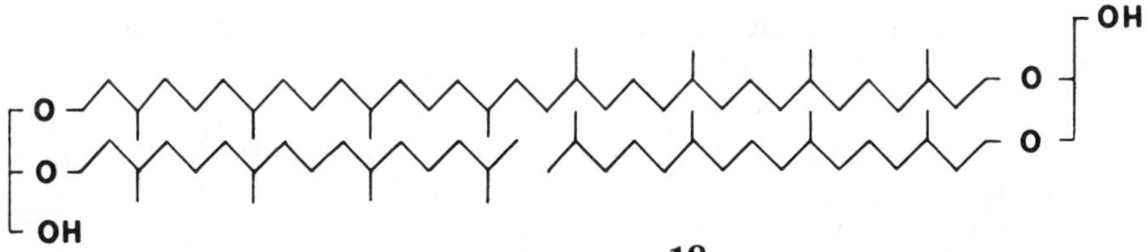

18

Whether this lipid might represent an intermediate in tetraether biosynthesis is not clear (see below). A fourth variation of the basic isopranoid glycerol ethers has also been recently established (81,82). It has been found to constitute about 85% of the total ethers of the thermophilic methanogen, *Methanococcus jannaschii*. This isopranoid glycerol ether can be viewed as a diether that has been covalently linked at the terminal end of the phytanyl chains. It is comprised of a single acyclic C_{40} biphytanyl diol which is ether-linked at the two terminal ends to a single glycerol molecule, resulting in a glycerol diether that contains a single C_{40} macrocyclic loop, **19**.

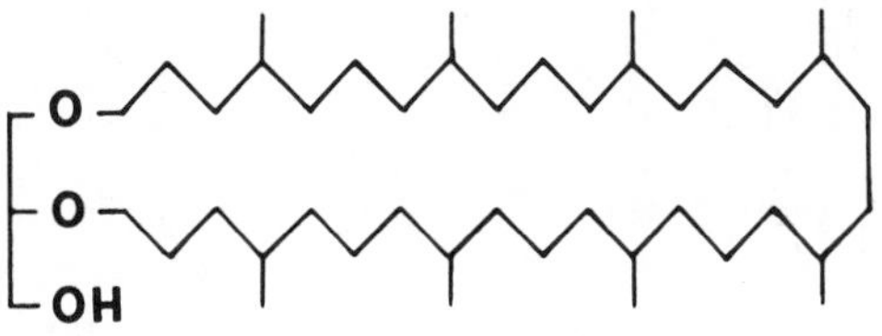

19

As a historical note, this structure is one of the initial two incorrectly proposed structures for what is now the characteristic archaebacterial tetraether (78,83,84).

Although a large number of new thermophilic archaebacteria have now been isolated, very few reports have yet appeared on the detailed nature and distribution of isopranyl ether lipids in these organisms. *Methanobacterium thermoautotrophicum* possesses about 44% diethers and 56% tetraethers (19,85), whereas *Methanothermus* species, which grow at much higher temperatures, contain about 13% diethers and 87% tetraethers (19). Among the nonmethanogenic thermophiles, including *Thermoproteus, Desulfurococcus,* and *Thermodiscus,* only the presence of C_{20} phytanyl and C_{40} biphytanyl chains have been noted as corroborating evidence that these organisms are indeed archaebacteria (19,29–31). Detailed studies are lacking on *Pyrodictium occultum,* the most thermophilic organism known, but preliminary evidence indicates the presence of phytanyl diethers (20%), and tetraethers containing only acyclic C_{40} biphytanyl chains (45%), with the remainder consisting of an as-yet-unidentified isopranoid ether lipid component (32, Langworthy and Stetter, in preparation). Detailed studies on the many new thermophilic archaebacteria (see Chapter 3, this volume) will prove challenging and should provide a number of new variations in isopranyl ether lipid structures.

Molecular Organization   The archaebacterial diethers are capable of providing a typical membrane lipid bilayer, but the tetraethers that occur in many archaebacteria can provide a different molecular organization (17–19). By virtue of their ability to extend from the inner face to the outer face of the membrane, the tetraethers, with a glycerol molecule exposed to the opposing membrane faces, provide a membrane assembly reflecting what could be considered a lipid "monolayer" rather than the normal bilayer of other organisms (Figure 2). Such a membrane, or regions of the membrane, possess the same dimensions as a normal bilayer, but the bilayer has in

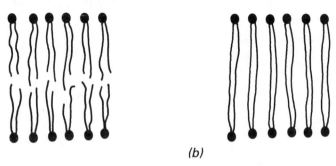

(a)   (b)

Figure 2. Schematic representation of a normal lipid bilayer (a) and of a covalently condensed lipid bilayer or "monolayer" afforded by the tetraethers of archaebacteria (b).

essence been covalently condensed at the center, joining together the inner and outer membrane faces. The inability of either *Thermoplasma* or *Sulfolobus* to freeze-fracture tangentially supports this view (17,19,22,71). Instead, cells characteristically are observed to cross-fracture perpendicularly through the membrane, as would be expected of a covalently cross-linked bilayer. Inner and outer membrane fracture faces are not observed. Moreover, Gliozzi et al. (86), employing artificial monolayer black membranes prepared from nonitolglycerol tetraethers as dispersions with squalene, have shown by current voltage measurements that the tetraethers are indeed organized as a covalently bound bilayer in which the molecules are completely stretched and span the membrane.

How diethers are integrated into the covalently bound bilayer membrane, whether in regions forming patches of normal bilayer or randomly inserted, and how membrane proteins interact and orient in such a hydrophobic matrix are unknown. Also unknown at this time is the nature of the membrane formed by the macrocyclic diether of *M. jannaschii*. Another interesting question is how the apolar chains, fixed at either 20 or 40 carbon atoms, can provide and maintain appropriate membrane fluidity, since there appears to be no way for adjusting chain length in response to temperature fluctuations as in eubacteria. Such a question might, however, be irrelevant. Recent studies on the physical properties of liposomes prepared from the polar lipids of *Thermoplasma* show that these lipids exhibit an extremely low permeability for low-molecular-weight solutes and that the lipids are in a liquid-crystalline state over a wide temperature range from 0 to 80°C (87). Differential scanning calorimetry shows no indication of a phase transition as compared with normal eubacterial membrane lipids. One report even indicates no transition up to 100°C (19). On the other hand, studies employing nonitolglycerol tetraethers from *Sulfolobus*, in which the polar head groups were removed, indicate the presence of conformational transitions and structural polymorphism in these lipids upon temperature variation when examined by X-ray diffraction (88), differential scanning calorimetry, or capacitance and conductance measurements (86,89,90). This latter observation may partly reflect the use of tetraethers lacking polar head groups. It would seem well worthwhile to establish the temperature at which the tetraethers do in fact melt into a completely liquid state. Since complete melting of the tetraethers would result in membrane disruption, this may give some indication of what the upper temperature for life really is.

In any case, tetraethers would seem to be well suited for life at high temperatures. By providing a covalently bound bilayer membrane, inner and outer halves of the bilayer cannot melt apart at high temperatures, thereby maintaining an intact hydrophobic core where normal bilayers would otherwise become disrupted. Furthermore, ether- rather than ester-linkages would protect the thermoacidophilic archaebacterial membrane from destruction due to hydrolysis in their hot acidic environment. However, the advantage of ether bonds for the nonacidophilic thermophiles is not clear.

A role for cyclization within the biphytanyl chains of the thermoacidophilic organisms could also be explained (17,19). Since the alkyl chains are fixed at 40 carbons and extend across the membrane, ring formation would increase internal membrane rigidity and simultaneously shorten tetraether length, thereby controlling membrane density, width, and fluidity. Such a trend, in which there is a corresponding increase in chain cyclization with increasing temperatures, has been observed in both *Sulfolobus* (89,91) and *Thermoplasma* (17). However, such a trend is not a general one in thermophilic archaebacteria, since most anaerobic thermophiles, even *Pyrodictium*, possess only acyclic biphytanyl chains with a trace to no cyclized chains present (Langworthy, Menath, and Stetter, in preparation). It has been suggested that this may simply reflect the fact that oxygen is necessary for ring biosynthesis (R. Thauer, pers. communication).

Biosynthesis   Within the eubacteria, the malonyl CoA pathway for fatty acid synthesis is the major route of apolar chain formation, while the mevalonate pathway for the synthesis of isoprenoids is a minor one. In archaebacteria, the relative importance of the pathways is reversed. The mevalonate route provides the isopranoid apolar chains, but fatty acid synthesis is only very minor (17–19). This represents a major point of divergence between archaebacteria and other groups of organisms. It is clear that the isopranyl C_{20} and C_{40} chains are synthesized via the mevalonate pathway, as evidenced by the incorporation of ^{14}C-mevalonic acid (71,74,75,80). What is not clear is how the glycerol diethers and tetraethers themselves are synthesized. The mevalonate pathway is operative, at least to the C_{20}geranylgeraniol pyrophosphate intermediate (19). Presumably this intermediate could be condensed to glycerol followed by a series of saturation steps to yield the phytanyl diethers (19,74,75). Perhaps the 2,3-cyclic glycerol phosphate, recently shown to occur in archaebacteria, may be involved (92). No clear evidence is available on the synthesis of tetraethers. However, based upon structural considerations (19,74,75), the fact that tetraethers are dimers of two "head-to-head" diethers condensed through the geminal ends of the phytanyl chains, suggests that they may be formed by covalent linkage of two diethers, or, in fact, of two diethers in possession of polar head groups (19). The fact that free C_{40} biphytanes or diols have not yet been detected in archaebacteria suggests such a possibility (17). How the diethers and tetraethers are biosynthesized presents a challenge and should prove to be unique.

3.6. Tetrahydroxybacteriohopane

Cyclized C_{30} triterpenes of the hopanoid type are known to be frequent constituents in many eubacteria (93–96). They are the structural equivalent of sterols and, because their biosynthesis from squalene is oxygen independent, they have been regarded as functional equivalents of sterols (94,96).

Among the thermophilic and acidophilic eubacteria, *Bacillus acidocaldarius* and related strains have been shown to possess large quantities (up to 1.5% of the cell dry weight) of a C_{35} hopanoid derivative, tetrahydroxybacterio-hopane (97,98). This fairly polar compound contains the C_{30} hopane nucleus, but it is substituted at carbon 29 with an *n*-1,2,3,4-tetrahydroxypentane chain, **20**.

20

In *B. acidocaldarius*, this compound serves as an apolar residue of the glycolipids of the organisms (see Section 4.1). *B. acidocaldarius* is thus far the only thermophile known to contain such a hopanoid.

A number of studies employing model membrane systems have shown that, like the sterols of eukaryotic cells, hopanoids are efficient in condensing artificial membranes (99), enhancing membrane viscosity (63), and decreasing phase transitions of lipids (100). Hopanoids can also serve as effective reinforcers of membranes (94,96). Although tetrahydroxybacteriohopanes occur in both mesophilic acidophiles and nonacidophiles (96,101), these compounds would seem of adaptive value to *B. acidocaldarius* since the membrane would likely become less stable and more permeable at high temperature in their absence (102).

4. COMPLEX LIPIDS

In comparison to the number of reports dealing with the nature and function of the apolar chains of thermophilic bacteria (facilitated no doubt by the ease of gas chromatography), little is really known about the structure or function of the complex lipids in these organisms. The extractable complex lipids consist of neutral lipids and polar lipids in which head groups such as sugars or phosphate radicals are attached to a hydrophobic glycerolipid. The polar lipids are therefore made up of glycolipids and acidic lipids, primarily phospholipids in eubacteria and phosphoglycolipids in archaebacteria (19,37). What will be described here are examples of the complex lipids of four different genera in which some of the lipid structures have been fairly well established. These include the two eubacterial genera *Bacillus* and *Thermus* and the two archaebacteria *Thermoplasma* and *Sulfo-*

lobus. Although studies are limited, it is clearly evident that thermophiles possess large amounts of glycolipids in comparison to mesophiles, suggesting a major role for these compounds in membrane thermostabilization.

4.1. *Bacillus*

Within the thermophilic and nonacidophilic *Bacillus* species, both *B. stearothermophilus* (103,104) and *B. caldotenax* (46) have been shown to contain mainly phospholipids as the principal complex lipid constituents. They are comprised of phospholipids common to mesophilic bacteria and include diphosphatidylglycerol (DPG), phosphatidylglycerol (PG) and phosphatidylethanolamine (PE). These phospholipids constitute 60 to 90% of the complex lipids. When the growth temperature of *B. caldotenax* was lowered from 65 to 40°C, Hasegawa et al. (46) found a decrease in the higher-melting point PE content and a concomitant increase in the lower-melting PG.

The lipids of the acidophile *B. acidocaldarius* have been more completely characterized (97,98). Extractable lipids represent about 8% of the cell dry weight and are made up of about 8% neutral lipids, 64% glycolipids and 20% acidic lipids. A number of the neutral lipids have been identified and include squalene, a precursor of pentacyclic triterpenes, along with free hopanoids consisting of about 86% hop-22(29)-ene and also including smaller amounts of hop-17(21)-ene and hopane (105). These three hopanes accounted for about 0.3% of the cell dry weight. Small amounts of free tetrahydroxybacteriohopanes are also present (98), as well as a series of polyprenols between 9 to 12 isoprene units long with the C_{50} and C_{55} compounds dominating (105). The lipoquinone, menaquinone-9, is also present (105).

Glycolipids are the major polar lipid class in *B. acidocaldarius* (98) and are comprised of two basic structural types: glucosyl-glucosamidyl-diacylglycerols and glucosamidyl-tetrahydroxy-bacteriohopanes. The diglycosyldiacylglycerols comprise about 70% of the total glycolipids and occur as glucosyl-(β-1→4)-N-acylglucosaminyl-(β-1→1)-diacylglycerol (25%), glucosyl-(β-1→4)-N-acylglucosaminyl-(β-1→1)-monoacylglycerol (41%) with trace amounts of glucosyl-(β-1→4)-N-acylglucosaminyl-(β-1→1)-glycerol. The amide-linked fatty acids in these glycolipids have a compositional distribution similar to the ester-linked acids. The remainder of the glycolipids consists of an unusual N-acylglucosamine in β-glycosidic linkage to the primary hydroxyl group of tetrahydroxybacteriohopane having the structure, 1-(O-β-N-acylglucosaminyl)-2,3,4-tetrahydroxypentane-29-hopane (98). It has been shown that this glycolipid produces a condensing effect in artificially prepared monolayer membranes, suggesting that, like cholesterol, the lipid may aid in limiting diffusion of H^+ ions through the membrane (99,106). On the other hand, as noted by these same investigators (102), direct measurement of the tetrahydroxybacteriohopane content by gas chromatography showed a significant increase with increasing temperatures up to the growth maximum of 65°C. The effect of pH did not appear to influence the tetrahydroxybacteriohopane content to any significant extent. It should be noted that by employing ^{14}C-mevalonate-labeled lipids, the tetrahydroxybacteriohopane glycolipid content does exhibit an increase at lower pH values, but seems uninfluenced by temperature (Ali and Langworthy, in preparation).

Among the acidic lipids, ordinary phospholipids, DPG, PG, PE, and traces of phosphatidic acid (PA) account for only 57% of the total charged lipids. The remaining 43% is a sulfonolipid (98). Sulfonolipids that contain a $C-SO_3^{2-}$ bond are common in photosynthetic cells, but are rare in non-photosynthetic bacteria (107). The sulfonolipid of *B. acidocaldarius* appears to be identical to the plant sulfonolipid, 6-sulfonoquinovosyl-(α-1→1)-di-acylglycerol. A function for this lipid is unknown, but would seem relevant to the acidophily of the organism. It has been suggested (98) that perhaps it functions in H^+ ion exclusion, or since it may remain ionized at all pH values, it may serve as a type of ion scavenger or exchanger. Preliminary evidence, however, suggests that the total content of the sulfonolipid is unaffected by alterations of pH values from 2 to 6 (Ali and Langworthy, in preparation).

4.2. Thermus

The complex lipids of several species of the genus *Thermus* have been identified, including *T. aquaticus, T. thermophilus*, and *T. flavus*, as well as several strains isolated from Iceland and from domestic hot water heaters (108–113). Total extractable lipids of the organisms amount to about 8 to 12% of the cell dry weight. Ray et al. (108) identified several of the neutral lipid components of *T. aquaticus*, including menaquinones-7, -8, and -9 and found the C_{40} carotenoids phytoene and Δ-carotene, along with large amounts of very polar carotenoids which were not identified. The neutral lipids, responsible for the yellow color of the organism, were reported to be about 60% of the total lipids, although this value would seem to be much lower in other species (113).

The glycolipids of *Thermus* are quite unusual and can occur in amounts of up to 50 to 70% of the total lipids in some species (110,113). The major glycolipid structure was first identified in *T. thermophilus* strain HB8 by Oshima and Yamakawa (109,112). It is a tetraglycosyldiacylglycerol which possesses four sugars as a tetrasaccharide glycosidically linked to the diac-ylglycerol residue. This is notable since the size of the hydrophobic moiety is double that normally found in mesophilic bacteria (114). Its structure has been established to be

galactosyl-(β-1→2)-galactosyl-(α-1→6)-N-acylglucosaminyl-(β-1→2)-
glucosyl-(α-1→1)-diacylglycerol.

The terminal galactose sugar is in the furanose form and *iso*-C_{17} is the only fatty acid in amide-linkage to glucosamine. Pask-Hughes and Shaw (113) have surveyed five different species of *Thermus* and have found all to contain similar tetraglycosyldiacylglycerols. Although the complete structures, in terms of types of glycosidic linkages or proving the presence of terminal furanose sugars was not established, the general structural features were elucidated. *T. aquaticus* strain YT1 contained glucosyl-glucosyl-glucosaminyl-glucosyl-diacylglycerol; *Thermus* species H, glucosyl-glycosyl-glucosaminyl-glucosyl-diacylglycerol; *Thermus* species NH, galactosyl-galactosyl-glucosaminyl-glucosyl-diacylglycerol; and *Thermus* species J galactosyl-glucosyl-glucosaminyl-glucosyl-diacylglycerol. Whether the galactose or glucose is the terminal sugar of the latter component was not established. *Iso*-C_{15} and *iso*-C_{17} fatty acids were the major ester- and amide-linked apolar chains.

The phospholipids of *Thermus* species are also unusual and perhaps unique. Ray et al. (108) identified DPG (3%), PA (1%), PG (3%) and phosphatidylinositol (PI) in the amount of 10% of the total phospholipids of *T. aquaticus*. The major phospholipid, however, which amounted to almost 80% of the lipid phosphorus, was only partially characterized. It was shown to contain one phosphate, three fatty acids, one glycerol, and a presumed long-chain amine containing a hydroxyl group. The minimum molecular weight of this compound was estimated to be 1,800. Among the *Thermus* species examined by Pask-Hughes and Shaw (113), all were found to possess this component in amounts of 35 to 42% of the total lipids in the strains examined. It was also found that the compound contained glucosamine which was N-acylated. The lipid therefore appears to be a glycophospholipid of some type. It would certainly be desirable to fully establish the structure of this phospholipid since it may prove relevant to the thermostability of the *Thermus* membrane.

Like *B. acidocaldarius*, most of the polar lipids of *Thermus* species contain carbohydrates which also include amino sugars. This suggests that sugar residues as polar head groups may have a special influence on the thermostability of membranes. It is of interest that spheroplasts prepared from *T. aquaticus* did not undergo thermal lysis even when boiled in distilled water (115). Furthermore, it has been shown (111,116) that by increasing glycolipid concentration or by the addition of glycolipids with an increasing number of sugars, phospholipid bilayers tend to become more rigid. The effect of growth temperature on the lipid composition should also be noted. When the growth temperature is increased from 50 to 75°C, there is a progressive increase in the total lipid content, and the amount of carotenoids increase 1.8-fold, phospholipids, 2-fold, and glycolipids 2- to 4-fold (108,110).

4.3. Thermoplasma

Thermoplasma is one of the more unusual thermophiles. Although only moderately thermophilic, growing optimally at 59°C, it is acidophilic (optimum, pH 2) and lacks a cell wall so that its membrane has no protection against the external conditions which consist essentially of an acid hydrolysis. Its uniqueness is also apparent by its obligatory requirement for H^+ions for cellular stability, as cells begin to lyse at pH values about 5 (117). There has been considerable interest in the structure of the *Thermoplasma* membrane (20,22,83), and historically it was the first archaebacterium to be shown to possess diglycerol tetraethers (71). In spite of such interest, the complex lipids are still ill defined.

The total lipids of *Thermoplasma* account for about 3% of the cell dry weight, or in isolated membranes, about 25%, while sugars make up about 10% and proteins constitute the remaining 65% (83,117). The extractable lipids contain about 17% neutral lipids, 25% glycolipids, and 57% phosphoglycolipids.

A variety of hydrocarbons have been identified in the neutral lipid fraction, which include characteristic archaebacterial isoprenoid constituents (19). These include the C_{30}isoprenoid squalene, a C_{25} hydropentaisoprene, C_{20} phytatetraene, and a series of C_{15} to C_{30} isoprenes, as well as a series of alkyl benzenes (17,19,118,119). Also found are lipoquinones, such as menaquinone-7 and small amounts of menaquinone-5 and -6, and a partially characterized quinone ($C_{47}H_{66}O_2$) designated thermoplasmaquinone that accounts for 25% of the total quinones (83,120).

The tetraether glycolipid structures of *Thermoplasma* are still unknown, although there are at least six different glycolipid species present (83). One component, however, which can be considered to be a glycolipid with an extended, nonbranched 25-sugar chain, has been identified (121,122). This lipoglycan, which is extractable by hot aqueous phenol but not the usual lipid solvents, accounts for about 3% of the cell dry weight. It has a molecular weight of 5,300 and is made up of one glucose and 24 mannose residues glycosidically linked to one side of a diglycerol tetraether molecule. It has the structure

[mannosyl-(α-1→2)-mannosyl-(α-1→4)-mannosyl-(α-1→3)]$_8$-glucosyl-
(α-1→1)-O-[diglyceroltetraether]-OH.

The molecule displays physical properties similar to Gram-negative bacterial lipopolysaccharides (123) and is located on the cell surface with the sugar chain extending into the external environment (124).

The phospholipids of *Thermoplasma* are not well characterized, but all appear to contain glycerolphosphate residues linked to diglycerol tetraethers (83). They also all contain an unusual unidentified sugar which is attached to the tetraether on the side opposite the glycerolphosphate radical (Langworthy, unpublished). They are all phosphoglycolipids and the major one, which can only be described as a glycerylphosphoryl-O-[diglyceroltetraether]-O-sugar, represents about 80% of the phospholipids and at least half of the total lipids of the organism. Four other minor phosphoglycolipids also contain free amine groups. Thus, even in this thermophilic archaebacterium, sugars again appear to dominate the polar lipid structure.

4.4. *Sulfolobus*

The complex lipids of *Sulfolobus* are the best defined of all the lipids of the thermophilic archaebacteria (78,80,125). Extractable lipids represent about 3 to 5% of the cell dry weight and consist of about 10% neutral lipids, 68% glycolipids, and 22% acidic lipids (78,80). Neutral lipids identified include the archaebacterial hydrocarbons, C_{30} octa- and decahydrosqualene, C_{25} fully saturated pentaisoprenes, C_{20} phytadiene and phytatriene, and alkyl benzenes (17,19,118,119). A tri-O-phytanyl glycerol is also present (126) which contains partially or fully saturated phytanyl chains. A new type of sulfur-containing quinone has also been identified (127), called caldariellaquinone, and shown to be benzo[b]thiophen-4,7-quinone.

The glycolipids are made up of two basic species, depending upon whether sugars are glycosidically bound to either the diglycerol tetraethers or nonitolglycerol tetraethers of the organism (see Section 3.5). Two sugars are attached to one side of the diglycerol tetraethers and one sugar to the nonitolglycerol tetraethers. They have been partially characterized (78,125) as the following:

glucosyl-($\beta\rightarrow$)-galactosyl-($\beta\rightarrow$)-O-[diglyceroltetraether]-OH and glucosyl-
($\beta\rightarrow$)-O-[nonitolglyceroltetraether]-OH.

Whereas the disaccharide unit is linked to glycerol, the monosaccharide is attached to one of the hydroxyl groups of the nonitol, but which one has not been established (125). In cells grown heterotrophically, the two glycolipids are present in about equal amounts, but in autotrophically grown cells, with sulfur as the energy source, the glucosyl-nonitolglycerol tetraethers predominate, representing close to 75% of the total glycolipids (80).

The acidic lipids contain one true phospholipid (20%), two phosphoglycolipids (75%), and one sulfolipid (5%) (78,80,125). The phospholipid is the diglycerol tetraether analogue of phosphatidylinositol in which inositol phosphate is linked to one side of the apolar tetraether as inositolphosphoryl-O-[diglyceroltetraether]-OH. The two phosphoglycolipids are the inositol phosphate derivatives of the two glycolipids of the organism, although where the inositol phosphate is attached to the glycolipids is not clear. A small amount of glycolipid sulfate is also present in which a single SO_4^{2-} radical is attached to the glucosyl-nonitolglycerol tetraether, but precisely where is not known. Two unidentified minor polar lipids have also been shown to be present in cells grown on sulfur (80). Thus, as with the other thermophiles examined, almost 90% of the complex lipids are polar lipids and all contain carbohydrates or a carbohydrate "surrogate", such as inositol.

5. CONCLUSION

Theoretical considerations on the molecular nature of thermophily by Brock (128) suggested that the plasma membrane would be one of the most critical and vulnerable cellular components and it must possess an inherent stability since it is the structure most directly exposed to the hot external environment. In terms of membrane lipids at least, we are far from an understanding of such stability. There is perhaps no single aspect characteristic of all thermostable membranes, as bacteria have probably evolved a variety of strategies to cope with—and to require—high temperatures. Analyses of thermophile lipids suggest some general trends, but often raise even more questions. For example, eubacterial as well as archaebacterial thermophiles possess methyl-branching as a major characteristic of the apolar chains. However, a number of physical studies—employing model membranes or natural membranes enriched in branched-chain fatty acids—have indicated that these act as membrane "fluidizers" rather than "stabilizers" as would be expected (58,129–132). Does this suggest that features other than the apolar chains are more important in thermal stabilization, or could methyl-branching be a means of allowing thermophiles to adjust to low temperature fluctuations? Studies are limited, but two features appear relevant. One is the nature of the polar lipid head groups. In all thermophiles examined, the polar lipids are enriched in carbohydrates, in some to the point that all contain a sugar residue. This implies a significant role in thermal stability. The second point that seems critical is the manner in which the apolar chains are linked to glycerol, and whether a lipid bilayer or covalently condensed bilayer can be formed. Eubacteria are not known to grow optimally above 85°C, presumably because a normal bilayer will begin to melt apart at some point. There are, however, a variety of archaebacteria growing above this temperature and these possess the covalently condensed bilayers provided by the tetraether glycerolipids, so that the membrane can not melt apart. Thus, it may be postulated that eubacteria, able to synthesize the analogous diglycerol tetraester lipids, may be found growing in the same temperature range as the extremely thermophilic archaebacteria. In any case, thermophilic membranes and lipids still present a unique and challenging problem leaving many unanswered questions about their structural and functional relationships.

A complete understanding of the molecular basis for thermostable membranes has not been achieved, but a large number of new and unique lipids have been uncovered and these are proving of interest in other respects. They provide a biological source for a number of lipids previously known only in sediments and deposits of shale and oil, and they can serve as molecular "signatures" for the existence of microbial populations in geological environments (17,19,133,134). Since they may have special physical properties, thermophile lipids may be of use in biotechnology as indicated by a preliminary report of the use of tetraether membrane lipids in desalinization (135). Finally, the unusual nature of many of the thermophile lipid structures is proving useful as a chemotaxonomic tool for assessing phylogenetic relationships (17,19,136). As the functions of these lipids become better defined, and as the new lipids which will surely be discovered are characterized, they should provide new insights into the evolution of prokaryotic cells from more primitive organisms.

ACKNOWLEDGMENTS

Portions of the work presented herein were supported in part by a Senior U.S. Scientist Prize from the Alexander-von-Humboldt-Stiftung to T.A.L. We thank C. Stadler for excellent editorial assistance.

REFERENCES

1. A. F. Esser, in M. Shilo, Ed., *Strategies of Microbial Life in Extreme Environments*, Dahlem Konferenzen, Verlag Chemie, Weinheim, 1979, p. 433.
2. M. B. Allen, *Bacteriol. Rev.* **17**, 125 (1953).
3. U. J. Heinen and W. Heinen, *Arch. Mikrobiol.* **72**, 199 (1972).
4. G. Darland and T. D. Brock, *J. Gen. Microbiol.* **67**, 9 (1971).
5. T. D. Brock and H. Freeze, *J. Bacteriol.* **98**, 289 (1969).
6. T. Saiki, R. Kimura, and K. Arima, *Agric. Biol. Chem.* **36**, 2357 (1972).
7. T. Oshima and K. Imahori, *Int. J. Syst. Bacteriol.* **24**, 102 (1974).
8. T. Jackson, R. F. Ramaley, and W. G. Meinschein, *Int. J. Syst. Bacteriol.* **23**, 28 (1973).
9. G. J. Merkel, D. R. Durham, and J. J. Perry, *Can. J. Microbiol.* **26**, 556 (1980).
10. W. E. Phillips, Jr. and J. J. Perry, *Int. J. Syst. Bacteriol.* **26**, 220 (1976).
11. M. Chan, R. H. Himes, and J. M. Akagi, *J. Bacteriol.* **106**, 876 (1971).
12. T. K. Ng, P. J. Weimer, and J. G. Zeikus, *Arch. Microbiol.* **114**, 1 (1977).
13. J. Wiegel, L. G. Ljungdahl, and J. R. Rawson, *J. Bacteriol.* **139**, 800 (1979).
14. B. Schink and J. G. Zeikus, *J. Gen. Microbiol.* **129**, 1149 (1983).
15. J. G. Zeikus, M. A. Dawson, T. E. Thompson, K. Ingvorsen, and E. C. Hatchikian, *J. Gen. Microbiol.* **129**, 1159 (1983).
16. C. R. Woese and R. S. Wolfe, Eds., *The Bacteria*, Vol. 8, Academic Press, New York, 1985.
17. T. A. Langworthy, T. G. Tornabene, and G. Holzer, *Zbl. Bakt. Hyg., I. Abt. Orig. C* **3**, 228 (1982).
18. T. A. Langworthy, *Curr. Top. Membr. Transp.* **17**, 45 (1982).
19. T. L. Langworthy, in C. R. Woese and R. S. Wolfe, Eds., *The Bacteria*, Vol. 8, Academic Press, New York, 1985, p. 459.
20. R. T. Belly, B. B. Bohlool, and T. D. Brock, *Ann. N.Y. Acad. Sci.* **225**, 94 (1973).
21. T. A. Langworthy and P. F. Smith, in N. R. Krieg, Ed., *Bergey's Manual of Systematic Bacteriology*, Vol. I, Williams and Wilkins, Baltimore, 1984, p. 790.
22. T. A. Langworthy, in M. F. Barile and S. Razin, Eds., *The Mycoplasmas*, Vol. I, Academic Press, New York, 1979, p. 495.
23. T. D. Brock, K. M. Brock, R. T. Belly, and R. L. Weiss, *Arch. Microbiol.* **84**, 54 (1972).
24. M. De Rosa, A. Gamborcorta, and J. D. Bu'Lock, *J. Gen. Microbiol.* **86**, 156 (1975).
25. W. Zillig, K. O. Stetter, S. Wunderl, W. Schulz, H. Priess, and I. Scholz, *Arch. Microbiol.* **125**, 259 (1980).

26. J. G. Zeikus and R. S. Wolfe, *J. Bacteriol.* **109**, 707 (1972).
27. K. O. Stetter, M. Thomm, J. Winter, G. Wildgruber, H. Huber, W. Zillig, D. Janecovic, H. König, P. Palm, and S. Wunderl, *Zbl. Bakt. Hyg. I. Abt. Orig. C* **2**, 166 (1981).
28. W. J. Jones, J. A. Leigh, F. Mayer, C. R. Woese, and R. S. Wolfe, *Arch. Microbiol.* **136**, 254 (1983).
29. W. Zillig, K. O. Stetter, W. Schäfer, D. Janekovic, S. Wunderl, I. Holz, and P. Palm, *Zbl. Bakt. Hyg. I. Abt. Orig. C* **2**, 205 (1981).
30. W. Zillig, K. O. Stetter, D. Prangishvilli, W. Schäfer, S. Wunderl, D. Janekovic, I. Holz, and P. Palm, *Zbl. Bakt. Hyg. I. Abt. Orig. C* **3**, 304 (1982).
31. F. Fischer, W. Zillig, K. Stetter, and G. Schreiber, *Nature* **301**, 511 (1983).
32. K. O. Stetter, H. König, and E. Stackebrandt, *System. Appl. Microbiol.* **4**, 535 (1983).
33. J. Israelachvili, in D. W. Deamer, Ed., *Light Transducing Membranes*, Academic Press, New York, 1978, p. 91.
34. A. G. Marr and J. L. Ingraham, *J. Bacteriol.* **84**, 1260 (1962).
35. T. A. Langworthy, *Methods Enzymol.* **88**, 396 (1982).
36. M. Kates, in T. S. Work and E. Work, Eds., *Laboratory Techniques in Biochemistry and Molecular Biology*, Vol. 3, Am. Elsevier, New York, 1972, p. 269.
37. H. Goldfine, *Curr. Top. Membr. Transp.* **17**, 1 (1982).
38. K. Y. Cho and M. R. J. Salton, *Biochim. Biophys. Acta* **116**, 73 (1966).
39. H. H. Daron, *J. Bacteriol.* **101, 145 (1970).**
40. M. Yao, H. H. Walker, and D. A. Lillard, *J. Bacteriol.* **102**, 877 (1970).
41. P. Y. Shen, E. Coles, J. L. Foote, and J. Stenesh, *J. Bacteriol.* **103**, 479 (1970).
42. K. C. Oo and K. L. Lee, *J. Gen. Microbiol.* **69**, 287 (1971).
43. M. Oshima and A. Miyagawa, *Lipids* **9**, 476 (1974).
44. U. J. Heinen and W. Heinen, *Arch. Mikrobiol.* **82**, 1 (1972).
45. A. Weerkamp and W. Heinen, *J. Bacteriol.* **109**, 443 (1972).
46. Y. Hasegawa, N. Kawada, and Y. Nosoh, *Arch. Microbiol.* **126**, 103 (1980).
47. W. Heinen, H. P. Klein, and C. M. Volkmann, *Arch. Mikrobiol.* **72**, 199 (1970).
48. P. H. Ray, D. C. White, and T. D. Brock, *J. Bacteriol.* **106**, 25 (1971).
49. G. J. Merkel and J. J. Perry, *Appl. Environ. Microbiol.* **34**, 626 (1977).
50. A. A. Herrero, R. F. Gomez, and M. F. Roberts, *Biochim. Biophys. Acta* **693**, 195 (1982).
51. J. L. Pond, T. A. Langworthy, and G. Holzer, *Science* **231**, 1134 (1986).
52. M. De Rosa, A. Gambacorta, L. Minale, and J. D. Bu'Lock, *Biochem. J.* **128**, 751 (1972).
53. M. Oshima and T. Ariga, *J. Biol. Chem.* **250**, 6963 (1975).
54. M. Oshima, Y. Sakaki, and T. Oshima, in S. M. Friedman, Ed., *Biochemistry of Thermophily*, Academic Press, New York, 1978, p. 31.
55. Y. Sakaki, M. Oshima, K. Yamada, and T. Oshima, *J. Biochem.* **82**, 1457 (1977).
56. M. De Rosa, A. Gambacorta, and J. D. Bu'Lock, *J. Bacteriol.* **117**, 212 (1974).
57. A. Blume, R. Dreher, and K. Poralla, *Biochim. Biophys. Acta* **512**, 489 (1978).
58. H. W. Beam and J. J. Perry, *J. Bacteriol.* **118**, 394 (1974).
59. K. I. Suzuki, K. Saito, A. Kawaguchi, S. Okuda, and K. Komagata, *J. Gen. Appl. Microbiol.* **27**, 261 (1981).
60. T. Endo, K. Inoue, S. Nojima, S. Terashima, and T. Oshima, *Chem. Phys. Lipids* **31**, 61 (1982).
61. E. Kannenberg, A. Blume, and K. Poralla, *FEBS Lett.* **172**, 331 (1984).
62. R. Benz, D. Hallman, K. Poralla, and H. Eibl, *Chem. Phys. Lipids* **34**, 7 (1983).
63. J. Sunamoto, K. Iwamoto, K. Inoue, T. Endo, and S. Nojima, *Biochim. Biophys. Acta* **685**, 283 (1982).
64. K. Poralla and W. A. König, *FEMS Microbiol. Letters* **16**, 303 (1983).
65. H. Allgaier, K. Poralla, and G. Jung, *Liebigs Ann. Chem.* **1985**, 378 (1985).
66. R. A. Klein, G. P. Hazelwood, P. Kemp, and R. M. C. Dawson, *Biochem. J.* **183**, 691 (1979).
67. G. P. Hazelwood, N. G. Clarke, and R. M. C. Dawson, *Biochem. J.* **191**, 555 (1980).

68. T. A. Langworthy, G. Holzer, J. G. Zeikus, and T. G. Tornabene, *System. Appl. Microbiol.* **4**, 1 (1983).
69. P. E. Kolattukudy, in P. K. Stumph and E. E. Conn, Eds., *The Biochemistry of Plants*, Vol. 4, Academic Press, New York, 1980, p. 571.
70. D. M. S. Horn and F. W. J. Hougen, *J. Chem. Soc.* **1953**, 3533 (1953).
71. T. A. Langworthy, *Biochim. Biophys. Acta* **487**, 37 (1977).
72. M. De Rosa, S. De Rosa, A. Gambacorta, L. Minale, and J. D. Bu'Lock, *Phytochemistry* **16**, 1961 (1977).
73. M. De Rosa, S. De Rosa, and A. Gambacorta, *Phytochemistry* **16**, 1909 (1977).
74. M. De Rosa, A. Gambacorta, B. Nicolaus, S. Sodano, and J. D. Bu'Lock, *Phytochemistry* **19**, 833 (1980).
75. M. De Rosa, A. Gambacorta, and B. Nicolaus, *Phytochemistry* **19**, 791 (1980).
76. M. Kates, *Prog. Chem. Fats Other Lipids*, **15**, 301 (1978).
77. M. De Rosa, A. Gambacorta, B. Nicolaus, B. Chappe, and P. Albrecht, *Biochim. Biophys. Acta* **753**, 249 (1983).
78. T. A. Langworthy, P. F. Smith, and W. R. Mayberry, *J. Bacteriol.* **119**, 106 (1974).
79. M. De Rosa, S. De Rosa, A. Gambacorta, and J. D. Bu'Lock, *Phytochemistry* **19**, 249 (1980).
80. T. A. Langworthy, *J. Bacteriol.* **130**, 1326 (1977).
81. P. B. Comita, R. B. Gagosian, H. Pang, and C. E. Costello, *J. Biol. Chem.* **259**, 15234 (1984).
82. P. B. Comita and R. B. Gagosian, *Science* **222**, 1329 (1983).
83. T. A. Langworthy, P. F. Smith, and W. R. Mayberry, *J. Bacteriol.* **112**, 1193 (1972).
84. M. De Rosa, A. Gambacorta, and J. D. Bu'Lock, *Phytochemistry* **15**, 143 (1976).
85. T. G. Tornabene and T. A. Langworthy, *Science* **203**, 51 (1979).
86. A. Gliozzi, R. Rolandi, M. De Rosa, and A. Gambacorta, *J. Membrane Biol.* **75**, 45 (1983).
87. D. Blöcher, R. Gutermann, B. Henkel, and K. Ring, *Biochim. Biophys. Acta* **778**, 74 (1984).
88. A. Gulik, V. Luzzati, M. De Rosa, and A. Gambacorta, *J. Molecular Biol.* **182**, 131 (1985).
89. A. Gliozzi, G. Paoli, M. De Rosa, and A. Gambacorta, *Biochim. Biophys. Acta* **735**, 234 (1983).
90. P. I. Lelkes, D. Goldenberg, A. Gliozzi, M. De Rosa, A. Gambacorta, and I. R. Miller, *Biochim. Biophys. Acta* **732**, 714 (1983).
91. M. De Rosa, A. Gambacorta, B. Nicolaus, and J. D. Bu'Lock, *Phytochemistry* **19**, 821 (1980).
92. R. J. Seely and D. E. Fahrney, *Current Microbiol.* **10**, 85 (1984).
93. M. Rohmer, P. Bouvier, and G. Ourisson, *Proc. Natl. Acad. Sci. U.S.A.* **76**, 847 (1979).
94. G. Ourisson and M. Rohmer, *Curr. Top. Membr. Transp.* **17**, 153 (1982).
95. K. Poralla, *FEMS Microbiol. Lett.* **13**, 131 (1982).
96. R. F. Taylor, *Microbiol. Rev.* **48**, 181 (1984).
97. T. A. Langworthy and W. R. Mayberry, *Biochim. Biophys. Acta* **431**, 570 (1976).
98. T. A. Langworthy, W. R. Mayberry, and P. F. Smith, *Biochim. Biophys. Acta* **432**, 550 (1976).
99. K. Poralla, E. Kannenberg, and A. Blume, *FEBS Lett.* **113**, 107 (1980).
100. E. Kannenberg, A. Blume, R. N. McElhaney, and K. Poralla, *Biochim. Biophys. Acta* **733**, 111 (1983).
101. S. Bringer, T. Härtner, K. Poralla, and H. Sahm, *Arch. Microbiol.* **140**, 312 (1985).
102. K. Poralla, T. Härtner, and E. Kanenberg, *FEMS Microbiol. Lett.* **23**, 253 (1984).
103. G. L. Card, C. E. Georgi, and W. E. Militzer, *J. Bacteriol.* **97**, 186 (1969).
104. G. L. Card, *J. Bacteriol.* **114**, 1125 (1973).
105. M. De Rosa, A. Gambacorta, L. Minale, and J. D. Bu'Lock, *Phytochemistry* **12**, 1117 (1973).

106. E. Kannenberg, K. Poralla, and A. Blume, *Naturwissenschaften* **67**, 458 (1980).
107. T. H. Haines, *Prog. Chem. Fats Other Lipids* **2**, 297 (1971).
108. P. H. Ray, D. C. White, and T. D.Brock, *J. Bacteriol.* **108**, 227 (1971).
109. M. Oshima and T. Yamakawa, *Biochemistry* **13**, 1140 (1974).
110. M. Oshima, in S. M. Friedman, Ed., *Biochemistry of Thermophily*, Academic Press, New York, 1978, p. 1.
111. R. A. Pask-Hughes, H. Mozaffary, and N. Shaw, *Biochem. Soc. Trans.* **5**, 1675 (1977).
112. M. Oshima and T. Ariga, *FEBS Lett.* **64**, 440 (1976).
113. R. A. Pask-Hughes and N. Shaw, *J. Bacteriol.* **149**, 54 (1982).
114. N. Shaw, *Adv. Microbiol. Physiol.* **12**, 141 (1975).
115. T. D. Brock, *Thermophilic Microorganisms and Life at High Temperatures*, Springer-Verlag, New York, 1978, p. 85.
116. F. Sharom, D. G. Barratt, A. E. Thede, and C. W. M. Grant, *Biochim. Biophys. Acta* **455**, 485 (1976).
117. P. F. Smith, T. A. Langworthy, W. R. Mayberry, and A. E. Houghland, *J. Bacteriol.* **116**, 1019 (1973).
118. T. G. Tornabene, T. A. Langworthy, G. Holzer, and J. Oró, *J. Mol. Evol.* **13**, 73 (1979).
119. G. Holzer, J. Oró, and T. G. Tornabene, *J. Chromatography* **186**, 795 (1979).
120. M. D. Collins and T. A. Langworthy, *System. Appl. Microbiol.* **4**, 295 (1983).
121. K. J. Mayberry-Carson, T. A. Langworthy, W. R. Mayberry, and P. F. Smith, *Biochim. Biophys. Acta* **360**, 217 (1974).
122. P. F. Smith, *Biochim. Biophys. Acta* **691**, 367 (1980).
123. K. J. Mayberry-Carson, I. L. Roth, and P. F. Smith, *J. Bacteriol.* **121**, 700 (1975).
124. K. J. Mayberry-Carson, M. J. Jewell, and P. F. Smith, *J. Bacteriol.* **133**, 1510 (1978).
125. M. De Rosa, A. Gambacorta, B. Nicolaus, and J. D. Bu'Lock, *Phytochemistry* **19**, 821 (1980).
126. M. De Rosa, S. De Rosa, A. Gambacorta, and J. D. Bu'Lock, *Phytochemistry* **15**, 1996 (1976).
127. M. De Rosa, S. De Rosa, A. Gambacorta, L. Minale, R. H. Thomson, and R. D. Worthington, *J. Chem. Soc. Perkin. Trans.* **1**, 653 (1977).
128. T.D. Brock, *Science* **158**, 1012 (1967).
129. J. R. Silvius and R. N. McElhaney, *Chem. Phys. Lipids* **24**, 287 (1979).
130. J. R. Silvius and R. N. McElhaney, *Chem. Phys. Lipids* **26**, 67 (1980).
131. R. N. McElhaney, *J. Mol. Biol.* **84**, 145 (1974).
132. E. Kannenberg, A. Blume, R. N. McElhaney, and K. Poralla, *Biochim. Biophys. Acta* **733**, 111 (1983).
133. G. Ourisson, P. Albrecht, and M. Rohmer, *Trends Biochem. Sci. (TIBS)—*, 236 (1982).
134. G. Holzer, *Colorado School of Mines Quarterly* **78**, 9 (1983).
135. S. Bauer, K. Heckmann, L. Six, C. Strobl, D.Blöcher, B. Henkel, T. Garbe, and K. Ring, *Desalination* **46**, 369 (1983).
136. T. A. Langworthy, *Yale J. Biol. Med.* **56**, 385 (1983).

THE GENES AND GENETIC APPARATUS OF EXTREME THERMOPHILES

TAIRO OSHIMA

Department of Life Science, Faculty of Science, Tokyo Institute of Technology, Ookayama, Tokyo 152, Japan

1. INTRODUCTION

To elucidate the molecular mechanism of thermophily, biochemical studies must be carried out on the genes and genetic apparatus of thermophiles as well as on enzymes and proteins.

DNA, genes, protein synthesis, and the related cellular components of thermophiles have been studied using a moderately thermophilic bacterium, *Bacillus stearothermophilus*, and several extremely thermophilic bacteria, such as *B. caldolyticus, Thermus thermophilus, T. aquaticus, T. flavus,* and *Sulfolobus acidocaldarius*. Early studies have been reviewed in some books and review papers (1–6).

For the writing of this article, some recent topics on biochemical studies of the genetic apparatus of extreme thermophiles have been selected. This paper does not present a comprehensive review of the field but is intended to promote further study on the molecular biology of thermophily.

2. CLONING AND THE STRUCTURE OF THERMOPHILE GENES

2.1. Molecular Cloning

DNA cloning and site-directed mutagenesis have opened a new era in the field of the biochemistry of thermophily. The role of an amino acid residue at a specific site in the conformational stability of a protein can be precisely examined by replacing that residue with other amino acids through change in the DNA sequence which codes for that protein. The implication of post-translational (or post-transcriptional, in the case of nucleic acids) modifications in regard to the stability of a biopolymer can be confirmed by introducing the modification enzyme(s) into a mesophilic organism. The gene cloning technique makes it possible to confer extra stability onto mesophilic proteins and nucleic acids. Eventually it may be possible to design enzyme proteins possessing even greater heat resistance than those in thermophiles.

Soon after DNA cloning had become an established technique, the structural gene for 3-isopropylmalate dehydrogenase of *T. thermophilus* was cloned into *Escherichia coli*. The gene corresponds to the *E. coli leu B* and *Neurospora crassa leu 2* genes. Nagahari et al. (7) cloned the gene from *T. thermophilus* strain HB27, and Tanaka et al. (8) used DNA extracted from strain HB8 for this purpose.

Cloning of thermophile genes presented a few complications. For instance, the restriction enzyme EcoR1 did not produce fragments of suitable size, probably because of the high G-C content of *T. thermophilus* DNA (the recognition site of EcoR1 is the A-T rich sequence GAATTC). To make a recombinant plasmid, 1 μg each of *Thermus* DNA and plasmid pBR322 were digested by HindIII. The restriction enzyme was then heat denatured.

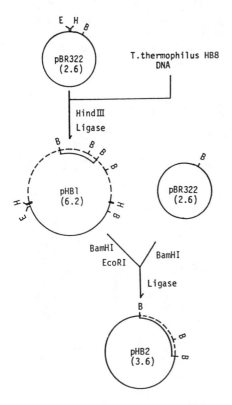

Figure 1. Recombinant plasmids pHB1 and pHB2. The hybrid plasmid pHB2 carries the *leu B* gene of *T. thermophilus* at the BamH1 site of pBR322. Modified from Tanaka et al. (8).

The DNA fragments produced from the thermophile DNA and the plasmid DNA were mixed and connected in the usual manner, using T4 DNA ligase.

Transformation with the recombinant DNA was carried out using an *E. coli leu B* mutant called C-600, and transformant colonies having Amp^r and 3-isopropylmalate dehydrogenase (Leu$^+$) genes were selected. The recombinant plasmid, pHB1, thus isolated contained thermophile genes for isopropylmalate dehydratase (*leu C* product) and 3-isopropylmalate dehydrogenase. As shown in Figure 1, the recombinant plasmid contained a 3.6 Md fragment derived from the thermophile DNA. To remove the dehydratase gene, Tanaka et al. (8) further digested the cloned thermophile DNA with BamH1. The recombinant plasmid, pHB2, thus obtained contained a 1.05 Md fragment of the thermophile DNA including the structural gene for 3-isopropylmalate dehydrogenase (see Figure 1). The chemical structure of pHB2 is schematically shown in Figure 3, along with the restriction sites.

E. coli cells harboring the recombinant plasmid pHB2 produced a heat-stable 3-isopropylmalate dehydrogenase. The heat resistance was similar to that of the dehydrogenase extracted from *T. thermophilus* HB8. This fact suggests that the thermophile enzyme is stabilized by an intrinsic mechanism without the binding of specific protector(s) or specific post-translational

a b

Figure 2. Gel electrophoretic patterns of the cell-free extract of *E. coli* harboring pHB2 before (a) and after (b) treatment at 70°C for 5 min. The arrow indicates the thermophile 3-isopropyl-malate dehydrogenase. Modified from Tanaka et al. (8).

modification(s). Likewise, heat stable isopropylmalate hydratase and 3-iso-propylmalate dehydrogenase were produced by *E. coli* harboring pHB1. Those observations and the related studies suggest that the evolution of thermophily is a multigenic process.

Gene cloning provides a new way to purify thermophile enzymes. When a cell-free extract of *E. coli* (pHB2) was heated at 70°C for 5 minutes, most of the proteins coded by *E. coli* genes were denatured and easily removed by centrifugation, whereas the heat stable dehydrogenase coded by the ther-mophile gene remained unchanged in the supernatant. In addition to the dehydrogenase, only a few other proteins were present in the supernatant, as indicated by gel electrophoresis (Figure 2). Eighty percent of the protein in the extract was removed by the heating procedure.

After the heat treatment, the supernatant was applied to a DEAE-cellulose (or DEAE-sephacel) column which was then eluted with a linear gradient of 0 to 0.4M KCl. This ion-exchange column chromatography procedure gave an almost homogeneous preparation of the enzyme. A combination of gene cloning and heat treatment can be applied to other thermophile enzymes and provides a general and simple procedure for enzyme purification.

The heat-stable isopropylmalate dehydrogenase activity in the crude extract of *E. coli* was about 7 times that of a cell-free extract of *T. thermophilus*, probably due to the high copy number of the plasmid in the transformed cells. This increased production is an additional advantage of gene cloning as a technique for enzyme purification.

The enzymatic properties of the thermostable 3-isopropylmalate dehydrogenase are similar to those of the same enzyme from *Salmonella typhimurium*, except for stability to heat. The enzyme consisted of two identical subunits, each having a molecular weight of about 36,000. The enzyme activity is specific for the threo-D_s isomer of 3-isopropylmalate. The Km values are about 1.5 mM for isopropylmalate and 0.5 mM for nicotinamide adenine dinucleotide (NAD). The presence of Mg^{2+} or Mn^{2+} is essential for activity, just as for the mesophile counterparts.

2.2. Structure of a Thermophile 3-Isopropylmalate Dehydrogenase Gene

The restriction map for pHB2 is shown in Figure 3. The 1.6 kb BamH1 fragment derived from the thermophile DNA was sequenced by Kagawa et al. (9). The base sequence was recently corrected (unpublished data), and the revised sequence is shown in Figure 4. The sequence was partially confirmed by amino acid sequencing of the enzyme. The enzyme subunit consists of 339 amino acid residues and has a molecular weight of 36,251.9.

One interesting observation was that the cloned DNA fragment contained a BamH1 site (-82 in Figure 4). For production of the enzyme in *E. coli*, two BamH1 fragments were necessary (8). Before the sequence was determined, the BamH1 site had been thought to be located in the *leu B* structural gene. However, our sequence study indicated that the BamH1 site was in the flanking region (noncoding region). It is not clear why the smaller BamH1

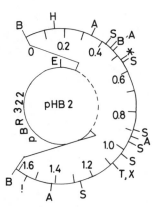

Figure 3.   Restriction map of the *leu B* gene and its flanking region. The figures indicate the distance (Kbp) from the BamH1 site. B=BamH1, E=EcoR1, H=Hinfl, S=SmaI, A=AluI, T=TaqI, X=XhoI, *=initiation site, !=termination site.

```
-530             GATCCTCCCCGCCCGCTTCATGAAGGTCCTCACCTTTGAGGGCCTGGGG

        RsaI
-481 CAGTACCTCTTCTACGACGAGCGCTTTGACGAGAAGGGGAACCCCAAGCCCCACCCCCTG

                                             HinfI
-421 AACGACCCCCGCTACGGGGCCACTGATCCTCTTGGTGGAGTCGGGCTTCGGCTCCGGCTC

-361 TAGCCGGAGACGCCCCCCAGGCCATCAAGCGGGCGGGGTTTAAGGCCATCATCGGGGAGG

             BglII
-301 CTTCAGCCGAGATCTTCTTCGGGAACGCCACCGCCGTCGGCCTCCCCTGCGTGAGCCTAG

                                                      AluI
-241 CCCCTGAGGACCTAGGCGTCCTCTTCCGCAGCGTGGAGGAGAACCCGGAGCTGGAGGTGG

-181 AGATTGACCTAGTGAACAAAGAGGAGACGCTTAGGGGAACCGACCGCTCCCCTCTTCATC

                AvaI SmaI                BamHI
-121 CGGGAAGAGGCCCGGGAGGCCCTAGTGGAGGGGCTTTGGGATCCCATCGGGGAGTCTTCT

                                                   Shine-Dalgarno
-61  GGAGGCCGGGGAGCTTTTGGACCAGTTTGACCGGAAGCTCCCCTACCCCAGGAGGACGGA

                     AvaI SmaI           AvaI BstEII
1    ATGAAGGTGGCCGTGCTCCCCGGGGACGGGATCGGCCCCGAGGTCACCGAGGCCGCCCTG
     MetLysValAlaValLeuProGlyAspGlyIleGlyProGluValThrGluAlaAlaLeu

61   AAGGTCCTGAGGGCCCTGGACGAGGCCGAGGGCGGGCCCACCTACGAGGTCTTCCCCTTC
     LysValLeuArgAlaLeuAspGluAlaGluGlyGlyProThrTyrGluValPheProPhe

                                         AvaI
121  GGCGGGGCGGCCATAGACGCTTCGGCGCCCTTCCCCGAGCCCACGCGAAAGGGCGTGGAG
     GlyGlyAlaAlaIleAspAlaSerAlaProPheProGluProThrArgLysGlyValGlu

181  GAGGCGGAGGCGGTGCTTCTGGGAAGCGTGGGGGGCCCAAGTGGGACGGCCTTCCCCGC
     GluAlaGluAlaValLeuLeuGlySerValGlyGlyProLysTrpAspGlyLeuProArg

                                      AflII
241  AAGATCCGCCCGGAGACGGGGCTTCTTTCCTTAAGGAAAAGCCAGGACCTCTTCGCCAAC
     LysIleArgProGluThrGlyLeuLeuSerLeuArgLysSerGlnAspLeuPheAlaAsn

301  CTCCGCCCGGCCAAGGTCTTCCCTGGGCTGGAAAGGCTTTCCCCCCTGAAGGAGGAGATC
     LeuArgProAlaLysValPheProGlyLeuGluArgLeuSerProLeuLysGluGluIle

     AvaI SmaI                            SacI SstI AluI
361  GCCCGGGGGGTGGACGTCCTCATCGTCCGGGAGCTCACCGGGGGGATCTACTTCGGGGAG
     AlaArgGlyValAspValLeuIleValArgGluLeuThrGlyGlyIleTyrPheGlyGlu

     AvaI SmaI                                           AvaI
421  CCCCGGGGGATGTCCGAGGCCGAGGCCTGGAACACGGAGCGCTACAGCAAGCCCGAGGTG
     ProArgGlyMetSerGluAlaGluAlaTrpAsnThrGluArgTyrSerLysProGluVal

             AvaI SmaI
481  GAGCGGGTGGCCCGGGTGGCCTTTGAGGCGGCGAGGAAGCGCAGGAAGCACGTGGTGAGC
     GlyArgValAlaArgValAlaPheGluAlaAlaArgLysArgArgLysHisValValSer

                      XhoI TaqI
541  GTGGACAAGGCGAACGTCCTCGAGGTGGGAGAGTTCCGGCGCAAGACCGTGGAGGAGGTG
     ValAspLysAlaAsnValLeuGluValGlyGluPheArgArgLysThrValGluGluVal

601  GGGCGGGGCTACCCCGACGTCGCCCTGGAGCACCAGTATGTGGACGCCATGGCCATGCAC
     GlyArgGlyTyrProAspValAlaLeuGluHisGlnTyrValAspAlaMetAlaMetHis

661  CTGGTCCGCTCCCCTGCCCGCTTTGACGTGGTGGTCACGGGGAACATCTTCGGGGACATC
     LeuValArgSerProAlaArgPheAspValValValThrGlyAsnIlePheGlyAspIle

                         AvaI SmaI
721  CTCGGGAACTTGCGAGCGGACCTCCCGGGCTCTCTAGGCCTCCTCCCCTCCGCCTCCTTG
     LeuGlyAsnLeuArgAlaAspLeuProGlySerLeuGlyLeuLeuProSerAlaSerLeu

781  GGAAGGGGCACCCCGGTCTTTGAGCCCGTGCACGGCTCCGCCCCCGACTACGCCGGCAAG
     GlyArgGlyThrProValPheGluProValHisGlySerAlaProAspTyrAlaGlyLys

                                                       AluI
841  GGGCGTAACCCCACGGCCGCCATCCTCTCCGCGGCCATGATGCTGGAGCAGCTTCGGCCT
     GlyArgAsnProThrAlaAlaIleLeuSerAlaAlaMetMetLeuGluGlnLeuArgPro

901  GGTGGGCTGGCGCGGAAGGTGGAAGACGCGGCCAAGGCCCTCCTGGAGACCCCGCCCCCG
     GlyGlyLeuAlaArgLysValGluAspAlaAlaLysAlaLeuLeuGluThrProProPro

961  GACCTCGGAGGAAGCGCGGCACGGGCCTTCACGGCCACGGTCCTCCGCCACCTCGCCTAA
     AspLeuGlyGlySerAlaAlaArgAlaPheThrAlaThrValLeuArgHisLeuAlaEND

1021 GATGGGGATATGATCCGGCACCGCTTCAGCGCCGAGGACTTCCACCGCATGGCCGAGGCG
              (MetIleArgHisArgPheSerAlaGluAspPheHisArgMetAlaGluAla

1081 GGGATC
     GlyIle)
```

Figure 4. Base sequence of *T. thermophilus* 3-isopropylmalate dehydrogenase gene. The sequence has recently been corrected. Amino acid sequence deduced from the base sequence is also shown.

fragment (from -529 to -82 in Figure 4) is necessary for expression of the coding region (1 to 1017) in *E. coli*.

The amino acid sequence deduced from the base sequence is also shown in Figure 4. The amino acid composition (Table 1) had several features characteristic of thermophilic proteins. No cysteine residue was found among the 339 amino acid residues. The cysteine content in some thermophile proteins is less than that in the corresponding proteins of mesophiles (1–5a). The high Glx/Asx and Arg/Lys ratios are also conspicuous in many thermophile proteins (1–5a). In the present enzyme (isopropylmalate dehydrogenase of *T. thermophilus*), Glx/Asx, Arg/Lys, and Ala/Gly were 33/21, 28/16, and 42/36, respectively. The ratios of Glx/Asx, Arg/Lys, and Ala/Gly are 32/35, 13/28, and 34/28, respectively, in the 3-isopropylmalate dehydrogenase from yeast. The thermophile dehydrogenase contained a large number of Pro (27), Val (32) and Leu (36) residues.

The most remarkable feature of the DNA structure is its usage of synonymous codons, as summarized in Table 1. The G-C content of DNA from *T. thermophilus* is about 70% and it has been speculated that the third letters in the degenerate codons are highly restricted to G or C (10). The sequencing results were found to agree with this speculation. In the coding region, the G-C content is 70.4%, but 64.8% in the noncoding region. The G-C content of the cloned fragment (whole sequence in Figure 4) is 68.3% and this is close to the G-C content of the whole chromosome (10).

The first letters have high G content; this fact leads to high Val, Ala, Glu, Gly contents in the protein. The four bases are almost equally used in the second letter, though the G content is slightly lower than others. The third letters are rich in G and C. The G-C content of the third letters is about 90%. These data are illustrated in Figure 5.

Since the codons used in the thermophile *leu B* gene are highly restricted to those whose third letters are G or C, the codon usage differs from those optimally used in *E. coli* (11). For instance, the optimal codons for Val in *E. coli*, GUU and GUA, are not used in the *T. thermophilus* gene whereas the nonoptimal codons in *E. coli*, GUC and GUG, are used 12 and 20 times, respectively, in the thermophile gene. The Ile codon AUA is one of the least used codons in *E. coli* and is known as the "modulator codon," since it represses the expression of the gene (11). This codon, AUA, is used once in the thermophile gene for 3-isopropylmalate dehydrogenase (Ile-45). With these divergent codons, it is not clear why the thermophile gene is efficiently expressed in *E. coli*. In fact, the thermophilic 3-isopropylmalate dehydrogenase was produced to a greater extent in *E. coli* harboring pHB2 than in *T. thermophilus* HB8. However, other genes from the thermophile might not be expressed in *E. coli* due to frequent use of the nonoptimal codons for *E. coli* translation.

In addition to the 3-isopropylmalate dehydrogenase, a gene from *T. thermophilus* involved in tryptophan synthesis (trpE) was transferred to and expressed in *E. coli* (S. Sato, unpublished data). Iijima et al. (12) cloned a

TABLE 1. Amino acid composition and codon usage of a thermophile 3-isopropylmalate dehydrogenase gene

Amino acid	No. of residues	Codon	No. used	Amino acid	No. of residues	Codon	No. used
Ala	42	GCT	1	Leu	36	CTT*	6
		GCC*	28			CTC*	15
		GCA	1			CTA*	1
		GCG	12			CTG	10
						TTA*	1
Arg	28	CGT	1			TTG*	3
		CGC	9				
		CGA*	2	Lys	16	AAA	1
		CGG*	10			AAG*	15
		AGA*	0				
		AGG*	6	Met	6	ATG	6
Asn	6	AAT*	0	Phe	12	TTT*	3
		AAC	6			TTC	9
Asp	15	GAT	0	Pro	27	CCT*	3
		GAC	15			CCC*	18
						CCA	0
Cys	0	TGT	0			CCG	6
		TGC	0				
				Ser	15	TCT	1
Gln	3	CAA*	0			TCC*	8
		CAG	3			TCA*	0
						TCG*	1
Glu	30	GAA	2			AGT	0
		GAG*	28			AGC	5
Gly	36	GGT*	0	Thr	13	ACT	0
		GGC	11			ACC	6
		GGA*	6			ACA*	0
		GGG*	19			ACG*	7
His	5	CAT	0	Trp	2	TGG	2
		CAC	5				
				Tyr	6	TAT*	1
Ile	9	ATT*	0			TAC	5
		ATC	8				
		ATA*	1	Val	32	GTT	0
						GTC*	12
						GTA	0
						GTG*	20

*Nonoptimal codons in *E. coli*

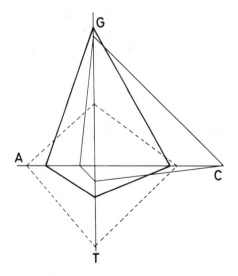

Figure 5. Base frequency used in the first (——), second (- - - -), and third (——) letters of the coding region of *T. thermophilus* 3-iso-propylmalate dehydrogenase gene.

gene for malate dehydrogenase into *E. coli* using DNA from *T. flavus*. A preliminary report on the partial sequence of the malate dehydrogenase gene has appeared (12), and the whole sequence will be published soon (personal communication). As described later, Erdman et al. (13) have cloned the 5S RNA gene from *T. thermophilus* into *E. coli*.

Molecular cloning of genes from a moderate thermophile, *B. stearother-mophilus*, has also been done, including genes coding for 3-isopropylmalate dehydrogenase. The gene for tyrosyl-tRNA synthetase of *B. stearothermo-philus* has also been cloned and the structure-function relationship of the enzyme was studied by site-directed mutagenesis (14). This is a pioneering work in a new field known as "protein engineering." The results showed that substitution of Thr-51 with Ala or Pro produced an enzyme with an affinity for ATP about 100 times higher than that of the native enzyme (15). In contrast, replacement of Cys-35 with Ser resulted in an enzyme with lesser affinity for ATP and a decrease in the activity of the enzyme.

Hoshino et al. (16) have studied genetic transformation in *T. thermo-philus*. However, the details of their research have not been published yet.

2.3. Cloning of Genes from Extreme Acido-Thermophiles

Recently, the isopropylmalate dehydrogenase gene from the extreme acido-thermophile *Sulfolobus acidocaldarius* was cloned and expressed in the author's laboratory (unpublished data). However, no report has appeared on the isolation of the structural genes of *S. acidocaldarius* which code for proteins. In this organism, only genes for rRNAs and tRNAs have so far been studied using molecular cloning techniques.

The most striking observation in regard to the genes of the acidophilic archaebacteria is the discovery of introns. Woese and his colleagues (17) isolated genes of *S. solfataricus* which may code for tRNA[Ser] and tRNA[Leu]. Each of these genes contained an extra oligonucleotide fragment immediately following the anticodon sequence which is supposedly excised when the tRNA molecules mature. S. Nishimura and co-workers, in collaboration with the present author's group, isolated several genes for tRNAs from *S. acidocaldarius* (unpublished data). However, no intron was found in some of these genes, indicating that not all the tRNA genes of *Sulfolobus* contain intervening sequences.

3. tRNA

Although *T. thermophilus* tRNA[mix] has a melting temperature of 87°C, which is about 8°C higher than that of *E. coli* tRNA[mix], the thermophile tRNA can be used instead of the mesophile tRNA in an *E. coli* cell-free protein synthesis system without increased miscoding (18). So far, *T. thermophilus* tRNAs are the only biopolymers whose unusual stability is clearly explicable based on molecular structure. The thermophile tRNA$_f^{Met}$ was different from its counterpart of *E. coli* in respect to the facts that: a) one G-U pair was replaced with a G-C pair in the T stem of the molecule, and b) the thermophile tRNA contained three modified bases, Gm-19, s^{2}T-55 and m^{1}A-59. Except for these posttranscriptional modifications, the nucleotide sequences of these tRNA$_f^{Met}$ are identical to each other (19).

Earlier studies (20–23) had shown that among the changes in chemical structure an additional hydrogen bond afforded by replacing a G-U pair with a G-C pair and thiolation at T-55 are important for the unusual stability of the tRNA molecule. It was speculated that the thiolation of position 2 of uracil (or thymine) strengthened the stacking interaction between neighboring bases in the helix (24). However, a more recent study has shown that the highly restricted conformation of the thiolated nucleotide also gives extra stability to the tRNA (25). Probably both mechanisms contribute to the unusual stability.

In the thermophile cells, some tRNA molecules are thiolated at position 55 and the rest are unmodified. The ratio of modified to unmodified molecules depends on the growth temperature. Two research groups (26,27) have succeeded in separating the thiolated and unthiolated tRNAs. As expected, the thiolated molecules had a higher melting temperature (Tm) than those which were unmodified. However, there was a small discrepancy regarding the extent to which thiolation contributed to the thermal stability; in the case of tRNA[mix], the difference in Tm was estimated to be about 8°C (22), whereas the difference was only 3°C in tRNA[Thr] (27).

According to a recent study, methylation at G-19 and at A-59 may also be responsible to some extent for the unusual stability. This is apparent from the ribonuclease resistance described below. The G-C content of the base-paired region of the tRNAs of *T. thermophilus* is generally 75 to 90%, whereas that of the mesophile tRNAs is 65 to 70%; this difference corresponds to one to three replacement(s) of A-U pair(s) with G-C pair(s). The empirical data (28) suggest that a 5% increment in G-C content in the base-paired region of a tRNA brings about a 1.5°C rise in the melting temperature. 2-Thiolated thymidine-free tRNA^{Thr} of the thermophile showed a higher melting temperature than that estimated from its G-C content in the base-paired region (28). This observation supports the possibility that modifications at positions 19 and 59 are of some importance to thermal stability, although the melting temperatures of *E. coli* tRNA$_f^{Met}$ and yeast tRNA^{Phe} are not raised by the in vitro methylation of G-19 and/or A-59 (28).

Ribose methylation at G-19 in the thermophile tRNA has been thought to confer on the ribonucleic acid an unusual resistance to ribonucleases. For confirmation of this, the specific enzyme, guanosine-2′-O-methylase, was extracted from the thermophile and purified (29). The enzyme catalyzes the ribose methylation of a specific guanosine in the D loop in the presence of S-adenosyl methionine as the methyl donor.

When tRNA from yeast was treated with the thermostable guanosine-2′-O-methylase in the presence of S-adenosyl methionine, only one methyl group was introduced into the corresponding ribose moiety of the guanosine residue in the D loop for each molecule of the yeast tRNA. The resulting methylated tRNA was as resistant to ribonucleases A and T$_1$ as the thermophile tRNAs (29). This is the first example of the artificial stabilization of biopolymers by the same means as those employed by the cell constituents of thermophiles.

The enzyme tRNA (adenine-1-)methyltransferase was also purified from *T. thermophilus* HB8 (30). This heat-stable enzyme specifically catalyzes the methylation of adenine-58 in the T loop.

The tRNAs from *S. acidocaldarius*, an extremely acidophilic, thermophilic archaebacterium, were also studied by several groups. The primary sequence of tRNA^{Met} was unique and was similar to some degree to those of eucaryotic tRNAs. For instance, the GTψCPu sequence characteristic of eubacterial tRNAs was replaced with a eucaryotic sequence GUUCPu (31). Similarly, the eucaryotic sequence was found in a tRNA from another acidothermophilic archaebacterium, *Thermoplasma acidophilum* (32). Woese et al. (17) and Kuchino et al. (33) found an intervening sequence to be present in tRNA genes. Introns were in the anticodon loop. The intervening sequence is not present in all tRNA genes of *Sulfolobus*: some tRNA genes contain no intron (33).

These biochemical properties supported the idea that the archaebacteria are phylogenetically unique and distant from eubacteria. This consideration has also been strongly supported by other biochemical studies on the protein-synthesizing machinery of *Sulfolobus, Thermoplasma*, and halophilic, methanogenic, and anaerobic archaebacteria (32,34–38).

4. 5S RNA AND ITS STRUCTURAL GENE

The nucleotide sequence of the 5S RNA of *T. aquaticus* was first reported by Nazar and Matheson (39). This sequence has recently been corrected (40). That of *T. thermophilus* was published in 1981 by Erdmann et al. (41) and minor corrections were made by Komiya et al. (42). These two sequences differ by only nine residues. The proposed secondary structures of these 5S RNAs from extremely thermophilic microorganisms are similar to the eubacterial ones, as shown in Figure 6.

The genetic organization of the rRNA genes of *T. thermophilus* was studied by Erdman and his co-workers (13), who found that the three rRNA structural genes were linked in the order 16S-5S-23S or 23S-5S-16S. Using the Southern hybridization procedure, they concluded that the thermophile carries at least two sets of three rRNA (5S, 16S, 23S) genes.

T. thermophilus 5S RNA was recently crystallized, and an X-ray analysis of the three-dimensional structure of the 5S RNA has been started (43). The results should help to elucidate the unusual stability of rRNAs and ribosomes of the thermophile.

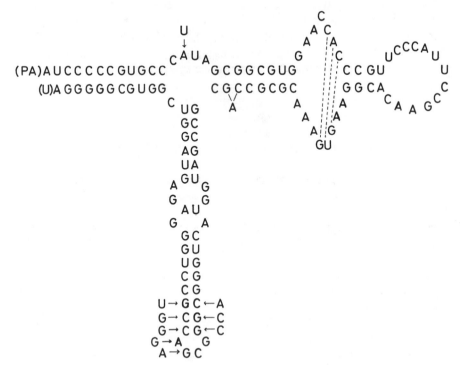

Figure 6. Secondary structure of 5S RNA of *T. thermophilus*. The base replacements found in *T. aquaticus* 5S RNA are shown by arrows.

5S RNAs from *S. acidocaldarius* and related species have also been studied in recent years and nucleotide sequences determined (44). The secondary structure deduced from the base sequence is unique and differs from those of both eubacterial and eucaryotic 5S RNAs, suggesting the unique phylogenetic status of the acido-thermophilic archaebacteria.

The organization of the rRNA genes of *Sulfolobus* and related species has been elucidated. In spite of the biochemical similarity in membrane lipids (45), rRNA sequences (46), and other cell components (47–50), each of the archaebacteria has its own particular rRNA gene organization. The 5S RNA gene of *Thermoplasma* is not linked to 16S and 23S genes as in the case of the 5S RNA genes of eucaryotes (51). The rRNA genes of *S. acidocaldarius* B12 are linked in the same order, 16S-23S-5S, as eubacterial genes (52). The close linkage of 16S-23S-5S rRNA genes was found in methanogenic archaebacteria (53). These data are summarized by Neumann et al. (52).

5. POLYAMINES

Polyamines are generally considered to be important cofactors in fundamental biochemical processes such as nucleic acid synthesis, protein synthesis, and cell division (54,55). Nucleic acids are stabilized by polyamines. A tRNA molecule appears to bind tightly with two molecules of polyamine (56). Polyamines are abundantly present in cells and tissues of rapidly growing organisms. Bacteria in the log phase have been found to contain large amounts of polyamines, whereas only small amounts of polyamines are present in the cells at the stationary phase.

In mesophilic bacteria, only putrescine (a diamine) and spermidine (a triamine) are known to be present. However, thermophiles contain polyamine(s) of longer chain length. The moderate thermophiles *B. stearothermophilus* and *B. acidocaldarius* produce spermine (a tetramine) in addition to spermidine and putrescine.

When in vitro protein synthesis catalyzed by a *T. thermophilus* cell-free extract was first studied, no reaction occurred at physiological temperatures of 60°C or higher (57). However, the addition of certain polyamines restored the activity at 65°C (57,58). Among the polyamines tested, spermine was the most active, while putrescine and other diamines did not support the reaction at high temperatures.

To identify the polyamine(s) acting as the essential cofactor(s) for in vivo protein synthesis under physiological conditions, the polyamine composition of *T. thermophilus* was analyzed. Cells of this organism contained more than twelve distinct polyamines. Eight of these twelve were new naturally occurring polyamines and two, norspermidine and homospermidine, were reported to be present in only a few organisms (59,60). The major components are two new tetramines, thermine and thermospermine. The bac-

terial cells also contain the hexamines, caldohexamine and homocaldohexamine, which are the longest-chain polyamines so far discovered in nature. Table 2 presents a list of the polyamines found in *T. thermophilus*. The polyamine composition of *T. aquaticus* is similar to that of *T. thermophilus*.

The in vitro protein synthesis catalyzed by a cell-free extract of *T. thermophilus* at high temperatures was restored by the addition of either of the new polyamines, thermine and thermospermine. Although spermine is rare in the thermophile cells, this amine supported cell-free protein synthesis, whereas the activity of spermidine in this regard was low. The reaction was more rapid in the presence of thermospermine than in the presence of thermine at 65°C. It was shown that polyamine stabilized a ternary complex between the ribosomes, the messenger, and aminoacyl-tRNA at high temperature. For the formation of an active complex, the presence of the polyamine was essential. Polyamines including putrescine also enhanced the polypeptide elongation reaction (57). However, the presence of polyamine at this later stage was not absolutely required.

These new polyamines in the thermophile may possibly regulate protein synthesis in vivo since the polyamine composition of the thermophile cells was found to depend on growth stage. The composition also depended on growth temperature and changed when the environmental temperature was changed (59,60).

The acido-thermophile *S. acidocaldarius* contained thermine in addition to putrescine, norspermidine, spermidine, and spermine. In vitro protein synthesis required the addition of thermine or spermine (61–61b). However, no polyamines of longer chain length than spermine were detected (60,61c). At high temperatures these unusual polyamines in thermophiles may play important roles in basic in vivo biochemical reactions other than translation. Careful attention should be given to these compounds to gain an understanding of DNA replication, transcription, and cell division at high temperature.

6. ELONGATION FACTORS AND AMINOACYL-tRNA SYNTHETASES

The polypeptide elongation factor Tu from *T. thermophilus* has been purified and characterized (62–64). This extremely heat-resistant elongation factor has been successfully studied by proton nuclear magnetic resonance measurements, and the involvement of a histidine residue at the GTP binding site was discovered by Miyazawa and his colleagues (65,66).

The polypeptide elongation factor G from the same microorganism was first studied by Arai et al. (62–64), who found the protein easily lost a short peptide (about 10–15 residues from its N-terminal end) by limited proteolysis. The modified protein still retained its activity. Recently, Reshetnikova and Garber (67) reported the growing of large crystals of this protein for X-ray analysis. Crystals of both native and trypsin-modified G factors from

TABLE 2. Unusual polyamines in *T. thermophilus*[a]

Trivial name	Systematic name	Chemical structure
Norspermidine	1,7-Diamino-4-azaheptane	$NH_2(CH_2)_3NH(CH_2)_3NH_2$
sym-Homospermidine	1,9-Diamino-5-azanonane	$NH_2(CH_2)_4NH(CH_2)_4NH_2$
Thermine	1,11-Diamino-4,8-diazaúndecane	$NH_2(CH_2)_3NH(CH_2)_3NH(CH_2)_3NH_2$
Thermospermine	1,12-Diamino-4,8-diazadodecane	$NH_2(CH_2)_3NH(CH_2)_3NH(CH_2)_4NH_2$
Homospermine	1,13-Diamino-4,9-diazatridecane	$NH_2(CH_2)_3NH(CH_2)_4NH(CH_2)_4NH_2$
Caldopentamine	1,15-Diamino-4,8,12-triazapentadecane	$NH_2(CH_2)_3NH(CH_2)_3NH(CH_2)_3NH(CH_2)_3NH_2$
Homocaldopentamine	1,16-Diamino-4,8,12-triazahexadecane	$NH_2(CH_2)_3NH(CH_2)_3NH(CH_2)_3NH(CH_2)_4NH_2$
Caldohexamine	1,19-Diamino-4,8,12,16-tetraazanonadecane	$NH_2(CH_2)_3NH(CH_2)_3NH(CH_2)_3NH(CH_2)_3NH(CH_2)_3NH_2$
Homocaldohexamine	1,20-Diamino-4,8,12,16-tetraazaeicosane	$NH_2(CH_2)_3NH(CH_2)_3NH(CH_2)_3NH(CH_2)_3NH(CH_2)_4NH_2$

[a]In addition to the unusual polyamines listed in the table, the thermophile cells contain two normal polyamines, spermidine and putrescine, and an unidentified compound.

T. thermophilus were grown for this purpose. The results of such an X-ray investigation would facilitate clarification of the molecular mechanisms of protein stabilization and of translation (68).

Miyazawa et al. (69,70) purified four aminoacyl-tRNA synthetases of *T. thermophilus*, methionyl-, valyl-, glutamyl-, and isoleucyl-tRNA synthetases, and initiated physicochemical studies on the reaction mechanisms of these enzymes. The methionyl-tRNA synthetase is a dimer enzyme and the other three are monomers. These enzymes resemble each other in amino acid composition, α-helix content, and the presence of two zinc atoms per enzyme protein. It was speculated that valyl- and isoleucyl-tRNA synthetases consist of two domains which correspond to the subunits of the methionyl-tRNA synthetase. Rivera et al. (71) reported the purification of prolyl-tRNA synthetase from *T. aquaticus* and stated that it consisted of two identical subunits as does the corresponding *E. coli* enzyme. The prolyl-tRNA synthesis catalyzed by the enzyme at elevated temperatures suggested that conformations of some parts of tRNA are essential for the enzyme reaction.

7. NUCLEASES

A site-specific endonuclease named Taq 1, which recognizes the four-base-pair sequence, 5'-TCGA-3'/3'-AGCT-5', was isolated from *T. aquaticus* YT-1 by Sato et al. (72). This enzyme cleaves the bond between T and C in the above sequence. An isoschizomer of this restriction enzyme was also found in *T. thermophilus* HB8 by Sato and Shinimiya (73) and was named Tth HB8 I. These heat-stable endonucleases are among the most useful enzymes for the sequencing and molecular cloning of DNA.

The counterpart of Tth HB8 I in the modification-restriction system, that is, a DNA methylase with the same specificity as Tth HB8 I, was identified and purified from *T. thermophilus* (74,75). The enzyme, with a molecular weight of 44,000, catalyzes the methylation at the N-6 position of adenine, using S-adenosyl-L-methionine as the methyl donor. A sequence study suggested that adenine in the sequence of 5'-TCGA-3' is methylated, thus protecting the chromosomal DNA from digestion by Tth HB8 I.

The search for restriction endonucleases in extreme thermophiles indicated the presence of other unique enzymes. A restriction endonuclease Tth 111 I was isolated and purified from *T. thermophilus* strain 111 (76). This enzyme recognized the following sequence:

$$\overset{\downarrow}{\text{GACN}}\text{NNGTC}$$

Under standard conditions, this enzyme cleaves DNA at the site indicated by the arrow in the above sequence.

Restriction endonuclease Tth 111 I is stable to heat, as would be expected, and is useful for generating relatively large DNA fragments, since the recognition sequences are few in DNA. There are two recognition sites in bacteriophage λ, and none in SV40 or ϕX-174 DNAs. This heat-stable endonuclease is now commercially available.

The extreme thermophiles produce two or more restriction endonucleases. The second enzyme in *T. aquaticus*, Taq II, has recently been purified and its sequence specificity (77) was determined as

$$5'\text{-GACCGANNNNNNNNNNN}^{\downarrow}\text{-3'/3'-CTGGCTNNNNNNNNN}_{\uparrow}\text{NN-5'}$$

The enzyme can thus be classified as a type II restriction endonuclease. The number of sites cleaved by this enzyme are one in SV40, five in pBR322, and more than 15 in λ DNAs. A third minor endonuclease is also present in the extreme thermophile. However, it is not clear yet whether this activity is due to the presence of the third enzyme or to a minor activity of Taq II (77).

The second enzyme in *T. thermophilus* strain 111 has also been purified (78,79). The specificity of this enzyme, Tth 111 II, is

$$5'\text{-CAAPuCAN}_{11}^{\downarrow}\text{-3'/5'-GTTPyGTN}_{9\uparrow}\text{-3'}$$

The number of recognition sites for this enzyme are 5 in pBR322, 12 in SV40 and more than 30 in λ DNAs.

Grachev et al. (80) reported the isolation of a new restriction endonuclease Taq XI from a strain of *T. aquaticus*. The enzyme recognizes the pentanucleotide sequence $CC^{\downarrow}(A/T)GG$ and cleaves it at the site indicated by the arrow. These workers reported that methylation of the C residue in the cleavage site does not protect the sequence from cleavage by this enzyme.

Takahashi et al. (81) isolated and purified a thermostable DNA ligase from *T. thermophilus* HB8. The enzyme consisted of a single polypeptide of molecular weight 79,000 and required the presence of NAD as a cofactor. Although the enzyme appeared to join cohesive-end but not blunt-end DNA molecules, its unusual stability will provide a useful tool in DNA recombination experiments.

Takahashi et al. (82) also studied a unique extracellular exonuclease of *T. thermophilus* HB8. The highly purified enzyme produces 5'-mononucleotides from either the 5' or the 3' terminus. This enzyme also possesses an endonuclease activity specific for superhelical, single-strand circular and covalently closed circular DNAs. Both endo- and exonuclease activities are sensitive to EDTA (83).

Watanabe et al. (84) reported the purification of adenosine 5'-triphosphate-dependent deoxyribonuclease from *T. thermophilus* HB8. Two forms of the enzyme were separated by affinity chromatography. The catalytic properties of these two enzyme proteins were similar to each other.

8. CONCLUDING REMARKS

DNA cloning and genetic manipulation techniques have brought a revolution to the biological sciences, and genetic studies on thermophiles using these recently advanced techniques seem to be one of the most promising ways for increasing our understanding of the molecular mechanisms of thermophily. Amino-acid replacements which are essential to make a normal protein thermostable will soon be clarified by in vitro mutagenesis of the protein.

Heat-stable thermophile enzymes can be easily purified by a combination of gene cloning and heat treatment, as described in this article. It can be expected that many thermophile enzymes will be highly purified and crystallized by this method in the near future, and eventually their three-dimensional structures will be elucidated by X-ray analysis.

Thermophiles are also interesting microorganisms in terms of biochemical evolution. Phylogenetic relationships of a thermophile with other thermophilic and mesophilic organisms will encourage comparisons of the chemical structures of various genes and genetic apparatus (85).

Thermophiles also provide useful experimental systems for studying basic biochemistry. Mutant proteins produced by site-directed mutagenesis of a mesophile protein would often be too unstable to carry out detailed studies because an amino acid replacement often perturbs the native conformation of the enzyme. In contrast, those produced by mutagenesis of a thermophile enzyme, in which the conformation is unusually stable, can be studied at lower temperatures. Thus cloned genes which code for thermophile enzymes would be useful materials for elucidating mechanisms of catalytic action. This line of study has already been reported for tyrosyl-tRNA synthetase of *B. stearothermophilus* (14,15).

Another aspect in which thermophiles contribute is in their production of unique biomolecules. Novel polyamines of thermophiles are promising compounds for studying the physiological roles of polyamines in living cells and tissues. Nucleases, especially site-specific restriction endonucleases of thermophiles, have been successfully used in molecular cloning and chemical analysis of genes.

REFERENCES

1. H. Zuber, Ed., *Enzymes and Proteins from Thermophilic Microorganisms*, Birkhäuser Verlag, Basel, 1976.
2. M. R. Heinrich, Ed., *Extreme Environments: Mechanisms of Microbial Adaptation*, Academic Press, New York, 1976.
3. D. J. Kushner, Ed., *Microbial Life in Extreme Environments*, Academic Press, New York, 1978.
4. R. E. Amelunxen and A. L. Murdock, *Crit. Rev. Microbiol.* **6**, 343 (1978).
5. S. M. Friedman, Ed., *Biochemistry of Thermophily*, Academic Press, New York, 1978.

5a. M. Shilo, Ed., *Strategies of Microbial Life in Extreme Environments*, Dahlem Konferenzen, Verlag Chemie, Weinheim, 1979.

6.. T. D. Brock, *Science* **230**, 132 (1985).

7. K. Nagahari, T. Koshikawa, and K. Sakaguchi, *Gene* **10**, 137 (1980).

8. T. Tanaka, N. Kawano, and T. Oshima, *J. Biochem.* **89**, 677 (1981).

9. Y. Kagawa, H. Nojima, N. Nukiwa, M. Ishizuka, T. Nakajima, T. Yasuhara, T. Tanaka, and T. Oshima, *J. Biol. Chem.* **259**, 2956 (1984).

10. T. Oshima and K. Imahori, *J. Biochem.* **75**, 179 (1974).

11. T. Ikemura, *J. Mol. Biol.* **151**, 389 (1981).

12. S. Iijima and T. Beppu, *Seikagaku* **53**, 817 (1981).

13. N. Ulbrich, I. Kumagai, and V. A. Erdmann, *Nucleic Acids Res.* **12**, 2055 (1984).

14. A. J. Wilkinson, A. R. Fersht, D. M. Blow, and G. Winter, *Biochemistry* **22**, 3581 (1983).

15. A. R. Fersht, J.-P. Shi, J. Knill-Jones, D. M. Lowe, A. J. Wilkinson, D. M. Blow, P. Brick, P. Carter, M. M. Y. Waye, and G. Winter, *Nature* **314**, 235 (1985).

16. T. Hoshino, personal communication.

17. B. P. Kaine, P. Gupta, and C. R. Woese, *Proc. Natl. Acad. Sci. U.S.A.* **80**, 3309 (1983).

18. Y. Ohno-Iwashita, T. Oshima, and K. Imahori, *Z. Allg. Mikrobiol.* **15**, 131 (1975).

19. K. Watanabe, Y. Kuchino, Z. Yamaizumi, M. Kato, T. Oshima, and S. Nishimura, *J. Biochem.* **86**, 893 (1979).

20. K. Watanabe, T. Oshima, M. Saneyoshi, and S. Nishimura, *FEBS Letters* **43**, 59 (1974).

21. K. Watanabe, T. Oshima, K. Iijima, Z. Yamaizumi, and S. Nishimura, *J. Biochem.* **87**, 1 (1980).

22. K.Watanabe, M. Shinma, T. Oshima, and S. Nishimura, *Biochem. Biophys. Res. Commun.* **72**, 1137 (1976).

23. P. Davanloo, M. Sprinzl, K. Watanabe, M. Albani, and H. Kersten, *Nucleic Acids Res.* **6**, 1571 (1979).

24. K. H. Scheit and P. Faerber, *Eur. J. Biochem.* **50**, 549 (1975).

25. Y. Yamamoto, S. Yokoyama, T. Miyazawa, K. Watanabe, and S. Higuchi, *FEBS Letters* **157**, 95 (1983).

26. K. Watanabe, T. Oshima, F. Hansske, and T. Ohta, *Biochemistry* **22**, 98 (1983).

27. N. Horie, S. Yokoyama, K. Watanabe, Y. Kuchino, S. Nishimura, and T. Miyazawa, *Biochemistry* **24**, 5711 (1985).

28. K. Watanabe, T. Oshima, and S. Nishimura, *Nucleic Acids Res.* **3**, 1703 (1976).

29. I. Kumagai, K. Watanabe, and T. Oshima, *J. Biol. Chem.* **257**, 7388 (1982).

30. I. A. Morozov, A. S. Gambaryan, T. N. Lvova, A. A. Nedospasov, and T. V. Venkstern, *Mol. Biol.* **18**, 1363 (1984).

31. Y. Kuchino, M. Ihara, Y. Yabusaki, and S. Nishimura, *Nature* **298**, 684 (1982).

32. M. W. Kilpatrick and R. T. Walker, *Nucleic Acids Res.* **9**, 4387 (1981).

33. Y. Kuchino, unpublished.

34. L. M. Van Valen and V. C. Mariorana, *Nature* **287**, 248 (1980).

35. D. G. Searcy, *Biochim. Biophys. Acta* **395**, 535 (1975).

36. M. Ohba and T. Oshima, in Y. Wolman, Ed., *Origin of Life*, D. Reidel Publishing, Dordrecht, 1981, p. 543.

37. M. Kessel and F. Klink, *Nature* **287**, 250 (1980).

38. A. T. Matheson, M. Yaguchi, W. E. Balch, and R. S. Wolfe, *Biochim. Biophys. Acta* **626**, 162 (1980).

39. R. N. Nazar, G. D. Sprott, A. T. Matheson, and N. T. Van, *Biochim. Biophys. Acta* **521**, 288 (1978).

40. E. Dams, P. Londei, P. Cammarano, A. Vandenberghe, and R. Dewachter, *Nucleic Acids Res.* **11**, 4667 (1983).

41. T. Kumagai, M. Digweed, V. A. Erdmann, K. Watanabe, and T. Oshima, *Nucleic Acids Res.* **9**, 5159 (1981).

42. H. Komiya, M. Kawakami, S. Takemura, I. Kumagai, and V. A. Erdmann, *Nucleic Acids Res.* **11**, 913 (1983).
43. K. Morikawa, M. Kawakami, and S. Takemura, *FEBS Letters* **145**, 194 (1982).
44. D. A. Stahl, K. R. Luehrsen, C. R. Woese, and N. R. Pace, *Nucleic Acids Res.* **9**, 6129 (1981).
45. T. A. Langworthy, *Curr. Top. Membr. Transp.* **17**, 45 (1982).
46. G. E. Fox, L. J. Magrum, W. E. Balch, R. S. Wolfe, and C. R. Woese, *Proc. Natl. Acad. Sci. U.S.A.* **74**, 4537 (1977).
47. W. Zillig, K. O. Stetter, and M. Tobien, *Eur. J. Biochem.* **91**, 193 (1978).
48. W. Zillig, K. O. Stetter, and D. Janekovic, *Eur. J. Biochem.* **96**, 597 (1979).
49. K. B. Searcy and D. G. Searcy, *Biochim. Biophys. Acta* **670**, 39 (1981).
50. C. R. Woese, *Sci. Am.* **244**, 94 (1981).
51. J. Tu and W. Zillig, *Nucleic Acids Res.* **10**, 7231 (1982).
52. H. Neumann, A. Gierl, J. Tu, J. Leibrock, D. Staiger, and W. Zillig, *Mol. Gen. Genet.* **192**, 66 (1983).
53. M. Jarsch, J. Altenbucher, and A. Boeck, *Mol. Gen. Genet.* **189**, 41 (1983).
54. S. S. Cohen, *Introduction to the Polyamines*, Prentice-Hall, New Jersey (1971).
55. C. W. Tabor and H. Tabor *Ann. Rev. Biochem.* **53**, 749 (1984).
56. S. S. Cohen, *Ann. N. Y. Acad. Sci.* **171**, 869 (1970).
57. Y. Ohno-Iwashita, T. Oshima, and K. Imahori, *Arch. Biochem. Biophys.* **171**, 490 (1975).
58. Y. Ohno-Iwashita, T. Oshima, and K. Imahori, in H. Zuber, Ed. *Enzymes and Proteins from Thermophilic Microorganisms*, Birkhäuser-Verlag, Basel, 1976, p. 333.
59. T. Oshima, *J. Biol. Chem.* **257**, 9913 (1982).
60. T. Oshima, in C. Tabor and H. Tabor, Eds., *Methods in Enzymology* **94**, Academic Press, New York, 1983, p. 401.
61. P. Cammarano, A. Teichner, G. Chinali, P. Londei, M. DeRosa, A. Gambacorta, and B. Nicolaus, *FEBS Letters* **148**, 225 (1982).
61a. T. Oshima, M. Ohba, and T. Wakagi, *Origins Life* **14**, 665 (1984).
61b. S. M. Friedman, *System. Appl. Microbiol.* **6**, 1 (1985).
61c. H. Kneifel, personal communication.
62. K. Arai, Y. Ota, N. Arai, S. Nakamura, C. Henneke, T. Oshima, and Y. Kaziro, *Eur. J. Biochem.* **92**, 509 (1978).
63. K. Arai, N. Arai, S. Nakamura, T. Oshima, and Y. Kaziro, *Eur. J. Biochem.* **92**, 521 (1978).
64. S. Nakamura, S. Ohta, K. Arai, N. Arai, T. Oshima, and Y. Kaziro, *Eur. J. Biochem.* **92**, 533 (1978).
65. A. Nakano, T. Miyazawa, S. Nakamura, and Y. Kaziro, *Biochemistry* **19**, 2209 (1980).
66. A. Nakano, T. Miyazawa, S. Nakamura, and Y. Kaziro, *FEBS Letters* **116**, 72 (1980).
67. L. S. Reshetnikova and M. B. Garber, *FEBS Letters* **154**, 149 (1983).
68. Y. N. Chirgadze, S. V. Nikonov, E. V. Brazhnikov, M. B. Garber, and L. S. Reshetnikova, *J. Mol. Biol.* **168**, *449 (1983)*.
69. D. Kohda, S. Yokoyama, and T. Miyazawa, *FEBS Letters* **174**, 20 (1984).
70. M. Hara-Yokoyama, S. Yokoyama, T. Miyazawa, *J. Biochem.* **96**, 1599 (1984).
71. J. A. Rivera, Q-S. Wang, and J. T-F. Wong, *Can. J. Biochem. Cell. Biol.* **62**, 507 (1984).
72. S. Sato, C. A. Hutchinson, III, and J. I. Harris, *Proc. Natl. Acad. Sci. U.S.A.* **74**, 542 (1977).
73. S. Sato and T. Shinomiya, *J. Biochem.* **84**, 1319 (1978).
74. S. Sato, K. Nakazawa, and T. Shinomiya, *J. Biochem.* **88**, 737 (1980).
75. M. McClelland, *Nucleic Acids Res.* **9**, 6795 (1981).
76. T. Shinomiya and S. Sato, *Nucleic Acids Res.* **8**, 43 (1980).
77. D. Barker, M. Hoff, A. Oliphant, and R. White, *Nucleic Acids Res.* **12**, 5567 (1984).
78. T. Shinomiya, M. Kobayashi, and S. Sato, *Nucleic Acids Res.* **8**, 3275 (1980).

79. T. Shinomiya, M. Kobayashi, S. Sato, and T. Uchida, *J. Biochem.* **92**, 1823 (1982).
80. S. A. Grachev, S. V. Mamaev, A. I. Gurevich, A. V. Igoshin, M. N. Kolosov, and A. G. Slyusarenko, *Bioorg. Khim.* **7**, 628 (1981).
81. M. Takahashi, E. Yamaguchi, and T. Uchida, *J. Biol. Chem.* **259**, 10041 (1984).
82. M. Takahashi, M. Kobayashi, and T. Uchida, *Nucleic Acids Res.* **8**, 5611 (1980).
83. M. Takahashi, M. Kobayashi, and T. Uchida, *J. Biochem.* **90**, 1521 (1981).
84. N. Watanabe, M. Umeno, and M. Anai, *J. Biochem.* **93**, 503 (1983).
85. H. Kuntzel, B. Piechulla, and U. Hahn, *Nucleic Acids Res.* **11**, 893 (1983).

APPLIED GENETICS OF AEROBIC THERMOPHILES

TADAYUKI IMANAKA AND SHUICHI AIBA

Department of Fermentation Technology, Faculty of Engineering, Osaka University, Yamada-oka, Suita-shi, Osaka 565, Japan

1. INTRODUCTION

Basic metabolic properties of thermophilic and mesophilic microorganisms are in general quite similar (1). Hence, the main difference between the two classes of microorganisms is in the optimum growth temperature. The phenomenon of thermophily (growth at temperatures higher than 55°C) has intrigued scientists for many years. Thermophiles have been used extensively in studies on the thermostability of enzymes and other constituents such as ribosome, tRNA, DNA, and membrane (2,3). Although some mechanisms contributing to thermostability have been presented (1,2), genetic analyses of thermophiles are scarce. If a method to permit an exchange of genetic information in thermophiles were established, and if genetic engineering techniques were made applicable to such thermophilic bacteria, the following advantages, by comparison with mesophiles, would arise:

1. The host-vector system should be quite safe from the standpoint of biohazards since thermophiles do not grow at room temperature (1).

2. The mode of expression of cloned genes from mesophiles and/or thermophiles can be compared, and the gene products can be comparatively examined at elevated temperatures in thermophiles. Accordingly, a genetic system could contribute to an elucidation of thermophily on molecular basis.

3. When the structural genes for thermostable enzymes are cloned in plasmid vectors which are efficiently expressed in thermophiles, productivity of the thermostable enzymes may be enhanced by virtue of the gene dosage effect.

4. The amount of cooling water required for cultivation of thermophiles in a large-scale fermenter is less than that for mesophiles.

5. The cultivation time is shortened because of accelerated growth rates of thermophiles (4).

6. The probability of contamination diminishes due to higher cultivation temperatures (>55°C).

In this chapter, an overall picture of the development of transformation systems for a moderate thermophile, *Bacillus stearothermophilus*, as well as for an extreme thermophile, *Thermus thermophilus*, will be discussed. Examples of gene cloning using these systems will also be presented.

2. HOST-VECTOR SYSTEMS IN *BACILLUS STEAROTHERMOPHILUS*

B. stearothermophilus is a thermophile that has been studied extensively (1). With respect to sporulation and protein secretion, *B. stearothermophilus* resembles the mesophile *Bacillus subtilis* (5), and a comparison between the thermophile and the mesophile would be useful.

Much effort has been exerted by many workers to develop a transformation system in *B. stearothermophilus*. In this context, Welker and his colleagues (6) established a transfection system which requires a helper phage.

As a consequence, they also found a restriction and modification system in the thermophile. In contrast, we established independently a transformation system in *B. stearothermophilus* using plasmids as vectors (7).

2.1. Characteristics of *Bacillus stearothermophilus* ATCC 12980

B. stearothermophilus ATCC 12980, the type strain (5), was subjected to spontaneous mutation (resistant to streptomycin (500 μg/ml)) and also cured spontaneously of the cryptic plasmid pBSO1. The mutant strain *B. stearothermophilus* CU21 thus obtained has been used throughout as the host for cloning (7). Since the strain CU21 is sensitive to many antibiotics (ampicillin, chloramphenicol, erythromycin, kanamycin, and tetracycline), drugresistant transformants can be easily detected (see later).

Since some amino acids and vitamins are required for the growth of *B. stearothermophilus* ATCC 12980, L broth (tryptone 10 g/l, yeast extract 5 g/l, NaCl 5 g/l, pH 7.3) and LG broth (L broth supplemented with 2.5 g glucose/l) are used. In the case of LG broth, the generation time is about 15 minutes and exponential growth continues up to $OD_{660} = 1$. Immediately after growth ceases, the viable cell number decreases drastically. This quick death might be due to a pH lower than the critical value because organic acids produced from glucose accumulate in the medium (5). On the other hand, when unsupplemented L broth is used, a slower growth rate is noted from the beginning of cultivation, but almost all cells are viable even in the stationary phase. In light of these results, L and LG broth are used for the preculture and main culture, respectively.

Since the maximum and minimum temperatures for growth of *B. stearothermophilus* are about 70 and 40°C, respectively, and since cells lose viability most rapidly at room temperatures, cells are preserved temporarily at 4°C and stored at −80°C.

2.2. Isolation of Drug-Resistance Plasmids from Thermophilic Bacilli

Vectors which are stable at temperatures up to 65°C are required before a system for the genetic manipulation of *B. stearothermophilus* can be established. Therefore, antibiotic-resistance plasmids were widely screened from thermophilic bacilli for the ability to transform competent *B. subtilis*. Bingham et al. (8) isolated the plasmid pAB124 (Tc^r (tetracycline resistance), 2.9 megadaltons (Md)). We could isolate plasmid pTB19 (Km^r (kanamycin resistance) Tc^r, 17.2 Md) and pTB20 (Tc^r, 2.8 Md) (9). Digestion of pTB19 with *Eco*RI followed by ligation, yielded deletion plasmids pTB51 (Km^r, 8.4 Md), pTB52 (Tc^r, 7.0 Md), and pTB53 (Km^{r}Tc^r, 11.2 Md) which were transferred to *B. subtilis* (9). Similarly, a deletion plasmid pTB90 (Km^{r}Tc^r, 6.7 Md) was constructed in *B. stearothermophilus* (7). It is interesting to note that restriction endonuclease cleavage maps of pTB20 and pAB124 are nearly identical, although these plasmids were independently isolated in

Japan and in England, respectively. In addition, two plasmids, pTHT15 (Tc^r, 2.9 Md) and pTHN1 (Km^r, 3.1 Md), were also isolated in Japan (10).

2.3. Transformation of *Bacillus stearothermophilus*

The procedure for transformation of *B. subtilis* with plasmid DNA using either competent cells or protoplasts has been well documented (11–13). Competent cells of *Bacillus licheniformis* could also be transformed in this manner with linear (chromosomal) DNA (14,15). This method has not been made applicable for the transformation of *B. licheniformis* with plasmid DNA (15), although protoplasts of *B. licheniformis* (15), *Bacillus megaterium* (16), and *Bacillus thuringiensis* (17) could be transformed with plasmid DNA. Hence the protoplast procedure was investigated for the transformation of *B. stearothermophilus*.

The procedure for transforming *B. stearothermophilus* protoplasts with plasmid DNA followed basically the procedure established earlier by Chang and Cohen (13) for the transformation of *B. subtilis* protoplasts with plasmid DNA. Accordingly, the description of this procedure will be minimized.

Since *B. stearothermophilus* ATCC 12980 died quickly at room temperature but was stably maintained at 4°C, the host cells were exposed to 48 and/or 4°C when required in the process of transformation.

The procedure for transformation of *B. stearothermophilus* ATCC 12980 consists of three steps: a) preparation of protoplasts with lysozyme, b) polyethylene glycol treatment to introduce plasmid DNA into protoplasts, and c) regeneration of protoplasts (7). In the regeneration agar, sodium succinate (used for *B. subtilis*) was replaced by sucrose, because *B. stearothermophilus* did not grow when sodium succinate was used.

It should be mentioned in this connection, that the use of an extremely low concentration of lysozyme, 1 μg/ml in contrast to 2 mg/ml for the protoplast formation of *B. subtilis* (13), was to enhance the regeneration frequency of protoplasts. Even under the extremely low concentration of the enzyme, the conversion ratio of intact cells to protoplasts was more than 99.99%. Regeneration frequencies of protoplasts were around 10% (7).

Transformation frequencies of *B. stearothermophilus* CU21 with various plasmids are shown in Table 1. The plasmids designated pUB110 (Km^r, 3.0 Md), pHV14 (Cm^r(chloramphenicol resistance), 4.6 Md), pHV11 (Cm^{r}Tc^r, 3.3 Md), pC194 (Cm^r, 2.0 Md), pTP4 (Cm^r, 2.9 Md), and pTP5 (Tc^r, 2.9 Md) are the vector plasmids for *B. subtilis*.

If the plasmid was prepared from *B. subtilis* instead of *B. stearothermophilus* CU21, transformation frequencies for plasmids pTB19, pTB90 and pTHT15 were reduced by two, three and two orders of magnitude, respectively. However, transformation frequencies of pUB110 remained almost unchanged despite the difference in the plasmid source. In fact, the difference in transformation frequencies, depending on the source of plasmid (either *B. subtilis* MI113 or *B. stearothermophilus* CU21) shows concrete

TABLE 1. Transformation of *B. stearothermophilus* CU21 with plasmid DNA

Plasmid	Source[a]	Antibiotic (μg/ml)[b]	Transformants per μg of DNA[c]	Transformation frequency per regenerant	Reference
pUB110	B. sub.MI113	Km (25)	1.6×10^5	1.6×10^{-3}	7
pUB110	B. stearo.CU21	Km (25)	5.9×10^5	5.9×10^{-3}	
pTB19	B. sub.MI113	Km (25)	4.9×10^3	4.9×10^{-5}	
pTB19	B. sub.MI113	Tc (5)	9.0×10^3	9.0×10^{-5}	
pTB19	B. stearo.CU21	Km (25)	1.0×10^5	1.0×10^{-3}	
pTB19	B. stearo.CU21	Tc (5)	5.5×10^5	5.5×10^{-3}	
pTB90	B. sub.MI113	Km (25)	2.2×10^4	2.2×10^{-4}	
pTB90	B. sub.MI113	Tc (5)	2.6×10^4	2.6×10^{-4}	
pTB90	B. stearo.CU21	Km (25)	1.3×10^7	1.3×10^{-1}	
pTB90	B. stearo.CU21	Tc (5)	2.0×10^7	2.0×10^{-1}	
pHV14	B. sub.MI113	Cm (20)	ND	$<1.0 \times 10^{-8}$	
pHV11	B. sub.MI113	Cm (20)	ND	$<1.0 \times 10^{-8}$	
pHV11	B. sub.MI113	Tc (5)	ND	$<1.0 \times 10^{-8}$	
pTB20	B. sub.MI113	Tc (5)	ND	$<1.0 \times 10^{-8}$	
pTB51	B. sub.MI113	Km (25)	ND	$<1.0 \times 10^{-8}$	
pTB52	B. sub.MI113	Tc (5)	ND	$<1.0 \times 10^{-8}$	
pTB53	B. sub.MI113	Km (25)	ND	$<1.0 \times 10^{-8}$	
pTB53	B. sub.MI113	Tc (5)	ND	$<1.0 \times 10^{-8}$	
pUB110	B. sub.RM125	Km (40)	1.3×10^5	1.0×10^{-2}	10
pTHN1	B. sub.RM125	Km (40)	5.4×10^5	4.3×10^{-2}	
pTHT15	B. sub.RM125	Tc (5)	3.1×10^3	2.5×10^{-4}	
pTHT15	B. stearo.CU21	Tc (5)	7.6×10^5	6.1×10^{-2}	
pC194	B. sub.RM125	Cm (7)	5.3×10^5	4.2×10^{-2}	
pTP4	B. sub.LMAH	Cm (7)	ND	$<5.0 \times 10^{-7}$	
pTP5	B. sub.LMAH	Tc (5)	ND	$<5.0 \times 10^{-8}$	

[a]B. sub., *Bacillus subtilis*; B. stearo., *Bacillus stearothermophilus*.
[b]Km, kanamycin; Tc, tetracycline; Cm, chloramphenicol
[c]ND, Not detected

evidence that both pTB19 and pTB90 from *B. subtilis* MI113 suffered restriction in transformation of *B. stearothermophilus*, whereas pUB110 was not restricted regardless of the plasmid source (7).

Although pC194 transformed *B. stearothermophilus*, its recombinant plasmids pHV14 and PHV11 could not transform the thermophile. The difference might be explained by the facts that pC194 was very unstable in the host bacterium, especially at elevated temperatures, and also that the cloned genes in the recombinant plasmids might have triggered "stress" in the host.

The high frequency and good efficiency of transformation in *B. stearothermophilus* protoplasts with pTB90 (20% transformants per regenerant and 2×10^7 transformants per microgram of plasmid DNA, respectively) are commensurable with those in *B. subtilis* with plasmid pC194 or pUB110 (13). This fact justifies the usefulness of pTB90 as a vector plasmid for molecular cloning of specific gene(s) in *B. stearothermophilus*.

2.4. Characteristics of Vector Plasmids for *Bacillus stearothermophilus*

The restriction endonuclease cleavage maps of plasmids pTB19 and pTB90 and the construction scheme of pTB90 in *B. stearothermophilus* are shown in Figure 1. The DNA sample of pTB19 when prepared from *B. stearothermophilus* CU21 might have contained a small amount of both pTB913 (a deletion plasmid from pTB19) and pBSO2 (a derivative of cryptic plasmid pBSO1) (18). pTB90 was obtained at a relatively low efficiency from the ligation mixture of the *Eco*RI digest of the DNA sample. The scheme of

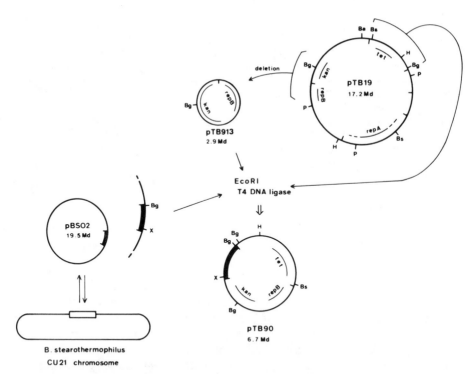

Figure 1. A schematic diagram of the construction of pTB90 in vitro (18). The three *Eco*RI fragments (2.9, 2.8, and 1.0 Md) composing pTB90 were from pTB913, pTB19, and pBSO2, respectively. Cleavage sites of *Bam*HI, *Bgl*II, *Bst*EII, *Pst*I, *Hin*dIII, and *Xba*I are indicated by Ba, Bg, Bs, P, H, and X, respectively. The bars inside the circles indicate the cleavage sites of *Eco*RI. The heavy line corresponds to the 1.0 Md *Eco*RI fragment. (Reproduced by permission of Society for General Microbiology.)

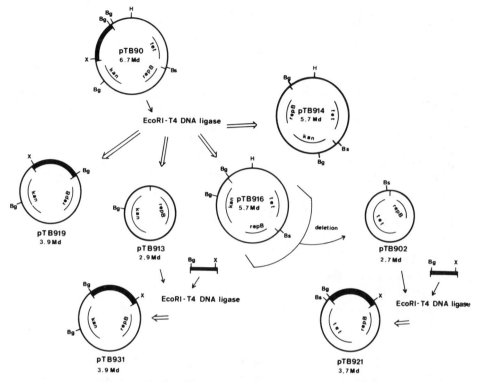

Figure 2. Restriction endonuclease cleavage maps of plasmids (18). Explanation of this figure is the same as in Figure 1. (Reproduced by permission of Society for General Microbiology.)

Figure 1 would account for the observation that pTB90 contained 2.9 Md and 1.0 Md *Eco*RI fragments which were not detected in the original plasmid pTB19.

Various deletion plasmids of pTB90 were constructed in order to examine the role of the specific 1.0 Md *Eco*RI fragment from pBSO2 in *B. stearothermophilus* (Figure 2). The presence of this 1.0 Md *Eco*RI fragment in the deletion plasmids from pTB90 increased transformation frequencies for *B. stearothermophilus* 10^3 to 10^4 times and lowered plasmid copy numbers in the host strain to about one-tenth of those found for plasmids lacking this fragment (Table 2). However, the detailed picture of the function of the 1.0 Md *Eco*RI fragment has not been revealed yet.

Two different replication determinants (RepA, RepB) were found on an antibiotic-resistance plasmid, pTB19, from a thermophilic bacillus (18). This finding is the first report of a specific plasmid from the genus *Bacillus* having two replication determinants on the same replicon. A deletion plasmid pTB913 (Km^r) was isolated from pTB19 (Km^{r}Tc^r) in *B. stearothermophilus* (see Figure 2). The same deletion plasmid (Km^r, 2.9 Md) was also obtained

TABLE 2. Properties of some *Bacillus* plasmids

Plasmid	Characteristics	Rep	1.0 Md EcoRI fragment	Copy no. in B. subtilis	Transformation of B. stearo-thermophilus	Copy no. in B. stearo-thermophilus
pTB19	Km^r Tc^r	A,B	−	1	+	1
pTB51	Km^r	A	−	8	−	−
pTB52	Tc^r	A	−	9	−	−
pTB53	Km^r Tc^r	A	−	8	−	−
pTB90	Km^r Tc^r	B	+	11	+	5
pTB914	Km^r Tc^r	B	−	13	+[a]	43
pTB916	Km^r Tc^r	B	−	15	+[a]	−[b]
pTB913	Km^r	B	−	25	+[a]	39
pTB919	Km^r	B	+	13	+	5
pTB931	Km^r	B	+	14	+	4
pTB902	Tc^r	B	−	24	+[a]	60
pTB921	Tc^r	B	+	13	+	7

[a]Low transformation frequency
[b]Deletion plasmid appears frequently

From Imanaka et al. (18)

from pTB19 in *B. subtilis*. In addition, another deletion plasmid (Tc^r, about 14 Md) was observed in *B. subtilis*. These results suggest that the two deletion plasmids emerged from pTB19 (Km^r Tc^r, 17.2 Md) by in vivo recombination and, conversely, pTB19 might have been constructed by in vivo fusion of two distinct plasmids specifying resistance to Km and Tc, respectively, in a host bacterium. It has been reported that many drug resistance plasmids (R plasmids) of *Enterobacteriaceae* are composite replicons (19,20).

Deletion plasmids containing RepA (pTB51, pTB52, and pTB53) derived from pTB19 could replicate only in *B. subtilis*, while other plasmids containing RepB (pTB19, pTB90, and its derivatives) could replicate in both *B. subtilis* and *B. stearothermophilus*, as shown in Table 2 (18). Further, copy numbers of RepA$^+$ plasmids in *B. subtilis* were somewhat lower than those of RepB$^+$ plasmids.

Gene expression and stability of pUB110, pTB19, and pTB90 in *B. stearothermophilus* were tested (Table 3) to check their validity as vector plasmids at elevated temperatures. pUB110 was stably maintained at 48 and 55°C, whereas the plasmid became unstable at 60 and 65°C. Since about 10% of the total population still carried pUB110 after about 20 generations at 60 or 65°C, the replication of the plasmid might not have been totally damaged at these higher temperatures. Plasmid pTB19 became unstable with an increase in temperature of cultivation. pTB90 was stably maintained at 48 and 55°C, whereas it became unstable at 60 and 65°C.

TABLE 3. Gene expression and stability of plasmids in *B. stearothermophilus*

Plasmid	Copy no. per chromosome	Growth temperature (°C)	No. of generations	Plasmid carrier (%)	Growth in LG broth plus drug (Km or Tc)
pUB110	~50	preculture		100	+
		48	18	100	+
		55	21	100	+
		60	19	8	−
		65	20	13	−
pTB19	~1	preculture		90	+
		48	19	68	+
		55	20	57	+
		60	19	54	+
		65	20	26	+
pTB90	5~18[a]	preculture		100	+
		48	23	98	+
		55	22	98	+
		60	20	58	+
		65	19	1	+

[a]Copy number $\simeq$ 5 in LG broth containing Km (5 μg/ml); copy number $\simeq$ 18 in LG broth containing Tc (5 μg/ml).

From Imanaka et al. (7)

    B. stearothermophilus strains carrying either pTB19 or pTB90 could grow in LG broth containing kanamycin or tetracycline even at 65°C. On the contrary, *B. stearothermophilus* (pUB110) could grow in LG broth plus kanamycin at 48 and 55°C, but not at 60 and 65°C. Supposing that the kanamycin-resistance gene of pUB110 was expressed normally even at temperatures higher than 60°C, the above-mentioned effect on the growth of *B. stearothermophilus* (pUB110) suggests that the protein product of the gene might have been thermolabile at the elevated temperatures.

    To make this point clear, we examined the Km[r] gene encoded by pTB913, a derivative of pTB19 (21). The product of a kanamycin resistance gene encoded by plasmid pTB913 isolated from a thermophilic bacillus was identified as kanamycin nucleotidyltransferase which is similar to that encoded by plasmid pUB110 from a mesophile, *Staphylococcus aureus*. The enzyme encoded by pTB913 was more thermostable than that encoded by pUB110 (Figure 3). In view of a close resemblance of restriction endonuclease cleavage maps around the *Bgl*II site in the structural genes of both enzymes, about 1,200 base pairs were sequenced, followed by amino-terminal amino acid sequencing of the enzyme. The two nucleotide sequences were found to be identical to each other except for only one base in the center of the structural gene. Each structural gene, initiating from GUG codon as me-

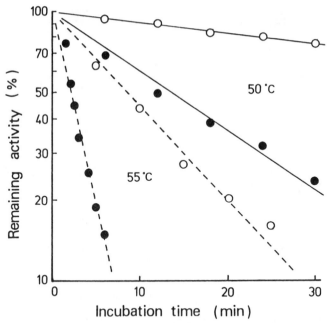

Figure 3. Thermostability of kanamycin nucleotidyltransferase (21). Both enzymes from pUB110 (●) and pTB913 (○) were used. Remaining activity after heating at 50°C (solid lines) or 55°C (broken lines) was expressed as percent of the original activity. (Reproduced by permission of American Society for Microbiology.)

thionine, was composed of 759 base pairs and 253 amino acid residues (molecular weight, about 29,000). The sole difference was transversion from a cytosine (pUB110) to an adenine (pTB913) at a position +389, counting the first base of the initiation codon as +1. That is, a threonine at position 130 for the pUB110-coded kanamycin nucleotidyltransferase was replaced by a lysine for the pTB913-coded enzyme. Thus, the difference in thermostability between the two enzymes is due to only a single amino acid replacement.

3. TRANSFORMATION OF *THERMUS THERMOPHILUS*

An extreme thermophile, *Thermus thermophilus* can grow at temperatures higher than 75°C (5). Although some plasmids have been isolated from several species of the genus *Thermus*, they are cryptic and hence selection of bacteria carrying the plasmid is difficult (22). Therefore, Hoshino et al. (10) attempted to develop a transformation system in *T. thermophilus* using chromosomal DNA.

For this purpose, several auxotrophic mutant strains were obtained from *T. thermophilus* HB27. These strains were precultured in TM medium (poly-

TABLE 4. Transformation of *Thermus thermophilus* with chromosomal DNA

Strain	No. of viable cells	No. of transformants	Transformation frequency (%)
K102 (Pro⁻)	2.1×10^8	2.5×10^7	11.9
K103 (Leu⁻)	2.0×10^8	1.5×10^6	0.75
K104 (Met⁻)	2.4×10^8	2.2×10^6	0.92
K105 (Lys⁻)	2.0×10^8	1.9×10^6	0.95
K106 (Trp⁻)	1.1×10^8	3.0×10^6	2.7

From Hoshino et al. (10)

peptone 4 g/l, yeast extract 2 g/l, NaCl 1 g/l, basal salts, pH 7.5). TM medium was inoculated with 1% seed culture and incubated with shaking at 70°C for 1 to 2 hours. At this time, the cells became competent without the $CaCl_2$ treatment needed for *E. coli*. The competent cells (0.36 ml) were mixed with chromosomal DNA (5 µg) of the prototroph and incubated with shaking at 70°C for 1 hour. The cells were plated on minimal agar medium to select the prototrophic transformants. The results are shown in Table 4. It was also reported that this procedure was applicable to other *Thermus* species such as *Thermus flavus* AT62, *Thermus caldophilus* GK24, and *Thermus aquaticus* YT1, although the transformation frequencies were lower (10).

This system shows considerable promise. If suitable vector plasmids carrying a drug resistance gene or other selectable markers were found and/or constructed, transformation of *Thermus* competent cells with the plasmids would become facile.

4. MOLECULAR CLONING IN *BACILLUS STEAROTHERMOPHILUS*

The well-documented transformation procedure of *B. subtilis* with plasmid DNA using either competent cells or protoplasts has permitted the cloning of penicillinase genes of *B. licheniformis* (15,23), α-amylase genes of *B. subtilis* (24), *Bacillus amyloliquefaciens* (25), and *B. licheniformis* (26), and protease genes of *B. subtilis* (27,28) and *B. amyloliquefaciens* (29,30) in *B. subtilis*. The cloning of specific gene(s) in the genus *Bacillus* would not only enable the breeding of industrially important strain(s), but also reveal mechanisms for secretion of extracellular enzymes. The above-mentioned transformation in the genus *Bacillus* has been concerned with mesophilic bacteria. Hence, the transformation system for a thermophile with plasmids was applied to the cloning of some specific enzyme genes.

B. stearothermophilus CU21 was used as the host strain, while pTB90 was used as the vector plasmid. Penicillinase genes from both the constitutive and wild-type strains of *B. licheniformis* 9945A have been cloned in *E. coli* with pMB9 (Tcʳ) as a vector plasmid (15). These recombinant plas-

TABLE 5. Penicillinase activity of *B. stearothermophilus*

Plasmid	Antibiotic (μg/ml)	Growth temperature (°C)	Penicillinase activity (U/mg of cells)	
			Total	Supernatant
None	None	48	ND[a]	—[b]
pTB90	Km(5)	48	ND	—
pLP21	Km(5)	48	90	7.4
pLP21	Km(5)	55	120	—
pLP21	Km(5)	60	17	—
pLP11	Km(5)	48	3,300	820
pLP11	Km(5)	55	610	—
pLP11	Km(5)	60	39	—
pLP11	Km(5) + Tc(1)	48	3,700	510
pLP11	Km(5) + Tc(2)	48	4,700	680
pLP11	Km(5) + Tc(3)	48	6,000	700
pLP11	Km(5) + Tc(4)	48	5,800	750

[a]ND, Not detectable (<0.01 U/mg of dry cell weight).

[b]—, Not tested

From Fujii et al. (31)

mids are designated pTTE11 (Tc^r *penP$^+$ penI10*) and pTTE21 (Tc^r *penP$^+$ penI$^+$*), respectively. *penP* and *penI* are the structural and repressor genes for penicillinase, respectively, and are coded on the same *Eco*RI fragment (2.8 Md).

The two *Eco*RI fragments (inducible and constitutive types) were cloned in pTB90 (31). *B. stearothermophilus* strains carrying the cloned penicillinase genes were grown at constant temperatures (48, 55, and 60°C) in LG broth to late exponential phase. Kanamycin (5 μg/ml) was added to the culture medium as a selective pressure to guarantee the presence of the recombinant plasmid. Penicillinase activities measured with and without the recombinant plasmid and in the absence of inducer (cephalosporin C) are shown in Table 5. No enzyme activities were detected from either *B. stearothermophilus* or the strain carrying only pTB90. *B. stearothermophilus* cells carrying pLP21 (*penP$^+$ penI$^+$*) produced nearly the same amount of penicillinase (90 to 120 units/mg of cells) at 48 and 55°C, whereas considerably less activity was detected by cultivation at 60°C (17 units/mg of cells). On the other hand, a large amount of penicillinase (3,300 units/mg of cells) was observed for the strain harboring pLP11 (*penP$^+$penI$^-$*).

It is evident from comparison of penicillinase activities between the strains carrying pLP21 and pLP11 in Table 5 that both genes, *penP* and *penI*, were expressed in *B. stearothermophilus*. The fact that the reduced penicillinase activity at higher temperatures of *B. stearothermophilus* (pLP21) was less marked than that of pLP11 carrier strain would suggest the thermal inactivation of the repressor. The deterioration of penicillinase activity at higher temperatures might be attributed to the thermal inactivation or rapid degradation of the enzyme by proteases from the host strain.

The copy number of pTB90 in *B. stearothermophilus* CU21 is amplified in the presence of tetracycline (7). The copy number of pLP11 should have therefore increased if tetracycline were added to the medium. As shown in Table 5, penicillinase activities of the strains harboring pLP11 were enhanced with increase in the amount of tetracycline added. The highest penicillinase activity (6,000 units/mg of cells) was observed when the bacterium was cultivated in the presence of 3 µg/ml of tetracycline; about a twofold increase of penicillinase compared with the absence of tetracycline (3,300 units/mg of cells) is noted. Lastly, it would be worthwhile to mention that about 10 to 20% of the total penicillinase was secreted into the culture medium of *B. stearothermophilus* CU21 at 48°C in contrast to about 30 and 60% secretion of the enzyme for *B. subtilis* and *B. licheniformis*, respectively.

Since the expression of penicillinase diminished remarkably at elevated temperatures, the study of the cloning of the structural genes for thermostable extracellular enzymes in *B. stearothermophilus* would be interesting.

The structural gene for a thermostable neutral protease from *B. stearothermophilus* was cloned in plasmid pTB90 (32). Recombinant plasmids pNP22 and pNP28 carried the protease gene, having the same 4.5 Md *Eco*RI-*Hin*dIII insert, and thermostable proteases were produced and secreted by recombinant-plasmid-carrier strains *B. stearothermophilus* MO-3 and *B. subtilis* MT-2, respectively (Table 6). *B. stearothermophilus* carrying the recombinant plasmid produced about 15-fold more protease (310 units/mg of cells) than did the wild-type strain of *B. stearothermophilus*.

TABLE 6. Production of extracellular protease activity by *B. stearothermophilus* and *B. subtilis*

Strain	Temperature (°C)	Protease (U/mg of cells)[a] at (h)		
		6	9	12
B. stearothermophilus CU21	55	19	17	17
B. stearothermophilus MO-3[b]	55	5.9	10	5.0
B. stearothermophilus MO-3 (pNP28)	55	210	290	310
B. subtilis MT-2[b]	37	4.2	7.0	6.2
B. subtilis MT-2 (pNP22)	37	90	210	180

[a]Samples were taken at the time indicated, and the protease in the culture supernatant was assayed. Maximum cell growth was observed at around 7 and 10 hours for *B. stearothermophilus* and *B. subtilis*, respectively.

[b]No halos were formed in plate assay by *B. stearothermophilus* MO-3 and *B. subtilis* MT-2; neither strain produced neutral protease on plates.

From Fujii et al. (32)

TABLE 7.   Production of α-amylase by *B. stearothermophilus* and *B. subtilis*

Strain	Temperature (°C)	α-Amylase (U/mg of cells)		
		late log	stationary	24h
B. stearothermophilus CU21	55	1.6	2.4	3.9
B. stearothermophilus AN174	55	ND[a]	ND	ND
B. stearothermophilus AN174 (pTB90)	55	ND	ND	ND
B. stearothermophilus AN174 (pAT9)	55	2.7	12.4	20.9
B. subtilis MI113	37	ND	0.1	0.3
B. subtilis TN106	37	ND	ND	ND
B. subtilis TN106 (pTB53)	37	ND	ND	ND
B. subtilis TN106 (pAT5)	37	0.1	0.3	1.1

[a]ND, Not detectable (<0.01 U/ml)

From Aiba et al. (33)

    The structural gene for a thermostable α-amylase from *B. stearothermophilus* was also cloned in plasmids pTB90 and pTB53 (33). Amylase activities of culture supernatant from both *B. stearothermophilus* and *B. subtilis* with and without a recombinant plasmid were examined (Table 7). The parental strain *B. stearothermophilus* CU21 produced 3.9 units/mg of cells. Neither the host strain nor the strain carrying vector plasmid pTB90 produced detectable amounts of amylase. In contrast, *B. stearothermophilus* AN174 carrying pAT9 produced five times more amylase (20.9 units/mg of cells) than did the original strain, *B. stearothermophilus* CU21.

    B. subtilis MI113 exhibited a low amylase activity (0.3 units/mg of cells), although the enzyme was substantially different from the thermostable amylase of *B. stearothermophilus*. Host strain *B. subtilis* TN106 with and without vector plasmid pTB53 did not produce amylase, whereas the strain carrying pAT5 did (1.1 units/mg of cells). These results show that the amylase gene encoded on the 4.8 Md *Hin*dIII fragment is expressed also in *B. subtilis*. However, the level of gene expression in *B. subtilis* was lower than that in *B. stearothermophilus*. This phenomenon might be due to the following reasons: a) low efficiency of the expression of the amylase gene from a thermophile in *B. subtilis*, b) low secretion efficiency of thermostable amylase in *B. subtilis*, and c) low copy number of the recombinant plasmid pAT5 in *B. subtilis*.

    The thermostable α-amylases were produced and secreted by the recombinant-plasmid-carrier strains of *B. stearothermophilus* AN174 and *B. subtilis* TN106. It has also been mentioned earlier that penicillinase and thermostable neutral protease are secreted by both *B. stearothermophilus* and *B. subtilis* carrying the specific recombinant plasmids, albeit the extent of secretion was different. These observations would suggest that some secretion mechanisms are shared by these mesophilic and thermophilic *Bacillus* species.

5. CONSTRUCTION OF MEROPOLYPLOID STRUCTURES IN HOST CELLS

Molecular cloning of specific gene(s) in either thermophiles or mesophiles permits the construction of meropolyploid structures in host cells, and the analysis such as *cis-trans* test of regulatory mechanism of gene expression is made possible. In this context, we have cloned penicillinase genes, that is, the structural gene (*penP*) and repressor gene (*penI*), from both wild-type and constitutive strains of *B. licheniformis* ATCC 9945A FD0120 in *E. coli*, *B. subtilis*, and *B. licheniformis* by using vector plasmids (15). Penicillinase genes from magnoconstitutive strain C01 could be cloned in *B. subtilis* with low-copy-number plasmid pTB53 (1 to 3 copies per chromosome), but not with high-copy-number plasmid pUB110 (~50 copies per chromosome). In contrast, penicillinase genes from wild-type strains could be cloned in *B. subtilis* even with the high-copy-number plasmid. These results suggest that if genes cloned on a plasmid were excessively expressed, such an expression would deter the normal activities of host cells. Hence transformants with such plasmids were rarely obtained.

Penicillinase was assayed in bacteria with and without a recombinant plasmid carrying *penP* (Table 8). No enzyme activities were detected for either *E. coli* C600-1 or the strain carrying only vector plasmid pMB9. *E. coli* C600-1 (pTTE11) produced nearly the same amount of penicillinase (about 7 units/mg of cells) as did *E. coli* C600-1 (pTTE21). Although the enzyme activity of the *penP*-carrying strains of *E. coli* was fairly low, both strains were resistant to ampicillin.

Very little enzyme activity was detected in *B. subtilis* carrying only vector plasmids pUB110 or pTB53. The penicillinase of *B. subtilis* MI112 was not induced by cephalosporin C. *B. subtilis* MI112 strains carrying either pTTB21 or pTTB22 produced nearly the same amount of penicillinase (83 to 100 units/mg of cells), although the cloned *Eco*RI fragments in pTTB21 and pTTB22 were inserted in the opposite direction. This fact suggests that the promoter of the penicillinase gene might have been on the cloned *Eco*RI fragment together with *penP*, and the promoter might have functioned normally in *B. subtilis*. In fact, the promoter has been identified in the upstream region of the structural gene (34).

B. subtilis carrying pTTB42 (low copy number) produced 26 units of penicillinase per mg of cells, about one-fourth of that in *B. subtilis* (pTTB21). This reduction in potency of penicillinase in *B. subtilis* (pTTB42) might be attributed most probably to the difference in copy number of plasmids. The induction ratio of the enzyme in *B. subtilis* (pTTB21 or pTTB42) was almost

TABLE 8. Penicillinase activity of plasmid-carrying strains

Strain	penP	penI	Anti-repressor gene	Plasmid	Copy no.[a]	penP	penI	Penicillinase activity[b] Uninduced	Induced	Induction ratio
E. coli										
C600-1	−	−	−	−		−	−	ND[c]		
C600-1	−	−	−	pMB9	high	−	−	ND		
C600-1	−	−	−	pTTE11	high	+	−	6.7		
C600-1	−	−	−	pTTE21	high	+	+	7.5		
B. subtilis										
MI112	−	−	−	−		−	−	1.0	0.3	0.3
MI112	−	−	−	pUB110	high	−	−	1.0		
MI112	−	−	−	pTB53	low	−	−	1.0		
MI112	−	−	−	pTTB21	high	+	+	100	87	0.9
MI112	−	−	−	pTTB22	high	+	+	83		
MI112	−	−	−	pTTB32	low	+	−	12,000		
MI112	−	−	−	pTTB42	low	+	+	26	20	0.8
B. licheniformis										
FD0120, R206	+	+	+	−				7.1	930	130
C01	+	−	+	−				3,000	3,200	1.1
M015-1	−	−	+	−				ND		
M015-1	−	−	+	pUB110	high	−	−	0.8		
M015-1	−	−	+	pTB53	low	−	−	ND		
M015-1	−	−	+	pTTB21	high	+	+	87		
M015-1	−	−	+	pTTB22	high	+	+	89	96	1.1
M015-1	−	−	+	pTTB32	low	+	−	10,000		
M015-1	−	−	+	pTTB42	low	+	+	22	88	4.0
R206	+	+	+	pTTB32	low	+	−	84	3,800	45

[a] High, 20–50 copies per chromosome; low, 1–3 copies per chromosome   From Imanaka (15)
[b] Activities are expressed in units of penicillinase per mg of cells
[c] ND, not detectable

174

unity. *B. subtilis* (pTTB32) produced the largest amount of enzyme (12,000 units/mg of cells).

Penicillinase activities were rarely detected in *B. licheniformis* M015-1 with or without vector plasmid (pUB110 or pTB53). Nearly the same activity of penicillinase (about 90 units/mg of cells) was manifested by *B. licheniformis* M015-1 carrying either pTTB21 or pTTB22. *B. licheniformis* M015-1 (pTTB42) produced 22 units of penicillinase per mg of cells. *B. licheniformis* M015-1 (pTTB32) produced 10,000 units of enzyme per mg of cells. The description of penicillinase activity, depending on the plasmid species, is quite similar to that with respect to the different host cells of *B. subtilis*.

It is evident from the data in Table 8 that the repressor of the penicillinase gene should have functioned in *B. subtilis*. In addition, neither *B. subtilis* (pTTB21) nor *B. subtilis* (pTTB42), both of which were *penP$^+$penI$^+$* and different in plasmid copy number, could be induced at all for penicillinase. These facts permit an inference that an antirepressor gene (or another regulatory gene), which should have performed as an effector in *B. licheniformis*, was not coded on the chromosome of *B. subtilis*.

Induction for penicillinase in *B. licheniformis*, however, occurred in a markedly different fashion than that in *B. subtilis*. The induction ratio in *B. licheniformis* M015-1 (*penP penI*) carrying pTTB22 (*penP$^+$penI$^+$*, high copy number) was 1.1, whereas that in strain M015-1 carrying pTTB42 (*penP$^+$penI$^+$*, low copy number) was 4.0. Furthermore, the induction ratio in wild-type strain FD0120 (*penP$^+$penI$^+$*, one copy for each on the chromosome) was 130. In other words, the greater the copy number for *penP$^+$ penI$^+$* per chromosome, the less the induction ratio.

Next, *B. licheniformis* R206, which was wild type with respect to the regulatory system of penicillinase, was used in place of strain M015-1 (the regulatory system was constitutive). *B. licheniformis* R206 (pTTB32) produced 84 units of penicillinase per mg of cells without the addition of inducer, whereas this strain produced 3,800 units of enzyme per mg of cells when induced. The induction ratio was 45.

The fact that the enzyme production (84 units/mg of cells) by *B. licheniformis* R206 (pTTB32) (uninduced) was repressed as compared with that (10,000 units/mg of cells) in *B. licheniformis* M015-1 (pTTB32) points out that the protein product (repressor) from the chromosomal origin (strain R206) must have functioned negatively on the penicillinase gene in plasmid pTTB32. These results make unlikely the possibility of operator constitutivity for plasmid pTTB32 and also indicate that plasmid pTTB32 was devoid of active repressor (*penI*).

As a matter of fact, when penicillinase production was induced, *B. licheniformis* R206 (pTTB32) produced much more enzyme (3,800 units/mg of cells) than did the host strain R206 (930 units/mg of cells). The difference between the two enzyme activities must be due to the derepression of penicillinase gene *penP* coded on the plasmid.

It is also shown that when induced, wild-type strain FD0120 produced much more penicillinase (930 units/mg of cells) than strain M015-1 carrying pTTB22 or pTTB42 (around 90 units/mg of cells). These facts could be explained as follows: repressor that could be titrated by antirepressor might have overwhelmed the function of antirepressor when the copy number of *penP*⁺ *penI*⁺ increased, and vice versa.

One of the points in this work is that recombinant plasmids were used to construct hetero- and meropolyploid structures in host cells for the examination of a regulatory mechanism of penicillinase synthesis. We have succeeded by the complementation test in confirming that penicillinase biosynthesis in *B. licheniformis* is, indeed, under the control of a repressor.

Using penicillinase markers in *B. licheniformis* strain 749 that were transferrable by transformation into strain 9945A, Sherratt and Collins (35) analyzed genetically the penicillinase locus and proposed a model for the regulatory mechanism of penicillinase biosynthesis. According to their genetic analysis, one regulatory gene (repressor gene, *penI*) is 90% linked to the structural gene (*penP*), and a second regulatory gene (effector or antirepressor gene) is 50% linked to the first regulatory gene. Our experimental data showed that the cloned *Eco*RI fragment (2.8 Md) from *B. licheniformis* 9945A coded penicillinase genes (*penP* and *penI*) and not the antirepressor gene. This result is in conformity with the genetic studies of Sherratt and Collins (35).

6. FINAL REMARKS

By using the host-vector system in *B. stearothermophilus*, we have cloned the structural genes of thermostable enzymes (neutral protease and α-amylase) of *B. stearothermophilus* (32,33), and determined the nucleotide sequences (36,37). According to the sequence determination, the guanine-cytosine (G-C) content of the coding region of neutral protease from *B. stearothermophilus* was 58%, while that of the third letter of the codons was 72%. In contrast, those values of the neutral protease gene from *B. subtilis* (mesophile) are 44 and 42%, respectively (28). In addition, the G-C content of the isopropylmalate dehydrogenase gene of *T. thermophilus* is 70%, whereas that of the third letter of the codons is extremely high at 89% (38). These observations point out that the larger value of optimum growth temperature of an organism is characterized by the greater G-C content in DNA. This tendency was remarkable, especially for the third letter of the codons of various coding regions.

Nucleotide sequence studies of Km^r genes of pUB110 and pTB913 showed that the substitution of a single amino acid residue stabilized the enzyme (21). Conversely, if subjected to site-directed mutagenesis such that a single replacement in the amino acid sequence of a given enzyme occurs, it would be possible to create more thermostable enzymes. Host-vector systems in thermophilic bacteria might offer an efficient screening means for such mutated genes in vivo.

A gene that controls the high-temperature regulon of *E. coli* has been cloned and analyzed (39), but the control mechanisms of gene expression in thermophiles at elevated temperatures are still obscure. We hope that the above-mentioned host-vector systems in thermophilic bacteria will contribute to the more profound and comprehensive view of thermophiles.

REFERENCES

1. L. G. Ljungdahl, in A. H. Rose and J. G. Morris, Eds., *Advances in Microbial Physiology*, Vol. 19, Academic Press, New York, 1979, p. 149.
2. R. E. Amelunxen and A. L. Murdock, in D. J. Kushner, Ed., *Microbial Life in Extreme Environments*, Academic Press, New York, 1978, p. 217.
3. J. Stenesh, in M. R. Henrich, Ed., *Extreme Environments: Mechanisms of Microbial Adaptation*, Academic Press, New York, 1976, p. 85.
4. S. Aiba, T. Imanaka, and J. Koizumi, *Ann. N. Y. Acad. Sci.* **413**, 57 (1983).
5. R. E. Buchanan and N. E. Gibbons, Eds., *Bergey's Manual of Determinative Bacteriology*, 8th ed., Williams & Wilkins, Baltimore, 1974.
6. J. F. Catterall, N. D. Lees, and N. E. Welker, in D. Schlessinger, Ed., *Microbiology—1976*, American Society for Microbiology, Washington, D.C., 1976, p. 358.
7. T. Imanaka, M. Fujii, I. Aramori, and S. Aiba, *J. Bacteriol.* **149**, 824 (1982).
8. A. H. A. Bingham, C. J. Bruton, and T. Atkinson, *J. Gen. Microbiol.* **114**, 401 (1979).
9. T. Imanaka, M. Fujii, and S. Aiba, *J. Bacteriol.* **146**, 1091 (1981).
10. T. Hoshino, Y. Koyama, and K. Furukawa, *Abstract of 2nd Symposium on Biotechnology*, Research Association for Biotechnology, Tokyo, 1984, p. 23 (in Japanese).
11. S. D. Ehrlich, *Proc. Natl. Acad. Sci. U.S.A.* **74**, 1680 (1977).
12. T. J. Gryczan, S. Contente, and D. Dubnau, *J. Bacteriol.* **134**, 318 (1978).
13. S. Chang and S. N. Cohen, *Mol. Gen. Genet.* **168**, 111 (1979).
14. C. B. Thorn and H. B. Stull, *J. Bacteriol.* **91**, 1012 (1966).
15. T. Imanaka, T. Tanaka, H. Tsunekawa, and S. Aiba, *J. Bacteriol.* **147**, 776 (1981).
16. B. J. Brown and B. C. Carlton, *J. Bacteriol.* **142**, 508 (1980).
17. S. I. Alikhanian, N. F. Ryabchenko, N. O. Bukanov, and V. A. Sakanyan, *J. Bacteriol.* **146**, 7 (1981).
18. T. Imanaka, T. Ano, M. Fujii, and S. Aiba, *J. Gen. Microbiol.* **130**, 1399 (1984).
19. D. Perlman and R. H. Rownd, *Nature* **259**, 281 (1976).
20. H. Danbara, J. K. Timmis, R. Lurz, and K. N. Timmis, *J. Bacteriol.* **144**, 1126 (1980).
21. M. Matsumura, Y. Katakura, T. Imanaka, and S. Aiba, *J. Bacteriol.* **160**, 413 (1984).
22. F. Hishinuma, T. Tanaka, and K. Sakaguchi, *J. Gen. Microbiol.* **104**, 193 (1978).
23. O. Gray and S. Chang, *J. Bacteriol.* **145**, 422 (1981).
24. Y. Takeichi, K. Ohmura, A. Nakayama, K. Otozai, and K. Yamane, *Agric. Biol. Chem.* **47**, 159 (1983).
25. I. Palva, *Gene* **19**, 81 (1982).
26. S. A. Ortlepp, J. F. Ollington, and D. J. McConnell, *Gene* **23**, 267 (1983).
27. M. L. Stahl and E. Ferrari, *J. Bacteriol.* **158**, 411 (1984).
28. M. Y. Yang, E. Ferrari, and D. J. Henner, *J. Bacteriol.* **160**, 15 (1984).
29. J. A. Wells, E. Ferrari, D. J. Henner, D. A. Estell, and E. Y. Chen, *Nucleic Acids Res.* **11**, 7911 (1983).
30. N. Vasantha, L. D. Thompson, C. Rhodes, C. Banner, J. Nagle, and D. Filpula, *J. Bacteriol.* **159**, 811 (1984).
31. M. Fujii, T. Imanaka, and S. Aiba, *J. Gen. Microbiol.* **128**, 2997 (1982).

32. M. Fujii, M. Takagi, T. Imanaka, and S. Aiba, *J. Bacteriol.* **154**, 831 (1983).
33. S. Aiba, K. Kitai, and T. Imanaka, *Appl. Environ. Microbiol.* **46**, 1059 (1983).
34. J. R. McLaughlin, S.-Y. Chang, and S. Chang, *Nucleic Acids Res.* **10**, 3905 (1982).
35. D. Sherratt and J. Collins, *J. Gen. Microbiol.* **76**, 217 (1973).
36. M. Takagi, T. Imanaka, and S. Aiba, *J. Bacteriol.* **163**, 824 (1985).
37. R. Nakajima, T. Imanaka, and S. Aiba, *J. Bacteriol.* **163**, 401 (1985).
38. Y. Kagawa, H. Nojima, N. Nukiwa, M. Ishizuka, T. Nakajima, T. Yasuhara, T. Tanaka, and T. Oshima, *J. Biol. Chem.* **259**, 2956 (1984).
39. F. C. Neidhardt, R. A. VanBogelen, and E. T. Lau, *J. Bacteriol.* **153**, 597 (1983).

APPLIED GENETICS OF
ANAEROBIC THERMOPHILES

PIERRE BÉGUIN AND JACQUELINE MILLET

Unit of Cellular Physiology, Department of Biochemistry and Molecular Genetics, Pasteur Institute, 28, rue du Dr Roux, 75724 Paris Cedex 15, France

1. INTRODUCTION

From a systematic point of view, anaerobic thermophiles are quite heterogenous since they belong to various taxonomic groups, and it is to be expected that each species is more closely related to other species of the same group, including nonthermophiles, than to anaerobic thermophiles belonging to other groups.

From a physiological point of view, two main categories may be distinguished among these organisms:

- Eubacterial microorganisms. Many can be used to ferment carbohydrates, yielding ethanol as the main product of interest.
- Archaebacterial methanogens. Besides producing methane, which can be used as a fuel, methanogens are an essential link in the metabolic chain leading to the anaerobic mineralization of organic wastes. By converting carbon dioxide, hydrogen, and low molecular weight acids into methane, they prevent the buildup of hydrogen or acid in concentrations that would inhibit fermenting organisms.

Anaerobic thermophiles grow relatively slowly in the laboratory. Even with modern anaerobic facilities, the large-scale screening of single colonies is far more difficult for these organisms than for aerobic bacteria such as *Escherichia coli* or *Bacillus subtilis*. Yet the potential advantages of high-temperature fermentations (for review see ref. 1 and Chapter 10, this volume) have stimulated a number of studies aiming at the improvement of thermophilic strains by random mutagenesis and selection. More recently, the ability of thermophiles to grow at temperatures at which most living organisms die and most enzymes are quickly inactivated has also received increased attention, and efforts have been made to understand the molecular basis of high temperature tolerance, especially at the level of gene expression. Similarly, the biology of methanogens, some of which are thermophiles, has also been studied with increased interest since it was discovered that they belong to the archaebacteria, a group so distantly related to both eubacteria and eukaryotes that it has been defined as a new kingdom (2).

The genes most amenable for improvement of methanogens are still ill defined. It is conceivable that more efficient methane-producing strains may be developed. For example, it would be of interest to increase their range of substrates or their tolerance to environmental conditions; similarly, methane superproducers would be of advantage. So far, however, very little is known about the means to isolate such mutants efficiently. Hence, most of the work dealing with genetics of methanogens involves the development of basic genetic tools.

In this chapter, we shall first review publications describing mutations affecting the efficiency with which anaerobic thermophiles convert carbohydrates into ethanol, which is usually the desired end product. We shall then discuss the characterization of genes from these organisms after cloning in foreign hosts. Finally, the prospects of developing genetic transfer systems in anaerobic thermophiles themselves will be reviewed.

2. APPLIED GENETICS OF THERMOPHILIC FERMENTATIONS

2.1. Carbohydrate Utilization

Most available plant carbohydrates, like starch, cellulose, hemicellulose, and pectin, are polymeric substrates and must be first hydrolyzed before being fermented. The biological depolymerization is slow and often limits growth and fermentation rates. Thus, attempts have been made to improve the efficiency of hydrolysis.

Derepression and Hyperproduction of Glucoamylase and Pullulanase   In a study of the regulation of glucoamylase and pullulanase production in *Clostridium thermohydrosulfuricum*, Hyun and Zeikus (J. Bacteriol., submitted) described the isolation of two regulatory mutants, Z21-109 and Z67-143. The mutants were obtained after mutagenesis by *N*-methyl-*N'*-nitro-*N*-nitrosoguanidine (NTG), followed by enrichment in the presence of starch as a sole carbon source, 2-deoxyglucose being added as a nonmetabolizable inducer of catabolic repression. For both mutants, catabolic repression by glucose was found to be abolished for both glucoamylase and pullulanase. Mutant Z21-109 produced 1.6-fold more glucoamylase and 1.8-fold more pullulanase than the wild type when grown on starch medium. Culture under growth-limiting conditions in a chemostat demonstrated that starch was required as an inducer for glucoamylase and pullulanase synthesis both in the wild type and in the mutant Z21-109. This strain also utilized starch more completely and produced more ethanol and less lactate than the wild type.

Cellulase Hyperproduction   A mutant of *Clostridium thermocellum* exhibiting hyperproduction of cellulolytic activity was isolated by Shinmyo et al. (3). Wild-type cells were treated with ultraviolet (UV) radiation and plated on agar containing cellobiose plus MN300 cellulose. Colonies having the largest clear zones of hydrolysis were chosen for assays of cellulolytic enzyme production in liquid medium. A mutant strain (AS39) was obtained by this procedure. The growth rates of the parent strain and the AS39 mutant were similar, but the mutant produced twice as much endo-β-glucanase and exo-β-glucanase as did the parent. Similar results were obtained using another medium (4). Strain AS39 appeared to be a regulatory mutant since the endo- and exo-β-glucanase activities are not due to the same enzyme (5).

Identification of a Cellulose-binding Factor   Cellulose degradation by *C. thermocellum* is effected by a set of enzymes with various specificities (5,6). In the culture medium, these proteins form a multienzyme complex, designated "cellulosome" by Lamed et al. (6). The role played in cellulose binding by one of the cellulosome components was demonstrated by a combined genetic-immunochemical approach (6–8). Part of the cellulolytic complex is located on the surface of the cells, allowing cells to bind to cellulose particles. A spontaneous adhesion-defective mutant (AD2) was isolated using repetitive cycles of enrichment which included growth on cellobiose and selective removal of bacteria adhering to fast-sedimenting cellulose particles (7). The absence of adhesion was only displayed by cellobiose-grown cells; after growth on cellulose, the mutant recovered the ability to bind to cellulose. An antiserum specific for the adhesion factor was prepared by immunizing rabbits against wild-type cells and adsorbing the serum on adhesion-defective cells. This serum was used to detect the binding factor in the culture medium, where it was found associated with the cellulosome. In immunoelectrophoresis experiments, the antiserum was found to react with a polypeptide with $M_r = 210,000$ having no cellulolytic activity.

2.2. Ethanol Production

Among chemicals that could be produced by fermentation of carbohydrates, ethanol is the compound for which there is the largest demand (9). The economics of the ethanol fermentation is strongly influenced by the yield of ethanol per mole of metabolized carbohydrate and by the ethanol tolerance of the fermenting organism, which limits the maximal end-product concentration obtainable. Hence, several groups have attempted to improve these two factors.

Improving the Fermentation Balance   Most thermophiles have a heterofermentative metabolism. As can be seen in Figure 1, glucose is converted into two moles of pyruvate via the Embden-Meyerhof pathway, with the concomitant reduction of two moles of NAD^+ and the synthesis of two moles of ATP. Pyruvate may either be reduced to lactate, thus recycling one equivalent of NADH, or it may be cleaved to yield acetyl-CoA. In anaerobic thermophiles, the latter reaction is generally achieved by pyruvate dehydrogenase (CoA-acetylating), together with the release of one mole of carbon dioxide and the reduction of two ferredoxin equivalents. Ferredoxin is then recycled by transferring electrons either to $NAD(P)^+$ or to hydrogenase, leading to the formation of H_2. In *Bacillus stearothermophilus*, which is also considered for thermophilic ethanol production, pyruvate cleavage to acetyl-CoA under anaerobic conditions is mainly achieved by pyruvate:formate lyase, which does not require reduction of a cofactor (10,11).

Besides being used in various anabolic reactions, acetyl-CoA can undergo two transformations. First, it can be converted into acetate and CoA by phosphotransacetylase and acetate kinase. This pathway leads to the phosphorylation of one mole of ADP and increases the recovery of metabolic energy. Second, it can be reduced by aldehyde dehydrogenase (CoA-acetylating) and alcohol dehydrogenase to ethanol and CoA. Unlike the acetate pathway, metabolism via the ethanol pathway does not increase the ATP yield; it is nevertheless essential since it is required for maintaining the

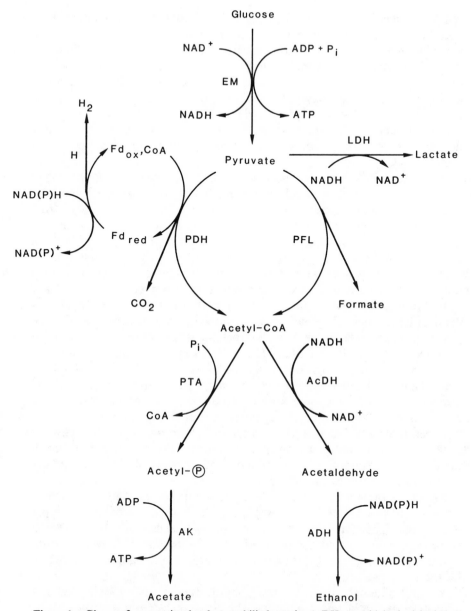

Figure 1. Glucose fermentation by thermophilic bacteria. AcDH, acetaldehyde dehydrogenase; ADH, alcohol dehydrogenase; AK, acetate kinase; EM, Embden-Meyerhof pathway; Fd_{ox}, Fd_{red}, oxidized and reduced forms of ferredoxin; H, hydrogenase; L, lactate dehydrogenase; PDH, pyruvate dehydrogenase; PFL, pyruvate:formate lyase. $NAD(P)^+$ and $NAD(P)H$ indicate that both dinucleotides can be involved in the reaction. Data from (10,68).

redox balance of the process. Thus, if no NAD(P)H is produced during the decarboxylation of pyruvate (i.e., if formate or hydrogen are formed in stoichiometric amounts), the redox balance requires that one mole of ethanol be produced per mole of acetate in order to recycle the two NADH produced during lycolysis; otherwise, the need of recycling excess NAD(P)H generated during pyruvate decarboxylation will lead to an increased ethanol:acetate ratio, with a correspondingly lower yield of ATP.

An obvious approach to increase ethanol yields is to search for mutants defective in pathways leading to undesired products, such as lactate and acetate.

A lactate-defective strain has been obtained in *C. thermocellum* through mutagenesis followed by a colony screening test based on the lactate dehydrogenase-catalyzed reduction of NAD$^+$ by lactate, coupled via phenazine methosulfate to the reduction of tetranitroblue tetrazolium (12). Although this mutant showed no detectable lactate production, its performance as an ethanol producer was not improved as compared to the parent strain, because it was no longer able to ferment its substrate completely. This may not necessarily be related to lactate production, however.

Fluoropyruvate can be used to select mutants defective in NAD$^+$-dependent lactate dehydrogenase, presumably because it is converted to fluorolactate, a toxic metabolite. Such fluorolactate-resistant mutants have been obtained in *B. stearothermophilus* (10), where lactate normally accounts for half of the fermentation products on a molar basis, and the yields of acetate and ethanol were increased correspondingly. In addition, for one of the mutants, it was possible to increase the ethanol:acetate ratio by growing the organism at elevated temperature and at neutral or slightly acidic pH. This has been interpreted as being due to the induction of a pyruvate dehydrogenase pathway in addition to the pyruvate:formate lyase pathway, although the authors recognized that the induction of an NAD$^+$-linked formate dehydrogenase would lead to the same result.

Another simple strategy for isolating fermentation mutants is to screen for low-acid-producing strains by plating them in the presence of a pH indicator such as methyl red. This technique was successfully used by Wang et al. (12), who isolated a mutant of *Clostridium thermosaccharolyticum* producing 0.35 g ethanol/g xylose, as compared to 0.23 g ethanol/g xylose for the parent strain.

While lactate dehydrogenase is an obvious target when looking for mutants defective in lactate production, the ethanol:acetate ratio may also be influenced by the activity of several enzymes, such as hydrogenase, ferredoxin:NAD$^+$-oxidoreductase or phosphotransacetylase, and the alterations leading to a modification of this ratio are generally unknown.

It is worth noting that the relative amount of the various fermentation products is subject to regulation. In *C. thermocellum*, lactate production increases at the end of exponential growth (12, H. Girard, personal communication). With *Thermoanaerobacter ethanolicus*, which grows between pH 5 and 10, good ethanol production was observed only when the pH was maintained between 5 and 8, when fermenting glucose, and between 5.5 and 7.5, when fermenting xylose. Moreover, the substrate concentration appears to be critical, since, in the wild-type strain, ethanol yields approaching the theoretical maximum required substrate concentrations to be maintained below 1% (13,14). In the mutant strain JW200 Fe(4), this limit has been raised to 4% glucose, 3% xylose, and 6% cellobiose or soluble cornstarch (15).

An unexpected fallout of the isolation of ethanol-tolerant mutants is that these mutants often display an improved ethanol yield as well. Wang et al. (12) isolated an ethanol-tolerant strain of *C. thermocellum* (ATCC 27405) yielding an ethanol:acetate ratio of 5:1, whereas the wild type produced an equimolar ratio. An ethanol-tolerant mutant of *C. thermosaccharolyticum* was also described in the same paper, with an ethanol:acetate ratio of 3:2 instead of 1:1 in the wild type. Similar observations were made in our laboratory with an ethanol-tolerant derivative of *C. thermocellum* NCIB 10682 (P. Tailliez et al., manuscript in preparation). Since methods for the genetic characterization of these ethanol-tolerant mutations do not presently exist, it is difficult to prove that ethanol tolerance and ethanol overproduction are due to the same mutation; however, the simultaneous occurrence of the two phenotypes in these different strains makes it likely.

Increasing Ethanol Tolerance The mechanisms of ethanol toxicity and tolerance are not understood, and they may vary according to the organism. Different bacteria may show different sensitivities to the various cellular targets of ethanol action so that different factors may be involved in growth limitation.

One aspect of ethanol toxicity seems to involve the perturbation of the membrane environment, presumably by modifying hydrophobic interactions between membrane components. In *E. coli*, ethanol induces the synthesis of membrane lipids enriched in unsaturated fatty acids and inhibits the cross-linking of peptidoglycans (16,17). An ethanol-tolerant mutant of *E. coli* was isolated by Fried and Novick (18) which showed altered properties of its lactose transport system, although the mutation was not linked to the *lacY* gene and probably involved an alteration of the environment of lactose permease. In *C. thermocellum*, ethanol induces growth arrest followed by a phase of resumed growth at a slower rate (19). During growth arrest, the composition of the membrane lipids changes, with a shift toward fatty acids having a lower melting point (20). This is reflected in the phase behavior of membrane lipids extracted from ethanol-adapted cultures, as measured by differential scanning calorimetry (21). With strain C9, an ethanol-tolerant mutant of *C. thermocellum*, the lipid composition was found similar to that of the wild type and underwent similar changes toward fatty acids of increased fluidity when challenged with inhibitory concentrations of ethanol (20). However, study of the phase change of membrane lipids showed that lipids extracted from the mutant had a T_m 7 to 10°C lower than the wild type and that the T_m did not change upon adaptation to ethanol. Finally, addition of ethanol to *C. thermocellum* membrane lipids had little effect on their phase behavior (21). It may thus be concluded that, in *C. thermocellum* as well as in *E. coli*, ethanol alters membrane properties by inducing changes in lipid composition. However, it is doubtful whether ethanol toxicity is due to an increase of membrane fluidity caused by these changes, as is sometimes surmised (20). Indeed, lipids extracted from the ethanol-tolerant mutant C9 had an even lower T_m than those of the wild type. Rather, ethanol probably causes a pleiotropic effect on the activity of membrane-bound enzymes, including enzymes involved in glycolysis (9) or possibly in other steps of fermentative metabolism. The latter hypothesis would explain why some ethanol-tolerant mutants also display alterations in their fermentation product ratios (see above).

In some cases, toxicity seems to be mediated by the accumulation of NADH caused by the oxidation of ethanol by an NAD^+-dependent ethanol dehydrogenase. An example is provided by the study of *Clostridium thermohydrosulfuricum* and its ethanol-tolerant derivative 39EA (22). The wild type showed temperature-independent growth inhibition about 20 g/l ethanol, whereas the mutant could tolerate up to 40 g/l at 60°C; this tolerance increased with decreasing temperature. The effect of several organic solvents was compared and it was found that the wild type was specifically sensitive to ethanol, whereas methanol and acetone were much less toxic. Growth stimulation was even observed with low concentrations of acetone, whose reduction to isopropanol provided the cells with the opportunity to increase acetate production. Acetone had approximately the same effect on the mutant, and its toxicity at high concentrations was about the same as for the wild type. Thus, in the wild type, the mechanism of inhibition by ethanol must be fairly specific and is probably not due to a general perturbation caused by the apolar character of this solvent, since acetone is even less polar and less toxic. In the mutant, specific sensitivity to ethanol was abolished and the toxicity observed at high concentrations of various solvents, including ethanol, could be due to general perturbations of hydrophobic interactions. The mechanism of inhibition by ethanol in the wild type was ascribed to oxidation of exogenous ethanol by NAD^+-dependent ethanol dehydrogenase, causing NADH to build up to the point where it inhibits glyceraldehyde-3-phosphate dehydrogenase and shuts off glycolysis. The mutant lacks NAD^+-dependent ethanol dehydrogenase and is no longer inhibited by NADH accumulation in the presence of ethanol. A similar mechanism has been identified in *C. thermohydrosulfuricum* for hydrogen sensitivity. Uptake hydrogenase causes electrons to flow from hydrogen to NAD^+ and a hydrogen-tolerant mutant lacking uptake hydrogenase has been isolated (J. G. Zeikus, personal communication).

Thus, possible causes of ethanol toxicity are numerous and often poorly defined. Therefore, ethanol-tolerant strains are still most efficiently isolated by random mutagenesis followed by selection in the presence of increased concentrations of ethanol.

3. SCREENING AND IDENTIFICATION OF SPECIFIC GENES

Cloning of DNA fragments from anaerobic thermophiles in *E. coli* presents no special difficulties and several specific genes have already been recognized in this way. These genes can then be analyzed with the molecular biology techniques available in *E. coli* and their nucleotide sequences can be determined. Although "surrogate genetics" is limited by the differences that may exist between the cloning host and the organism from which the gene was originally isolated, specific DNA fragments can be used as probes to characterize mRNA from the donor organism, permitting a definition of the structure of the regions controlling initiation and termination of transcription. If needed, gene products can be synthesized in large quantities after cloning in appropriate expression vectors. Finally, selective markers, like genes conferring antibiotic resistance or amino acid prototrophy, can be incorporated into otherwise cryptic plasmids isolated from anaerobic thermophiles, making them more suitable for use as shuttle vectors (see Section 4.1).

Several approaches may be used to identify specific genes after cloning in a foreign host:

1. In some cases, it is possible to use hybridization probes homologous to the desired gene. This approach has been used to characterize rRNA and tRNA genes of methanogens (23,24, L. Sibold, personal communication). Homology has also been detected between hybridization probes derived from the nitrogen-fixation (*nif*) genes of *Klebsiella pneumoniae* and *Anabaena* strain 7120 and the DNA restriction fragments of various methanogens, including *Methanobacterium thermoautotrophicum* (strain ΔH) (25). These DNA fragments correspond to genuine *nif* genes, as shown by the recent discovery that at least two methanogens, *Methanosarcina barkeri* (strain 227) and *Methanococcus thermolithotrophicus*, can fix N_2 (27,27).

2. The cloned genes may complement a defective function of the host, such as an auxotrophic mutation. With *C. thermocellum*, two cosmids were found that were able to complement either *leuB* and *leuC* or *trpE* mutations of *E. coli* (28). No such complementation has yet been described for thermophilic methanogens, but the *argG* function of *E. coli* can be supplied by appropriate DNA fragments from *M. barkeri* and *Methanococcus voltae* (29,30). Similarly, clones bearing fragments of *M. voltae* and *Methanobrevibacter smithii* were found to complement *hisA* and *purE* mutations of *E. coli*, respectively (31).

    Restoration of a prototrophic phenotype may sometimes occur

owing to a mutator effect of the cloned insert. Gomez et al. (32) have isolated a recombinant plasmid containing *C. thermocellum* DNA that increases the reversion frequency of *E. coli* HB101 to Pro[+] and Leu[+]. The cloned DNA appeared to contain an insertion sequence, termed IS120, which behaved like a transposing element in *E. coli* and occurred at a frequency of about 10 copies/cell in *C. thermocellum* (33). The nucleotide sequence of IS120 includes a 7 bp inverted repeat at both its termini and a 3 bp direct duplication at the site of insertion. Three open reading frames were identified, coding for two polypeptides of about 11,000 daltons and one of about 38,000 daltons. The 38,000 dalton protein and at least one of the 11,000 dalton proteins were synthesized in *E. coli* maxicells (33).

3. The cloned gene may direct the synthesis of a polypeptide which can be identified by enzymatic or immunological tests. This approach was chosen in our laboratory to clone cellulase genes from *C. thermocellum*. Screening of gene banks for endoglucanase activity was performed by using carboxymethylcellulose, either in viscosimetric assays (28) or in the Congo red plate test devised by Teather and Wood (34,35). Six clones bearing different endoglucanase genes were isolated (28, Pétré et al., manuscript in preparation). The enzyme expressed by one of the clones was found to be immunologically identical with the $M_r = 56,000$ endoglucanase, termed endoglucanase A (EGA), which had been previously purified in our laboratory from a *C. thermocellum* culture supernatant (36); the corresponding gene was termed *celA*. The endoglucanases expressed by the other cloned had not been characterized previously. They were termed endoglucanase B, C, (EGB, EGC), etc., and the corresponding genes *celB, celC*, etc. Three clones, bearing the genes *celA, celB*, and *celC*, are now well characterized (35,37,38, Pétré et al., manuscript in preparation). The three genes have been subcloned, yielding fragments of 2.8, 2.7, and 2.7 kb, respectively. Endoglucanase expression in *E. coli* was found to be independent of cloning orientation, indicating that transcription was initiated within the DNA insert. The cloned fragments do not cross-hybridize and are not adjacent.

The enzymes EGB and EGC have been purified from extracts of *E. coli* bearing appropriate plasmids (37, Pétré et al., manuscript in preparation). Antiserum directed against *E. coli*-synthesized EGB was used to demonstrate that the corresponding protein was indeed secreted by *C. thermocellum* into the culture medium, where it was identified as a 66,000 dalton polypeptide by the "Western blot" technique (37). Since the molecular weight of *E. coli*-synthesized EGC is about 38,000, it appears that EGA, EGB, and EGC are all different from the $M_r = 83,000$ to 94,000 endoglucanase purified from *C. thermocellum* culture supernatant by Ng and Zeikus (39).

The enzymes EGA, EGB, and EGC display different enzymatic specificities toward cellodextrins of various degrees of polymerization. While EGC hydrolyzes cellotriose, cellotetraose, and cellopentaose, EGB has no activity toward cellotriose but hydrolyzes cellotetraose and cellopentaose, and EGA reacts with cellopentaose but only poorly with cellotetraose and not at all with cellotriose (Longin et al., manuscript in preparation).

Efficient hydrolysis of crystalline cellulose usually requires the synergistic action of endoglucanases and cellobiohydrolases (40,41). The latter enzymes have little or not activity toward carboxymethylcellulose (41,42) and cannot be detected using the Congo red test (43). However, methylumbelliferyl-β-cellobioside (44) and p-nitrophenyl-β-cellobioside (45) have recently been described, respectively, as fluorogenic and chromogenic substrates for cellobiohydrolases, and it can be expected that genes coding for these enzymes will soon be cloned and characterized in *C. thermocellum*. A cellobiohydrolase gene has already been isolated from *Cellulomonas*, a cellulolytic mesophile (43).

The finding that genes originating from organisms only distantly related to *E. coli* can be expressed in this bacterium with a detectable efficiency suggests that gene expression mechanisms show a remarkably broad specificity. A study of the DNA sequence of the *celA* gene of *C. thermocellum* (38) indicates that the regions controlling protein secretion and gene expression bear close similarity to the corresponding sequences found in *E. coli* and *B. subtilis*. The amino-terminal sequence determined for the mature protein purified from *C. thermocellum* culture supernatant is preceded by a signal peptide of 32 amino acids which is closely related to signal peptides described for other secreted proteins from Gram-positive bacteria. A putative ribosome-binding site which closely matches the canonical Shine-Dalgarno sequence (46,47) is found upstream from the start of the translated sequence. As to transcription control sites, analysis of *C. thermocellum* mRNA indicates that transcription is monocistronic and that promoter and termination sites are closely related to those described for *E. coli* and *B. subtilis* (P. Béguin, unpublished data).

The expression of genes from methanogens in *E. coli* was more unexpected, since transcription and translation systems are quite different in archaebacteria and eubacteria. Translation is not affected by the same antibiotics (48) and the structure of DNA-dependent RNA polymerases is quite different (for a review see 29). Thus, it will be very interesting to see whether genes from methanogens that are expressed in *E. coli* use the same promoters in both organisms.

4. DEVELOPMENT OF GENE TRANSFER SYSTEMS

The study of the genetics of anaerobic thermophiles is severely hampered by the lack of DNA transfer systems. Hence, some of the most important genetic tools are not available for genetic studies. To mention just a few examples, it is impossible to investigate the function of specific genes, particularly those involved in regulation, by constructing appropriate merodiploids. Likewise, site-directed mutagenesis of anaerobic thermophiles by recombination with a DNA fragment bearing a transposon cannot be performed. As to the localization of genetic markers, no extended genetic map can be established, except by successive cloning and identification of overlapping DNA segments ("chromosome walking"), clearly a cumbersome procedure for long stretches of DNA.

Concerning the construction of strains with improved performances for industrial purposes, it would be highly desirable to be able to introduce amplifiable expression vectors bearing genes of interest originating either from the same or from foreign organisms.

The development of gene transfer systems requires, first, that adequate vectors be devised for the cloning and expression of genes in the desired organism, and second, that means should be found to introduce plasmid DNA bearing the appropriate genes into the organism of interest and to select transformed clones.

4.1. Plasmids

Despite the need for appropriate vectors for genetic engineering, there are few reports of the presence of plasmids in anaerobic thermophiles. Nine strains of *C. thermocellum*, including five new isolates and four strains from culture collections, were tested in our laboratory by the "miniprep" method of Birnboim and Doly (49), and no plasmid was detected. However it is not yet clear whether this reflects the scarcity of plasmid-bearing strains or the failure to reveal plasmids by conventional methods. Indeed, plasmid identification and characterization in various bacteria often requires modification of standard isolation procedures. Critical steps appear to be adequate cell lysis, inactivation of nucleases, and elimination of DNA-associated proteins (50–55).

Cryptic plasmids were identified by Weimer et al. (55) in thermophilic anaerobic bacteria possibly related to the genera *Thermobacteroides, Thermoanaerobacter*, or *Thermoanaerobium*, which ferment hemicellulose. The isolation procedure involved overnight treatment with sodium dodecyl sulfate and centrifugation for 4 days in a CsCl gradient in order to remove associated protein material. One of the plasmids, pM24D (pDP5009), with a length of 2.3 kb, was studied in more detail. It has unique sites for *Acc*I, *Bal*I, *Sau*3A1, and *Sau*96I and codes for at least one polypeptide with an M_r of 8,000. Cloning of pM24D into pBR322 yielded a hybrid plasmid, pKW17, which could possibly be developed as a shuttle vector (56).

A 4.5 kb cryptic plasmid (pME2001) was also characterized in the Marburg strain of *Methanobacterium thermoautotrophicum* (54). A restriction map was constructed, with unique sites for seven restriction endonucleases.

Analysis by agarose gel electrophoresis and electron microscopy of a purified preparation showed the formation of multimers; it will be interesting to explore the type of mechanism involved. Hybridization experiments revealed the presence in another strain of *M. thermoautotrophicum* (strain ΔH) of a plasmid with sequence homology to plasmid pME2001. While plasmid pME2001 was apparently produced in high copy number and represented a significant portion of the total DNA, the plasmid of the ΔH strain was present in extremely low concentration. It is not clear whether the low yield of the ΔH plasmid was due to loss during the isolation procedure or to its absence in some of the host cells. Recently, Meile and Reeve (57) inserted pME2001 into various replicons having different host ranges, with a view to obtaining shuttle vectors able to replicate in methanogens and in various hosts, such as *E. coli, B. subtilis, Staphylococcus aureus*, and the yeast *Saccharomyces cerevisiae*. Selection in eubacterial hosts was made possible by the presence of genes conferring resistance to chloramphenicol and ampicillin, and in yeast by the *ura-3* gene complementing uracil auxotrophy. To provide a selective marker for transformation of methanogen auxotrophs, a DNA fragment from *Methanococcus vannielii* complementing *E. coli hisA* mutations was also cloned in some of the hybrid plasmids.

Although endogenous plasmids are good candidates for the development of cloning systems, another promising alternative lies in the availability of broad host-range vectors. Among thermophiles, *B. stearothermophilus* has been successfully transformed by pTB19, a plasmid originating from another thermophilic *Bacillus* (58), and by pUB110, which was first isolated from *S. aureus* (59). Even though *S. aureus* is a mesophile, pUB110 can be stably maintained at temperatures up to 55°C (60); it has also been recently introduced into the mesophilic anaerobe *Clostridium acetobutylicum* (52). Since this plasmid contains a selectable marker (resistance to kanamycin), it should be a candidate of choice to develop a genetic engineering system in thermophilic clostridia and possibly other Gram-positive thermophiles.

4.2. Transformation

To our knowledge, the successful introduction of foreign genetic material into anaerobic thermophiles has not yet been reported. An essential requirement of cloning systems is the availability of markers enabling the selection of hybrid clones. Hence efforts have been made to isolate auxotrophs and antibiotic-resistant mutants. Gomez et al. (32) have tested the effect on *C. thermocellum* of various mutagenic agents including nitrosoguanidine, NTG, ultraviolet (UV) radiation and gamma radiation. Ultraviolet radiation was found most effective in generating mutants resistant to 5-fluorouracil and rifampicin. *C. thermocellum* auxotrophs can be efficiently enriched using penicillin and D-cycloserine (61); two spontaneous Ade⁻, one spontaneous Leu⁻, and one UV-induced Ile⁻ mutants were isolated after enrichment using these antibiotics.

More recently Kiener et al. (62) described the isolation of mutants of *M. thermoautotrophicum* (strain Marburg) resistant to antibiotics and to amino acid analogues. The following inhibitors were used: bacitracin, lasalocid, chloramphenicol, and 2-bromoethanesulfonic acid (2-BES) which is a structural analogue of coenzyme M (2-mercaptoethanolsulfonic acid (63)). Twenty amino acid analogues were tested and only DL-ethionine was found to act as an inhibitor. Several independent experiments were performed using NTG as mutagenic agent, but only mutants resistant to DL-ethionine or to 2-BES were found. Since these mutants were altered in cell morphology, it was suggested that the analogue resistance may be due to changes in the cell envelope rather than in specific target sites in the cytoplasm. Unfortunately, the plating efficiency of the three analogue mutants was deemed too low to use them for genetic experiments. In the same publication, six auxotrophic mutants were described. They were obtained after NTG mutagenesis and each of them required only a single growth factor (leucine, phenylalanine, adenosine, thiamine). As the frequency of revertants to prototrophy was below the level of detection, these mutants may be useful to develop a system of transfer of genetic material in *M. thermoautotrophicum*.

Development of successful transformation protocols is highly empirical and differs from species to species. Modifications of the protocol established by Chang and Cohen (64) for the PEG-induced transformation of *B. subtilis* protoplasts have been used to transform *B. stearothermophilus* (65) and *C. acetobutylicum* (52), suggesting that the method may be adapted to other Gram-positive strains. Some of the technical problems appear to arise from nuclease activity and/or from the difficulty of obtaining regeneration of protoplasts. Mild heating of the bacteria seemed to inactivate the nuclease in the case of *C. acetobutylicum*, and modifying the lysozyme concentration used to prepare protoplasts appeared to be useful in the case of *B. stearothermophilus* (65). The isolation of strains yielding regeneration-proficient protoplasts may also be helpful (66).

5. CONCLUSION

Most of the interest in eubacterial anaerobic thermophiles stems from their potential use in high-temperature fermentations. Empirical attempts at improving their properties have already led to the isolation of significantly better strains, particularly with respect to ethanol production. Cloning and characterization of specific genes should allow the definition of consensus regions controlling gene expression and possibly the identification of specific features responsible for thermotolerance. Available data (38,67, Chapters 6 and 7, this volume) indicate that, at least for eubacteria, such features are rather subtle. At the same time, the cloning of genes coding for thermophilic enzymes such as cellulases, amylases, or xylanases in appropriate expression vectors may facilitate the mass production of these enzymes. The pace at which anaerobic thermophiles will be further improved will no doubt be accelerated once genetic engineering becomes possible, a development which is likely to occur in the near future.

Besides their use for methane production, much of the interest in methanogens arises from the fact that they belong to the archaebacteria. Since it was recognized that these bacteria display features that set them apart from other forms of life, their basic molecular biology has been studied with increased attention. While much of the basic research on methanogen gene expression may not lead to immediate applications, it is clear that a better understanding of the basic biology of these organisms and the development of genetic tools will be profitable to applied research in the long term.

ACKNOWLEDGMENTS

The authors wish to thank Drs. T. K. Ng, P. J. Weimer, J. G. Zeikus, T. Leisinger, B. Snedecor, L. Sibold, P. Tailliez, R. Longin, and H. Girard for reprints and for communicating results prior to publication. They are also grateful to Drs. J.-P. Aubert and L. Sibold for critical reading of the manuscript. Work in our laboratory was supported by a research contract from Solvay et Cie., Brussels, Belgium and Rhône-Poulenc, Paris, France, and by research funds from University of Paris 7.

REFERENCES

1. B. Sonnleitner and A. Fiechter, *Trends Biotechnol.* **1**, 74 (1983).
2. C. R. Woese and G. E. Fox, *Proc. Natl. Acad. Sci. USA* **74**, 5088 (1977).
3. A. Shinmyo, D. V. Garcia-Martinez, and A. L. Demain, *J. Appl. Biochem.* **1**, 202 (1979).
4. D. V. Garcia-Martinez, A. Shinmyo, A. Madia, and A. L. Demain, *Eur. J. Appl. Microbiol.* **9**, 189 (1980).
5. E. A. Johnson and A. L. Demain, *Arch. Microbiol.* **137**, 135 (1984).
6. R. Lamed, E. Setter, R. Kenig, and E. A. Bayer, *Biotechnology and Bioengineering Symposium*, No. 13, Wiley, New York, 1983, p. 163.
7. E. A. Bayer, R. Kenig, and R. Lamed, *J. Bacteriol.* **156**, 818 (1983).
8. R. Lamed, E. Setter, and E. A. Bayer, *J. Bacteriol.* **156**, 828 (1983).
9. A. A. Herrero, *Trends Biotechnol.* **1**, 49 (1983).
10. B. S. Hartley and M. A. Payton, *Biochem. Soc. Symp.* **48**, 133 (1983).
11. M. A. Payton, *Trends Biotechnol.* **2**, 153 (1984).
12. D. I. C. Wang, G. C. Avgerinos, I. Biocic, S.-D. Wang, and H.-Y. Fang, *Phil. Trans. R. Soc. Lond.* **B300**, 323 (1983).
13. L. G. Ljungdahl, F. Bryant, L. Carreira, T. Saiki, and J. Wiegel, in A. Hollaender, Ed., *Trends in the Biology of Fermentations for Fuels and Biochemicals*, Plenum, New York, 1981, p. 397.
14. L. H. Carreira, L. G. Ljungdahl, F. Bryant, M. Szulczinski, and J. Wiegel, *Proceedings of the IVth International Symposium on Genetics of Industrial Microorganisms*, Tokyo, 1982, p. 351.
15. L. H. Carreira, J. Wiegel, and L. G. Ljungdahl, *Biotechnology and Bioengineering Symposium*, No. 13, Wiley, New York, 1983, p. 183.

16. T. M. Buttke and L. O. Ingram, *Biochemistry* **17**, 637 (1978).
17. L. O. Ingram, *J. Bacteriol.* **146**, 331 (1981).
18. V. A. Fried and A. Novick, *J. Bacteriol.* **114**, 239 (1973).
19. A. A. Herrero and R. F. Gomez, *Appl. Environ. Microbiol.* **40**, 571 (1980).
20. A. A. Herrero, R. F. Gomez, and M. F. Roberts, *Biochim. Biophys. Acta* **693**, 195 (1982).
21. W. Curatolo, S. Kanodia, and M. F. Roberts, *Biochem. Biophys. Acta* **734**, 336 (1983).
22. R. W. Lovitt, R. Longin, and J. G. Zeikus, *Appl. Environ. Microbiol.* **48**, 171 (1984).
23. M. Jarsch, J. Altenbuchner, and A. Böck, *Mol. Gen. Genet.* **189**, 41 (1983).
24. H. Neumann, A. Gierl, J. Tu, J. Leibrock, D. Staiger, and W. Zillig, *Mol. Gen. Genet.* **192**, 66 (1983).
25. L. Sibold, D. Pariot, L. Bhatnagar, M. Henriquet, and J.-P. Aubert, *Molec. Gen. Genet.* **200**, 40 (1985).
26. P. A. Murray and S. H. Zinder, *Nature* **312**, 284 (1984).
27. N. Belay, R. Sparling, and L. Daniels, *Nature* **312**, 286 (1984).
28. P. Cornet, D. Tronik, J. Millet, and J.-P. Aubert, *FEMS Microbiol. Letters* **16**, 137 (1983).
29. J. N. Reeve, N. M. Trun, and P. T. Hamilton, in A. Hollaender, R. D. DeMoss, S. Kaplan, J. Koniski, D. Savage, and R. S. Wolfe, Eds., *Genetic Engineering of Microorganisms for Chemicals*, Plenum, New York, 1982, p. 233.
30. A. G. Wood, A. H. Redborg, D. R. Cue, W. B. Whitman, and J. Koninski, *J. Bacteriol.* **156**, 19 (1983).
31. P. T. Hamilton and J. N. Reeve, in W. R. Strohl and O. H. Tuovinen, Eds., *Microbial Chemoautotrophy*, Ohio State University Press, Columbus, 1984, p. 291.
32. R. F. Gómez, B. Snedecor, and B. Méndez, *Developments in Industrial Microbiol.*, Vol. 22, Society for Industrial Microbiology, Arlington, Va., 1981, p. 87.
33. B. Snedecor, E. Chen, and R. F. Gómez, *Proceedings of the IVth Symposium on Genetics of Industrial Microorganisms*, Kyoto, 1982, p. 356.
34. R. M. Teather and P. J. Wood, *Appl. Environ. Microbiol.* **43**, 777 (1982).
35. P. Cornet, J. Millet, P. Béguin, and J.-P. Aubert, *Bio/Technology* **1**, 589 (1983).
36. J. Pètre, R. Longin, and J. Millet, *Biochimie* **63**, 629 (1981).
37. P. Béguin, P. Cornet, and J. Millet, *Biochimie* **65**, 495 (1983).
38. P. Béguin, P. Cornet, and J.-P. Aubert, *J. Bacteriol.* **162**, 102 (1985).
39. T. K. Ng and J. G. Zeikus, *Biochem. J.* **199**, 341 (1981).
40. V. S. Bisaria and T. K. Ghose, *Enzyme Microbiol. Technol.* **3**, 90 (1981).
41. N. Creuzet, J.-F. Bérenger, and C. Frixon, *FEMS Microbiol. Letters* **20**, 347 (1983).
42. T. D. Bartley, K. Murphy-Holland, and D. E. Eveleigh, *Anal. Biochem.* **140**, 157 (1984).
43. N. R. Gilkes, M. L. Langsford, D. G. Kilburn, R. C. Miller, and R. A. J. Warren, *J. Biol. Chem.* **259**, 10455 (1984).
44. H. van Tilbeurgh, M. Claeyssens, and C. K. de Bruyne, *FEBS Letters* **149**, 152 (1982).
45. M. V. Deshpande, K.-E. Eriksson, and L. G. Pettersson, *Anal. Biochem.* **138**, 481 (1984).
46. J. Shine and L. Dalgarno, *Nature* **254**, 34 (1975).
47. G. D. Stormo, T. D. Schneider, and L. M. Gold, *Nucl. Acids Res.* **10**, 2971 (1982).
48. R. Hilpert, J. Winter, W. Hammes, and O. Kandler, *Zbl. Bakt. Hyg. I. Abt. Orig.* **C2**, 11 (1981).
49. H. C. Birnboim and J. Doly, *Nucl. Acids Res.* **7**, 1513 (1979).
50. M. S. Strom, M.W. Eklund, and F. T. Poysky, *Appl. Environ. Microbiol.* **48**, 956 (1984).
51. H. P. Blaschek and M. A. Klacik, *Appl. Environ. Microbiol.* **48**, 178 (1984).
52. Y.-L. Lin and H. P. Blaschek, *Appl. Environ. Microbiol.* **48**, 737 (1984).
53. M. Thomm, J. Altenbuchner, and K. O. Stetter, *J. Bacteriol.* **153**, 1060 (1983).
54. L. Meile, A. Kiener, and T. Leisinger, *Mol. Gen. Genet.* **191**, 480 (1983).
55. P. J. Weimer, L. W. Wagner, S. Knowlton, and T. K. Ng, *Arch. Microbiol.* **138**, 31 (1984).
56. S. Knowlton and L. W. Wagner, *Abstracts, American Society for Microbiology*, 85th Annual Meeting, Las Vegas, 1985.
57. L. Meile and J. N. Reeve, *Bio/Technology* **3**, 69 (1985).
58. T. Imanaka, M. Fujii, and S. Aiba, *J. Bacteriol.* **146**, 1091 (1981).

59. R. W. Lacey and I. Chopra, *J. Med. Microbiol.* **7**, 285 (1974).
60. T. Imanaka, *Trends Biotechnol.* **1**, 139 (1983).
61. B. S. Méndez and R. F. Gomez, *Appl. Environ. Microbiol.* **43**, 49 (1982).
62. A. Kiener, C. Holliger, and T. Leisinger, *Arch. Microbiol.* **139**, 87 (1984).
63. W. E. Balch and R. S. Wolfe, *J. Bacteriol.* **137**, 256 (1979).
64. S. Chang and S. N. Cohen, *Mol. Gen. Genet.* **168**, 111 (1979).
65. T. Imanaka, M. Fujii, I. Aramori, and S. Aiba, *J. Bacteriol.* **149**, 824 (1982).
66. S. Knowlton, J. D. Ferchak, and J. K. Alexander, *Appl. Environ. Microbiol.* **48**, 1246 (1984).
67. M. Matsumura, Y. Katakura, T. Imanaka, and S. Aiba, *J. Bacteriol.* **160**, 413 (1984).
68. R. Lamed and J. G. Zeikus, *J. Bacteriol.* **144**, 569 (1980).

9

INDUSTRIAL APPLICATIONS OF THERMOSTABLE ENZYMES

THOMAS K. NG AND WILLIAM R. KENEALY

Central Research and Development Department, E. I. du Pont de Nemours and Company, Experimental Station, Wilmington, Delaware 19898

1. INTRODUCTION

The recent interest in "biotechnology," coupled with the discovery of new, novel thermophiles, has prompted studies on the utilization of thermophiles and their enzymes for industrial purposes. Several programs directed toward understanding the mechanism of thermophily and isolating industrially useful microorganisms from extreme environments have already been funded by the government and private sectors (1). There are suggestions that enzymes could be genetically engineered to function at 100°C with faster rates and that bioreactors (fermenters) using thermophiles could operate with no external cooling (1). Additional advantages for the use of thermophiles in biotechnology have also been proposed (2–5).

Thermostable enzymes are now utilized extensively in industrial processing (6,7). However, before we are overwhelmed by such enthusiasm, we should examine what constitutes a thermostable enzyme and what is the rationale for their utility in industry. Most importantly, we need to understand the mechanism of thermophily so that enzymes with enhanced thermostability can be engineered. Once the decision for using thermostable enzymes is made, a good source of that particular enzyme is required. In this chapter, we will discuss the pros and cons of obtaining thermostable enzymes from thermophiles as opposed to obtaining such enzymes from mesophiles. In addition, the alternative approach, the enhancement of thermostability by physical/chemical techniques, will also be included. A review of industrial applications will serve to illustrate the impacts and limitations of thermostable enzymes. Finally, we shall consider the future of thermostable enzymes and their industrial applications.

Thermostable enzymes not involved in industrial applications are not included in this chapter. There are several excellent review articles on this subject (8–10).

2. THERMOSTABLE ENZYMES

2.1. Definition

The term "thermostable" has been given a multitude of meanings in the literature. Most of the definitions of thermostability relate to the inherent nature of the enzyme and take into consideration the source of the enzyme (11–13). Thus, the proposed classification of enzymes from thermophiles by Campbell and Pace (11) relied on the relative stability of proteins when compared to the growth temperature of the organism. In this respect, a *thermostable enzyme* was defined as one having a maximal reaction temperature above that of the optimal growth temperature for the microorganism. Clearly, such a definition is invalid in view of the fact that enzymes from microorganisms growing at temperatures above 90°C are "thermostable" regardless of whether their optimum activity temperature is below or above the optimum growth temperature. It becomes obvious that thermostability is a relative term.

A temperature stability curve is often used by enzyme manufacturers for their product specifications (6). In this procedure, thermostability is defined as the retention of activity after heating to 60°C or above for a prolonged period (6). The drawback of such a procedure is that it only measures how well an enzyme tolerates high temperature and does not take into consideration the ability of the enzyme to renature during the cooling period. Thus, enzymes that are thermostable by this definition should actually be termed *thermotolerant* because the retention of enzyme activity after heating is measured at a mesophilic temperature. Unfortunately, as we will discuss later, many studies on the mechanism of thermophily have been done using such a definition.

Perhaps the most apprpriate and practical way to express thermostability is by measuring the half-life of enzyme activity at several elevated temperatures. The half-life is defined as the reaction time for the enzyme activity to drop to half the initial value under specific conditions (6). If the enzyme activity is measured specifically at that elevated temperature, then thermostability becomes an operational definition, encompassing the function of time and temperature and is defined independently of the source. The term "thermostable" can be expressed on only a relative scale. That is to say, a truly thermostable enzyme will have a half-life at 50°C considerably longer than its thermolabile counterpart. With half-life data: 1) a plant operator can decide when to recharge an immobilized enzyme bed; 2) a process engineer can determine whether certain commercial enzyme preparations can withstand the required operating temperature; 3) the enzyme manufacturer can specify the shelf life of the product; and 4) most importantly, a researcher can compare under identical conditions the thermostability of an enzyme from different sources or from different variants. This definition is useful since it brings several basic and applied parameters into consideration.

2.2. Rationale for Use in Industry

The above considerations bring us to examine the rationale for employing thermostable enzymes in industry. At first inspection, one might think of applying thermostable enzymes for faster reactions at higher temperature. While it is true that reaction rate increases with increasing temperature, the catalytic rates of thermostable enzymes from thermophiles are no different from their mesophilic counterparts (8–10,14). The fact is that thermophiles do not require or necessarily possess enzymes with higher *activities* when growing optimally at higher temperature. On the other hand, thermostable enzymes from mesophiles do exhibit increased reaction rate when utilized at high temperatures, although they simultaneously experience a higher denaturation rate. This distinction between thermostable enzymes from mesophiles and from thermophiles is of prime importance when we discuss the utility of thermostable enzymes later.

Primarily, thermostable enzymes are selected because of process constraints and/or cost effectiveness. In the two major areas of industrial applications, namely starch processing and detergent formulation, high- to moderately high-temperature operating conditions dictate the use of thermostable enzymes. Besides process requirements, reactions at high temperature also improve the mass transfer rate, lower the viscosity, minimize the risk of contamination (5), and in a few cases shift the equilibrium of the reactions to favor formation of products (14). In fact, the majority of industrial enzymatic reactions operate at temperature above 50°C (6). Additional cost savings in production of enzymes from thermophiles have been suggested (2–5,15). However, the significance of such factors in production-scale facilities has not yet been clarified.

The predominant reasons for selecting thermostable enzymes are the major cost savings resulting from longer storage stability and the higher activity at high temperature (6).

Enzymes, like all proteins, are gradually denatured upon storage, and since enzymes are sold normally in activity units rather than as weight of enzyme, the manufacturers will add approximately ten percent more activity to compensate for inactivation between the time of production and the expiration time of the product (6). Thermostable enzymes because of their enhanced thermostability are also more resistant to most chemical denaturants (16,17). Thus, a thermostable enzyme would have a longer shelf life than its mesophilic counterpart to use thus resulting in major cost savings. The second reason for using thermostable enzymes as stated above, is the benefit from having a higher reaction rate with higher temperature. This means higher productivity and lower capital investment. Since most industrial enzyme reactions are operated between 50 and 100°C (2), enzymes from thermophiles will have less of an increase in activity as compared to that from mesophiles between this temperature range. Thus, thermostable enzymes from extreme thermophiles may have little value as far as increase in activity is concerned for moderately thermophilic processes.

Despite such limitations, industrial enzymes with higher chemical and thermal stabilities are often desired. The ability to achieve these via genetic or chemical engineering depends on a better understanding of the mechanism of thermophily. This will be discussed in the next section.

2.3. Mechanism of Thermophily

Although microorganisms can regulate environmental factors such as pH and osmotic pressure, they cannot regulate their internal temperature. Consequently, thermophilic organisms must possess intrinsically thermostable cellular components. Other than their thermostability, enzymes from these thermophiles appear to be catalytically indistinguishable from their mesophilic counterparts. They have similar reactivity and catalytic sites (8–10). There is no evidence to suggest that thermostable enzymes have compensated for their higher stability by having higher turnover rates (5,8,9). Such a misconception might arise from the observation that most commercial thermostable enzymes which are of mesophilic origin display higher activities at their high operating temperature than thermolabile enzymes. However, this increase is generally due to the effect of temperature on the catalytic function rather than the intrinsic properties of the enzymes.

What is the molecular basis for the enhanced stability of thermophilic enzymes? Such information is of great scientific and commercial importance. It can address the fundamental question of the evolution of thermophily and yield basic knowledge for the design and engineering of an enzyme so that it can function at extreme temperatures. A number of reports have been written on this subject (8–10), and only some generalized conclusions are presented here.

In the first place, the thermostability of an enzyme is a function of the enzyme's stabilizing forces (18). These include hydrogen bonding, hydrophobic bonding, ionic interactions, metal binding, and disulfide bridges. In their absence, destabilizing forces arise from the conformational entropy of the protein; in other words, the energy required to maintain and define tertiary structure will denature the enzyme. Naturally, each of these stabilizing forces, either by itself or in combination, has been suggested as a possibility for enhanced thermostability (8). Although it is generally true that such stabilizing forces contribute to overall stability, the means by which this is achieved varies among different enzymes. For example, the replacement of glutamate at position 49 by methionine increased the thermostability of tryptophan synthetase in *E. coli* by increasing hydrophobic bonding (19). However, the identical amino acid replacement in another enzyme might be totally ineffective or actually destabilize the enzyme. Enhanced thermostability is not due to a single attribute or mechanism but to a combination of stabilizing effects derived from the various interactions mentioned above (16,20). The extent of these interactions depends greatly on the protein/solvent environment.

Second, the net energy of stabilization (ΔG) is small, often between 40 and 80 kcal per mole (21,22). This represents the difference between the two large opposing forces, that is, between the stabilizing and the destabilizing forces which by themselves are large (>400 kJ/mole) (23). On one hand, minor perturbations like temperature fluctuation could easily disrupt the balance of forces by disproportionately increasing the conformational entropy. On the other hand, enhanced thermostability could be achieved by minor changes in the primary structure which might result in higher hydrophobic interactions, metal binding, and so on. Often such changes are minor and might involve a single amino acid substitution (19,20,24). For instance, the single amino acid substitution mentioned earlier for *E. coli* tryptophan synthetase increased the net energy of stabilization by only 3.14 kJ/mole but this was enough to offset the increased instability induced by heating the enzyme at 58°C for 20 minutes (19). Coincidentally, this extra

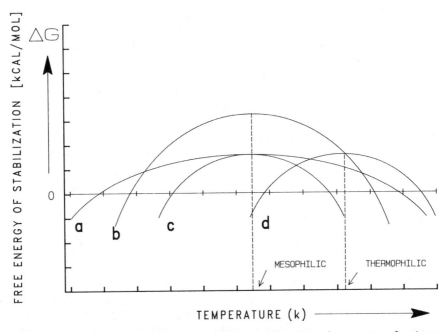

Figure 1. Hypothetical free energy of stabilization as a function of temperature for a) mesophilic enzyme with extended temperature stability range, b) mesophilic enzyme with extended temperature stability and higher energy of stabilization, c) normal mesophilic enzyme as reference, and d) normal thermophilic enzyme from a thermophile.

destabilizing force also equaled 3.14 kJ/mole. Such subtle changes normally have little effect on the overall tertiary structure, and, as a result, structural or amino acid sequence comparisons fail to pinpoint the molecular basis for enhanced thermostability.

Third, there are two different modes of thermostability just as there are two different types of thermostable enzymes. 1) Thermostable enzymes that are of mesophilic origin can be regarded as enzymes that exhibit extended temperature range and higher stability. In comparison to their mesophilic counterparts (Curve c, Figure 1), they are likely to have higher net energy of stabilization (ΔG) (Curve b) or an extended range of ΔG (Curve a). This extra energy is derived from various interactions mentioned earlier. In the case of thermolysin, the bound calcium ions provide extra stabilizing energy through ionic interactions (25), while intrinsic difference in amino acid composition might be responsible for higher stability in α-amyloglucosidase from *Aspergillus niger* (6). Because of this, and because enzyme activity increases proportionately with temperature, thermostable mesophile enzymes are utilized extensively in industry. 2) Thermostable enzymes from thermophiles do not necessarily have higher net energy of stabilization or wider temperature range for activity. Instead, their intrinsic primary structure, (i.e., amino acid sequence) provides extra stabilizing energy through various interactions which counterbalance the higher entropy involved in the maintenance of the tertiary structure at thermophilic temperature. In this regard, enzymes from thermophiles show a curve for the net energy of stabilization which is identical to that of mesophiles except that it is shifted forward along the temperature scale (Curve d, Figure 1). It should be emphasized that not all enzymes, whether from thermophilic or mesophilic sources, can be distinctly classified as such. Variations among the ΔG curves might exist.

Finally, the thermostability of an enzyme is highly dependent on its environment. This conclusion is inferred from the fact that the majority of the stabilizing forces for an enzyme are derived via interactions with the solvent. Therefore, the observed thermostability of an enzyme in a dilute buffered solution is very much different than what might be found inside the cell. At the same time, the location of a particular amino acid within the protein tertiary structure is just as important as its chemical nature. Therefore, statistical comparisons of amino acid compositions between thermophilic and mesophilic enzymes to determine the stabilizing effect of one particular amino acid are often misleading (26). In this respect, it is not surprising that an enzyme like porcine lipase is stable and active at 100°C in an organic solvent (27). In a later section, we will discuss enhanced thermostability induced artificially through interactions with the enzyme environment.

3. SOURCES OF THERMOSTABLE ENZYMES

Thermostable enzymes can be categorized into two groups according to their sources, those derived by natural selection and those created by physical/chemical modifications of thermolabile enzymes.

3.1. Natural Sources

It is a general belief that industrial thermostable enzymes are obtained from thermophiles and that the higher the growth temperature of the organism, the more active the enzyme at elevated temperature. On the contrary, the majority of industrial thermostable enzymes are produced by mesophiles (6). Normally, this type of thermostable enzyme is obtained by extensive, nonselective screening and subsequent strain improvement. The failure of thermophiles to gain a higher share of the market is twofold. One is the fact mentioned earlier that thermostable enzymes from thermophiles have less activity than their mesophilic counterparts under similar conditions at elevated temperatures. The other is that thermostability is not the only criterion for an "effective" industrial enzyme. The enzyme has to be produced in large quantity, be easily recoverable (preferably extracellular), and be produced by a microorganism with low cultivation cost. A classical example is α-amylase produced by *Bacillus licheniformis* strain T5. The enzyme is significantly more stable than most α-amylases from other cultures or strains. However, its relatively low activity makes it uneconomical for commercial production. In this case, the productivity is increased two to three times by selecting strain T316 from strain T5 through conventional genetic techniques (28).

Despite such drawbacks, thermophiles might still be good sources of thermostable enzymes since the yield and the cost of cultivation can be lowered by strain manipulation. The thermophilic α-amylase gene from *Bacillus stearothermophilus* has been cloned and expressed in *Escherichia coli* (29,30, see also Chapter 7).

Moreover, the availability of thermostable enzymes from thermophiles will continue to expand as more new novel thermophiles are isolated. Unlike mesophilic sources for thermostable enzymes, high-temperature environments are generally found in only a few restricted locations. These include geothermal areas like Yellowstone National Park and deep-sea thermal vents. By and large, most of these have not been fully explored (see Chapters 2 and 3). The potential for expanding our present understanding of thermophile enzymes seems tremendous and it may be only a matter of time before enzymes from these sources will be studied industrially.

3.2. Induced Thermostability

Thermostability can be induced by association with nonprotein stabilizing compounds or physical-chemical modifications. Such enhancement includes interactions with substrates, solvents, and salts as well as chemical modifications (18).

Enzymes are generally more stable when they are complexed with substrates or cofactors than in the free form (31). Thus, glutamate synthetase from *B. stearothermophilus*, which normally loses its activity at 65°C, is stable in the presence of NH_4^+, glutamate, Mg^{2+}, and ATP. It is even stabilized by feedback inhibitors such as alanine and histidine (32,33). The use of the substrate as an enzyme stabilizer is fully exploited by the starch industry (6). Typically, high substrate concentrations are employed throughout the starch hydrolysis process. This, together with the stabilizing effect of calcium, permits α-amylase from *B. licheniformis* (6) to function above 100°C. The stability of α-glucoamylase is also significantly enhanced in the presence of glucose or its analogues (34).

As discussed earlier, the forces that stabilize a protein are derived from the interaction of the protein with the solvent. Therefore, changes in the solvent system will affect thermostability. Unfortunately, most organic solvents cause structural instability and irreversible denaturation, but enhanced thermostability has been observed with polyhydroxyl alcohols (18). Glycerol, sucrose, and ethylene glycol are generally used to stabilize most enzymes during storage. The stabilizing effect of these polyhydroxyl alcohols has been reviewed by Myers and Jakoby (35). Most interestingly, porcine pancreatic lipase, which normally functions at mesophilic temperature, is stable and catalytically more active at 100°C upon replacement of the water with organic solvents (27). In this case, the solvent is 1M alcohol in tributyrin, which also serves as the substrate for the enzyme. It is, therefore, unclear whether the stabilizing effect is due to substrate binding or solvent-protein interactions. The dehydrated enzyme also exhibits a different substrate specificity, suggesting a change in the conformation of the active site.

Most salts have deleterious effects on enzyme stability. Yet, it is also well known that enzymes in 3M ammonium sulfate are very stable. This difference represents the free energy lost or gained during the "salting out" process, as the nonpolar groups are exposed on the protein surface (18).

Of all cations, calcium seems to have the most profound effect on thermostability. Calcium stabilizes most α-amylases at high temperature (36). Removal of calcium results in lower stability or even inactivation. Other cations such as magnesium, strontium, or barium can mimic the effect of calcium on α-amylase from *Aspergillus orzyae* (37), but α-amylase from *Bacillus caldolyticus* has an absolute requirement for calcium (38).

The important role of calcium is best illustrated in the case of *thermolysin*. Thermolysin, a zinc-containing protease from *Bacillus stearothermophilus*, has four bound calcium ions. Three of the four calcium ions can be removed without affecting the activity of the enzyme, but thermostability is markedly lowered (39). This indicates that at least one calcium is required for activity (40). Because of the dependence on calcium for protection against autolysis (41) and from x-ray crystallographic studies (25), it was concluded that the two calcium ions located on the surface prevent the enzyme from being hydrolyzed by other thermolysin molecules while the other two calcium ions cross-linked the two identical subunits to provide a rigid polypeptide backbone with extra stability (42).

Another common approach to enhance thermostability is by chemical modification of the enzyme. This involves alteration of the protein surface by chemical reagents or by covalent linkage to polymeric materials. Enzymes are modified either by monofunctional or bifunctional reagents. In general, monofunctional reagents like acid anhydride or aldehyde/sodium borohydride alter surface charges, whereas bifunctional reagents like glutaraldehyde cross-link amino acid groups. The general effect of such modification is to provide extra energy of stabilization. However, the chemical basis for such stability varies among different enzymes and has to be determined empirically in each case. Enzymes that have been modified in such a manner and exhibited increased thermostability have been reviewed by Schmid (18).

At present, the major way to obtain industrial thermostable enzymes is by direct isolation from nature. This source of enzymes will continue to expand as more novel microorganisms from extreme environments are discovered (see Chapter 3). On the other hand, as techniques for enhanced stability by chemical/physical modification improve, an increasing number of industrial enzymes will be stabilized via this route (6).

4. INDUSTRIAL APPLICATION OF THERMOSTABLE ENZYMES

Most of the industrial enzymes are utilized at temperatures above 50°C (6). These thermostable enzymes are added to detergents, employed for the production of natural sweeteners, and used in the manufacture of pharmaceuticals. Of the $400 million of enzymes sold worldwide in 1984 (6,43), 90% were thermostable enzymes used by the detergent and starch industries. A recent study suggested that these two areas have the greatest growth potential in the near term (44). Table 1 lists some of the key industrial thermostable enzymes and their use. To illustrate their importance, two major classes of thermostable enzymes are discussed in the following section.

4.1. Saccharification Enzymes

Glucosidases   Presently, thermostable enzymes are utilized extensively throughout the starch processing industry. As shown in Table 2, thermostable carbohydrases are directly involved in the manufacture of all starch-derived products.

TABLE 1. Applications of key industrial thermostable enzymes

Enzyme	Operating temperature (°C)	Major applications
Carbohydrases		
α-Amylase (bacterial)	90–110	Starch hydrolysis, brewing, baking, detergents
Glucoamylase	50–60	Maltodextrin hydrolysis
α-Amylase (fungal)	50–60	Maltose
Pullulanase	50–60	High glucose syrups
Xylose isomerase	45–55	High fructose syrups
Pectinase	20–50	Clarification of juices/wine
Cellulase	45–55	Cellulose hydrolysis
Lactase	30–50	Lactose hydrolysis, food processing
Proteases		
Acid protease	30–50	Food processing
Fungal protease (neutral protease)	40–60	Baking, brewing, food processing
Alkaline protease	40–60	Detergent
Lipases	30–70	Detergent, food processing

TABLE 2. Enzymatic starch processing

Substrate	Process	Enzymes	Products
Gelatinized starch	Liquifaction (thinning + dextrinization)	α-Amylase	Maltodextrins
Maltodextrins	Saccharification	Glucoamylase Fungal α-amylase Glucoamylase + pullulanase Glucoamylase + fungal α-amylase	Glucose syrups Maltose syrups High glucose syrups Mixed syrups
High glucose syrups	Isomerization	Xylose isomerase	High fructose syrups

α-Amylase is utilized for the liquefaction of starch. Its functions are to lower the viscosity of the gelatinized starch and hydrolyze it to maltodextrins so as to obtain a degree of polymerization of 10 to 15 (Table 2). Bacterial α-amylase (1,4-α-glucan glucanohydrolase) is an endoglucanase that preferentially hydrolyzes internal α-1,4-glucosidic bonds. Before the availability of thermostable α-amylase, enzymes from *Bacillus subtilis* and *B. amyloliquefaciens* had to be added before and after the high-temperature gelatinization step. This requirement has essentially been eliminated by the introduction of *B. licheniformis* α-amylase. This enzyme retains most of its activity at 110°C for a short period. Its thermostability is dependent on calcium and substrate concentration. In a typical process, 100 ppm calcium in the form of lime water is added together with the enzyme to a 40% starch slurry. The temperature is raised to 110°C for 5 to 10 minutes and then rapidly lowered to 95°C for 1 to 2 hours. The enzyme is then inactivated by heating the mixture at 125°C and pH 4 for a brief period to prevent further dextrinization. The major product is maltodextrins. Only a limited amount of this product is used as a viscosity modifier and filler in the food industry with the majority being used for the production of sugar syrups or fermentation substrates.

Fungal α-amyloglucosidase (1,4-α-D-glucan glucohydrolase or glucoamylase) is used in the production of glucose syrup with dextrose equivalents of 95 to 97%. α-Amyloglycosidase is an exo-acting enzyme that releases glucose from the reducing ends of maltodextrins. Commercial preparations from *Aspergillus* spp. and *Rhizopus* spp. also contain debranching activity that cleaves α-1,6 linkages from amylopectin. In contrast to the liquefaction step, saccharification is a considerably slower process, with typical run times of between 48 and 72 hours. Saccharification also operates at lower temperature and pH, due to the unavailability of a more suitable enzyme. The enzyme from *Aspergillus niger* and *A. orzyae* operates optimally at 55°C and pH 4 to 5. This incompatibility between the operating conditions of liquefaction and saccharification hampers the development of a continuous starch-to-glucose process.

In order to obtain glucose syrup with greater than 98% dextrose equivalents, pullulanase is usually added. This is because α-amylase generally produces a series of nonreducing cyclic glucosyl polymers called cyclodextrins at high glucose concentration and thus reduces the total dextrose equivalents. In addition, pullulanase debranches the α-1,6 linkage of amylopectin forming more linear malto-oligosaccharides. Bacterial pullulanase from *Klebsiella aerogenes* is most commonly used. However, extensive application of pullulanase in the starch industry is limited by the instability of the enzyme since the *Klebsiella* pullulanase rapidly loses activity at pH 4.5 and 60°C.

For the production of maltose syrup, thermostable fungal α-amylase is used. The enzymes from *Aspergillus niger* and *A. orzyae* differ from bacterial α-amylase by producing maltose as the major end product of prolonged incubation. The enzyme also produces limited dextrins and glucose. In this fashion, fungal α-amylase is similar to cereal or microbial β-amylase. Consequently, fungal α-amylase has essentially replaced cereal α-amylase in the commercial processes. However, it is important to note that fungal α-amylase (1,4-α-D-glucan maltohydrolase) is mechanistically different from microbial α-amylase (1,4-α-D-glucan glucanohydrolase).

For economic reasons, fungal α-amylase is often used in conjunction with amyloglucosidase to produce a mixed syrup with a dextrose equivalent of 60%. This mixed syrup, which is often termed high-conversion syrup, is used as an inexpensive substrate for the brewing and fermentation industries.

Isomerases Perhaps the most significant impact in the industrial application of enzymes is in the production of high-fructose corn syrup (HFCS). The present United States annual market is around $1.5 billion (45). Fructose, an isomer of D-glucose, is the sweetest of all natural sugars. Its use as a natural sweetener was previously limited by its availability. Glucose isomerase has not been found to occur in nature and chemical isomerization is uneconomical. It was then discovered that xylose isomerase also catalyzes the isomerization reaction. At present, the annual United States production of HFCS amounts to 6 billion pounds with a market value of $1.5 billion. The HFCS industry consumes $40 million worth of xylose isomerase. The modern process utilizes immobilized xylose isomerase in a fixed-bed reactor. High-purity glucose syrup is pumped continuously through the reactor at 60°C and pH 8, and a syrup with approximately 52% fructose and 48% glucose is obtained. The xylose isomerases from *Bacillus coagulans*, *Streptomyces albus*, *Arthrobacter* spp., and *Actinoplanes missouriensis* are most frequently used (46).

Other Carbohydrases Two other important industrial carbohydrases are the pectolytic enzymes and lactase. Pectolytic enzymes, which are a mixture of pectin esterase and pectinase, are used for the clarification of fruit juices (47). Lactase is used in the dairy industry for the breakdown of lactose to glucose and galactose (48). Even though some of these processes operate above 50°C, enzymes with high thermostability are generally not used. Most of the lactase and pectolytic enzymes obtained from mesophilic microorganisms lose a considerable amount of activity at temperatures above 50°C. In this respect, enzymes with higher stability would be desirable.

A recent development in the industrial enzyme area is the suggested use of cellulase for the production of glucose from cellulose (49). This in turn would provide an abundant source of substrate for the fermentation industry. Cellulase is a collective term for several enzymes that hydrolyze native cellulose to a mixture of cellobiose and glucose (50). The most commonly studied cellulases are those from *Trichoderma* spp. (49), *Thermonospora* spp. (51,52), and *Clostridium thermocellum* (53). All of these enzymes exhibit optimum activities between 50 and 60°C. In comparison to the enzymatic starch hydrolytic process, cellulose hydrolysis by cellulase for the production of glucose is very slow. Clearly, a cellulase with much improved activities is required before an economical enzyme-based process is possible. Moreover, cellulases produce high concentrations of cellobiose, a 1,4-β-linked disaccharide which, unlike maltose, is not metabolized by most industrial microorganisms. Consequently, a thermostable cellulase will have little value unless it has significantly improved activity and an active β-glucosidase component that could increase the glucose content by hydrolyzing the excess cellobiose.

4.2. Proteolytic Enzymes

As a single class, proteolytic enzymes constitute about two-thirds of the total industrial enzyme market. Of these, 25 percent are thermostable, alkaline proteases utilized in the detergent industry (6). Despite such impressive numbers, the use of enzymes in detergents had to overcome several obstacles. In the first place, the advantage of using proteases in laundering clothes that contain little protein is not obvious. For this reason, the first enzyme detergent was marketed primarily for cleaning work clothes used in the food industry (54). Subsequent tests revealed that even in minute quantity protein will bind most inert particles (e.g., soil), making dirt difficult to remove. Such finding affirms the validity of incorporating protease into detergent formulations. Moreover, very low levels of enzyme (usually at 1 ppm) are enough to disrupt the binding action of the proteinaceous materials. Such excellent cleaning ability helped enzyme-based detergents capture 50 percent of the United States and European household detergent markets in the early 1970s (55). However, problems were then encountered when some detergent packaging workers developed allergies to the powdered enzyme. Although the allergy problem was resolved by encapsulating the enzyme preparations and making them essentially dust free (48), putative health and safety considerations prompted the removal of all enzymes from United States formulations. Now that the European experience with enzyme-containing detergents has clearly demonstrated the safety of the product (which coincidentally agreed with the conclusion of the National Academy of Sciences investigation in 1971 (56)), there is quite a resurgence of enzymes in the United States detergent marketplace.

A protease suitable for use in detergents requires more than just good enzyme activity. Besides being able to maintain activity under various stringent washing conditions (e.g., 50°C, pH 10), it must be resistant to other additives like surfactants and sequestering agents. As indicated earlier, thermostable enzymes generally are also more resistant to chemical denaturation. Consequently, thermostable alkalophilic proteases are suitable for detergent applications from both physical and chemical standpoints. So far, only proteases from alkalophilic *Bacillus* spp. are satisfactory in detergent applications. The protease from *Bacillus licheniformis* is most commonly used. This enzyme has temperature and pH optima at 60°C and 10, respectively (54). Proteases with higher tolerance to pH and chemical denaturants are available from other *Bacillus* spp. (54,56). Some of these like Esperase® from Novo and Milezyme® 8X from Miles are for liquid formulations and are used for dehairing in the tanning industry. On the other hand, calcium-dependent proteases like caldolysin from *Thermus* sp. T-351 (57) or thermolysin from *Bacillus thermoproteolyticus* (56) are not suitable because most detergents contain calcium-binding sequestering agents such as tripolyphosphate (54).

The future development of thermostable proteases in detergent enzymes will be determined by two counteracting factors. On the one hand, there is a need for proteases to be used in liquid formulations. Such enzymes should be highly stable so that they have a reasonable shelf life. Proteases that are resistant to laundry and dishwashing bleaches like chlorine or peroxide are also highly desirable. Subtilisin from *Bacillus amyloliquefaciens* was modified via genetic engineering to have improved resistance to hydrogen peroxide (58). On the other hand, energy considerations encourage the practice of presoaking laundry and the use of low-temperature washing. This will stifle the future development of thermostable proteases that are not as active as mesophile proteases at room temperature. Likewise, the popularity of the prespotting practice (i.e., adding concentrated enzyme solution to particularly dirty spots) will encourage the use of active mesophile proteases.

Although proteolytic enzymes are used primarily in the detergent industry, significant quantities, amounting to ten percent of the total enzyme market, consist of acid rennet used in cheese manufacturing (42). Although calf rennet is most commonly used, the rising cost of calf rennet has encouraged the use of fungal rennet as replacement. The first commercial microbial rennet was derived from a thermophilic fungus, *Mucor pusillus*

(56). Ironically, this thermostable protease marketed by Meito Sangyo Co. was used at mesophile temperatures (56) and its failure in the market was because the temperature required to inactivate this highly thermostable protease resulted in the production of undesirable compounds in the cheese. This enzyme was soon replaced by mesophile protease from *Mucor miehei* (48,56). This example helps to reemphasize the point that thermostable enzymes do not necessarily have intrinsic advantages for industrial applications.

In this section, we have examined the importance of thermostable enzymes in two key industrial processes, namely detergent formulation and starch processing. These processes together consume approximately 49% of the total market share for industrial enzymes. Clearly, there are other important application areas, such as textiles, leather tanning, diagnostics, and biotransformation, where industrial enzymes play an important role. Yet, thermostable enzymes have so far had a lesser impact in these areas than the two that we have discussed (6,59).

5. FUTURE PROSPECTS

At present, the industrial applications of thermostable enzymes are limited to the few areas which have been examined in this chapter. Moreover, the majority of industrial thermostable enzymes are not derived from thermophilic microorganisms. Perhaps it is time to assess the industrial potential of thermostable enzymes realistically, in particular those derived from thermophiles. Hopefully, this will shed some light on the future research needs in this area.

With the discovery of many new novel thermophiles (see 60 and Chapter 3, this volume), the sources of thermostable enzymes have expanded significantly beyond conventional thermophiles. Yet beyond those from the genera *Bacillus* and *Thermus* (8–10,61), relatively little is known about enzymes from other thermophiles. Some of these enzymes may possess unique properties because of their extreme thermostability. For instance, the DNA-dependent RNA polymerase from *Thermoproteus tenax* has a half-life at 100°C of 135 minutes. Most interestingly, this enzyme, like those from the thermoacidophilic branch of archaebacteria, resembles that of yeast RNA polymerase A(I) (62). Secondary alcohol dehydrogenase from *Thermoanaerobium brockii* is highly stereospecific and has a broad substrate specificity (63). An 80% optical purity was obtained in the reduction of 2-pentanone (64).

How do enzymes function at temperatures above 100°C? Thermodynamic calculations suggest that the peptide bonds of protein are less stable at higher temperatures (65). Nevertheless, thermostable enzymes do not exhibit lower stability or specific activity at their optimal temperature when compared to their mesophilic counterparts. There is also no reason to suggest that thermostable enzymes have higher turnover rates. As a matter of fact, the specific growth rates of thermophiles do not differ significantly from that of the mesophiles (5). These and many other questions will continue to provide an enlightening and challenging area for us to examine.

The future industrial growth of thermostable enzymes will probably be determined by economic factors. The development of new processes or the use of thermostable enzymes instead of thermolabile enzymes should be primarily a matter of cost effectiveness. In most cases, predictions for growth in industrial enzymes are conflicting (43,66,67).

In new process development, enzymes have to compete with existing chemical catalysts. Enzymes are clearly superior only in the cases where stereospecific reactions are involved. So far, thermophilic microorganisms have not been a satisfactory source of thermostable enzymes. This may be because relatively little is known about enzymes from extreme thermophiles. Notwithstanding, carbohydrases from extreme thermophiles have been reported (68). The α-amylase from *Clostridium thermosulfurogenes* (69,70) and the pullulanase and glucoamylase from *Clostridium thermohydrosulfuricum* (71) may hold great industrial potential. A new class of thermophile pullanases from *Bacillus acidopullulyticus* has recently been characterized (72). New applications and process improvements should be possible as we gain a better understanding of enzymes from these novel thermophiles. The use of thermophile enzymes for organic synthesis is particularly appealing. Because of their enhanced thermostability, thermophile enzymes also are generally more stable to organic solvents (57,73). Biotransformations with several thermophilic microorganisms have been reported (see Chapter 10, this volume). Particularly noteworthy, is the stereoselective reduction of carbonyl compounds by *Thermoanaerobium brockii* (74) where some of the reduction reactions proceed stereochemically in a reverse fashion relative to yeast (75).

A promising approach to obtain "tailor-made" enzymes is through genetic engineering (76). This is achieved by altering the amino acid composition of an enzyme specifically so as to increase the catalytic function or structural stability. For example, the thermostability of bacteriophage T4 lysozyme was increased by inserting a disulfide bond at a specific site in the protein molecule (77). Furthermore, human α_1-antitrypsin has been modified by similar technique to be more resistant to oxidation (78,79). A different approach was taken by researchers at Synergen (Boulder, CO). By inserting the gene for a thermolabile kanamycin nucleotidyltransferase into the thermophilic *Bacillus stearothermophilus* and looking for resistant mutants under nonpermissive growth temperature (kanamycin resistance at 60°C), they were able to isolate a mutant producing a thermostable kanamycin nucleotidyltransferase (80). Unfortunately, the construction of thermostable enzymes is still highly empirical, since so little is known about the molecular basis of thermostability and mechanism of catalysis in enzymes, particularly industrial enzymes.

Where the enzyme cost is only a minor portion of the total process cost (less than 1%) and/or the enzyme has a limited market volume, the cost of extensive research and development on one single enzyme is prohibitive if the major incentive is an economic one (58). From this we might conclude that the future of thermostable enzymes and their industrial applications are interdependent. Advancement in the scientific area should result in process improvement and open new avenues for industrial applications, while profitability in the market place should provide incentives for basic research on the study of thermophilic microorganisms.

ACKNOWLEDGMENTS

We wish to thank P. J. Weimer for his valuable suggestions and L. Savage for typing the manuscript.

REFERENCES

1. M. E. Curtin, *Biotechnology* **3**, 36 (1985).
2. J. G. Zeikus, *Enzyme Microb. Technol.* **1**, 243 (1979).
3. J. Wiegel, *Experientia* **36**, 1434 (1980).
4. B. Sonnleitner, in A. Fleichter, Ed., *Advances in Biochemical Engineering*, Vol. 28, Springer-Verlag, New York, 1983, p. 69.
5. A. Fontana, in *European Congress on Biotechnology*, 3rd ed., 1984, p. I221.
6. T. Godfrey and J. Reichelt, Eds., *Industrial Enzymology*, Nature Press, London, 1983.
7. B. P. Wasserman, *Food Technol.* **38**, 78 (1984).
8. R. E. Amelunxen and A. L. Murdoch, in D. J. Kushner, Ed., *Microbial Life in Extreme Environments*, Academic Press, New York, 1978, p. 217.
9. S. M. Friedman, Ed., *Biochemistry of Thermophily*, Academic Press, New York, 1976.
10. H. Zuber, Ed., *Enzymes and Proteins from Thermophic Microorganisms*, Birkhauser Verlag, Basel, 1976.
11. L. L. Campbell and B. Pace, *J. Appl. Bacteriol.* **31**, 24 (1968).
12. R. Singleton, Jr. and R. E. Amelunxen, *Bacteriol. Rev.* **37**, 320 (1973).
13. H. Zuber, *Colloq. Ges. Biol. Chem.* **32**, 114 (1981).
14. A. M. Klibanov, *Adv. Appl. Microbiol.* **29**, 1 (1983).
15. B. M. Daniel, D. A. Cowan, and H. W. Morgan, *Chem. New Zealand* **45**, 94 (1981).
16. T. Oshima, *Enzyme Eng.* **4**, 41 (1978).
17. E. Stellwagen and H. Wilgus, in S. M. Friedman, Ed., *Biochemistry of Thermophily*, Academic Press, New York, 1978, p. 223.
18. R. D. Schmid, in T. K. Ghose, A. Fleichter, and N. Blakebrough, Eds., *Advances in Biochemical Engineering*, Vol. 12, Springer-Verlag, New York, 1979, p. 41.
19. K. Yutani, K. Ogasahara, Y. Sugino, and A. Matsushiro, *Nature* **267**, 274 (1977).
20. M. A. Holmes and B. W. Matthews, *J. Mol. Biol.* **160**, 623 (1982).
21. N. C. Pace, *CRC Crit. Rev. Biochem.* **1**, 1 (1975).
22. C. Tanford, *Adv. Protein Chem.* **24**, 1 (1970).
23. J. F. Brandts, in S. N. Timasheff and G. D. Fasman, Eds., *Structure and Stability of Biological Molecules*, Marcel Dekker, New York, 1969, p. 213.
24. M. Matsumura, Y. Katakura, T. Imanaka, and S. Aiba, *J. Bacteriol.* **100**, 413 (1984).
25. B. W. Matthews, L. M.Weaver, and W. R. Kester, *J. Biol. Chem.* **24**, 8030 (1974).

26. P. Argos, M. G. Rossmann, U. M. Grau, H. Zuber, G. Frank, and J. P. Tratschin, *Biochemistry* **18**, 5698 (1977).
27. A. Zaks and A. M. Klibanov, *Science* **224**, 1249 (1984).
28. J. Sanders, ASM Presentation on Genetic Improvement of an Amylase-Producing *Bacillus*, Las Vegas, 1985.
29. *Genetic Technology* **2**, 6 (March 1982).
30. N. Tsukagoshi, H. Ihara, H. Yamagata, and S. Udaka, *Mol. Gen. Genet.* **193**, 58 (1984).
31. N. Citri, *Adv. Enzymol.* **37**, 397 (1973).
32. A. Hachimori, A. Matsunaga, M. Shimizu, T. Sarnejima, and Y. Nosoh, *Biochem. Biophys. Acta* **350**, 461 (1974).
33. F. C. Wedler and F. M. Hoffman, *Biochemistry* **13**, 3215 (1974).
34. S. Moriyama, R. Matsuno, and T. Kamikubo, *Agric. Biol. Chem.* **41**, 1985 (1977).
35. J. S. Myers and W. B. Jakoby, *Biochem. Biophys. Res. Commun.* **51**, 631 (1973).
36. E. H. Fisher and E. A. Stein, in P. D. Boyer, H. Hardy, and K. Myrbach, Eds., *The Enzymes*, Vol. 4, Academic Press, New York, 1960, p. 313.
37. H. Todda and K. Narita, *J. Biochem. (Tokyo)* **62**, 767 (1967).
38. W. Heinen and A. M. Lauwers, in H. Zuber, Ed., *Enzymes and Proteins from Thermophilic Microorganisms*, Birkhauser Verlag, Basel, 1976, p. 77.
39. J. Feder, L. K. Garrett, and B. W. Wildi, *Biochemistry* **10**, 4552 (1971).
40. G. Voordouw and R. S. Roche, *Biochemistry* **14**, 4659 (1975).
41. G. Voordouw and R. S. Roche, *Biochemistry* **14**, 4667 (1975).
42. F. W. Dahlquist, L. W. Long, and W. L. Bigbee, *Biochemistry* **15**, 1103 (1976).
43. *Chemical Week* **133**, 30 (Nov. 30, 1983).
44. *Biotechnology News* **5**, 8 (May 22 1985).
45. *Chemical Week* **135**, 27 (Nov. 7, 1984).
46. J. Reichelt, in T. Godfrey and J. Reichelt, Eds., *Industrial Enzymology*, Nature Press, New York, 1983, p. 375.
47. F. M. Bombouts and W. Pilnik, in A. H. Rose, Ed., *Microbial Enzymes and Bioconversions*, Academic Press, New York, 1980, p. 269.
48. K. Burgess and M. Shaw, in T. Godfrey and J. Reichelt, Eds., *Industrial Enzymology*, Nature Press, New York, 1983, p. 260.
49. M. Mandels, in G. T. Tsao, Ed., *Annual Reports on Fermentation Processes*, Vol. 5, Academic Press, 1982, p. 35.
50. Y. H. Lee and L. T. Fan, in T. K. Ghose, A. Fleichter, and N. Blakebrough, Eds., *Advances in Biochemical Engineering*, Vol. 17, Springer-Verlag, New York, 1983, p. 101.
51. D. L. Crawford and E. McCoy, *Appl. Microbiol.* **24**, 150 (1972).
52. F. J. Stutzenberger, *Biotechnol. Bioeng.* **21**, 909 (1979).
53. J. G. Zeikus and T. K. Ng, in G. T. Tsao, Ed., *Annual Reports on Fermentation Processes*, Vol. 5, Academic Press, New York, 1980, p. 263.
54. H. C. Barfoed, in T. Godfrey and J. Reichelt, Eds., *Industrial Enzymology*, Nature Press, New York, 1983, p. 284.
55. G. A. Starace, *J. Amer. Oil Chem.* **60**, 1025 (1983).
56. K. Aunstrup, in A. H. Rose, Ed., *Microbial Enzymes and Bioconversions*, Academic Press, New York, 1980, p. 49.
57. H. W. Morgan, R. M. Daniel, D. A. Cowan, and C. W. Hickey, Eur. Patent Appl. No. 80302743.2, Publ. No. 0 024 182 (1981).
58. *Biotechnology News* **5**, 3 (July 18, 1985).
59. A. Wiseman, Ed., *Handbook of Enzyme Biotechnology*, Halsted Press, New York, 1975.
60. T. D. Brock, *Thermophilic Microorganisms and Life at High Temperature*, Springer-Verlag, New York, 1978.
61. O. Prangishvilli, W. Zillig, A. Gierl, L. Biesert, and I. Holz, *Eur. J. Biochem* **122**, 471 (1982).
62. M. R. Heinrich, Ed., *Extreme Environments: Mechanisms of Microbial Adaptation*, Academic Press, New York, 1976.

63. R. J. Lamed and J. G. Zeikus, *Biochem. J.* **195**, 183 (1981).
64. J. G. Zeikus and R. J. Lamed, U.S. Patent No. 4,352,885 (1982).
65. R. H. White, *Nature* **310**, 430 (1984).
66. *Chemical Week* **133**, 30 (Nov. 30, 1983).
67. *Chemical and Eng. News* **61**, 11 (Sept. 12, 1983).
68. H. H. Hyun, Ph.D. Dissertation, Dept. of Bacteriol., Univ. Wisconsin, 1984.
69. H. H. Hyun and J. G. Zeikus, *Appl. Environ. Microbiol.* **49**, 1162 (1985).
70. H. H. Hyun and J. G. Zeikus, *Appl. Environ. Microbiol.* **49**, 1174 (1985).
71. H. H. Hyun and J. G. Zeikus, *Appl. Environ. Microbiol.* **49**, 1168 (1985).
72. M. Schulein and B. Hojer-Pedersen, in A. I. Laskin, G. T. Tsao, and L. B. Wingard, Jr., Eds., *Enzyme Engineering 7: Annals of the New York Academy of Sciences*, Vol. 434, Wiley, New York, 1984.
73. O. Cowan, R. Daniel and H. Morgan, *Trends Biochem.* **3**, 68 (1985).
74. D. Seebach, M. F. Zuger, F. Giovannini, B. Sonnleitner, and A. Fleichter, *Agnew. Chem. Int. Ed. Engl.* **23**, 151 (1984).
75. R. MacLeod, H. Prosser, L. Fikentscher, J. Lanyi, and H. S. Mosher, *Biochemistry* **3**, 838 (1964).
76. G. Winter and A. R. Fersht, *Trends Biotechnol.* **2**, 5 (1984).
77. L. J. Perry and R. Wetzel, *Science* **226**, 557 (1984).
78. M. Courtney, S. Jallat, L-H. Tessier, A. Benavente, and R. G. Crystal, *Nature* **313**, 149 (1985).
79. S. Rosenberg, P. J. Barr, R. C. Najarian, and R. A. Hallewell, *Nature* **312**, 77 (1984).
80. D. Hirsch, *Presentation at Genex-UCLA Symposium on Protein Structure and Design*, Keystone, CO, April 1985.

10

USE OF THERMOPHILES FOR THE PRODUCTION OF FUELS AND CHEMICALS

PAUL J. WEIMER

Central Research and Development Department, Bldg. 402, E. I. du Pont de Nemours and Company, Inc., Experimental Station, Wilmington, Delaware 19898

1. INTRODUCTION, SCOPE AND DEFINITIONS

Thermophilic microorganisms have provided biologists and biochemists with fascinating and useful model systems for investigating physiological,

217

ecological, and biochemical processes at both the gross and molecular level (1–3). Nevertheless, exploitation of these organisms by the industrial microbiologist to provide products useful to human society has been almost insignificant. It is the purpose of this report to review the actual and potential production of fuels, community chemicals, and specialty chemicals by thermophilic microorganisms.

For present purposes, "obligate" thermophiles will be considered to be those organisms that have temperature optima $\geq 50°C$ and are incapable of growth at 30°C (see Chapter 1). Such obligate thermophiles are exclusively bacteria. Where appropriate, summaries will also be given of certain fermentations by facultative thermophiles, organisms capable of growth at both 30 and 55°C and which generally display temperature optima in the 45 to 50°C range. Representatives of this latter group include certain bacteria as well as several "thermophilic" fungi (3,4). The so-called thermophilic or thermotolerant yeasts are excluded from discussion, owing to the lack of demonstrated ability of any true yeast to grow at temperatures above 48°C (5).

Although a major intent of this review is to supply an overview of the commonly envisioned fuel and bulk chemical fermentations which employ obligately thermophilic microorganisms (e.g., for ethanol production), it is hoped that a review of the dispersed and often obscure literature on production of potential specialty chemicals by thermophiles might stimulate further research in this generally unexplored area.

Several other applications of thermophilic microorganisms which relate to fuels and chemical production are dealt with in Chapters 9, 11, and 12 of this volume and are not discussed here.

2. GENERAL SUBSTRATE CONSIDERATIONS

Commercial production of fuels and commodity chemicals by thermophilic fermentation must be almost exclusively based on carbohydrate available from plant biomass, as only these materials are convertible to the products of interest at reasonable rates without toxicity problems and/or are utilizable by thermophiles for bulk chemical fermentations. A detailed discussion of the source and availability of biomass and the economics of biomass utilization is beyond the scope of this review (for details, see 6–12). Here, only an outline is given of the more prominent biomass sources in terms of their utilizability by thermophilic bacteria.

Cellulosic biomass is available in the form of wood chips, residues from agricultural and forest products (e.g., lumber milling and pulping) operations, municipal refuse, and animal waste (6–8). The amount, localization, and collectability of these wastes are strong determinants of raw material cost (7,8). Although cellulosic biomass is several times cheaper than purified, easily fermentable carbohydrate such as corn starch or corn syrup, much of this cost advantage is lost because of the pretreatment processes necessary to render the carbohydrate fraction of the biomass utilizable at reasonable rates (9). Furthermore, the cost of cellulosic wastes may be expected to rise considerably as uses are found for them. Despite its higher costs, starch is a more useful substrate for fermentative production of fuels and bulk chemicals due to its easier fermentability and the presence of an existing technology for its recovery from plant material, particularly from corn (10,11). Corn syrup (hydrolyzed corn starch) is only slightly more expensive than corn starch and is produced in enormous amounts by the corn wet-milling industry. Because of its depolymerized nature, it is especially easily fermented by microorganisms. Corn syrup (and thus the parent material corn starch) is also the basis for many of the existing bulk chemical fermentations (e.g., to ethanol) employing mesophilic microorganisms. As will be demonstrated below, thermophilic fermentations to such products as ethanol or acetic acid are characterized by final product concentrations well below those of analogous mesophilic fermentations and thus would be unlikely to be competitive with these processes where easily utilizable carbohydrate (starch or glucose) serve as substrates. Whey may also be useful as a substrate (12), but only on a local basis, to obviate transportation costs, and only after further addition of other carbohydrate to increase its own low carbohydrate content (~5 percent, as lactose). Exploitation of thermophilic carbohydrate fermentations for bulk chemical production, if it is to occur at all, will most likely involve thermophilic systems which display clear superiority to analogous mesophile systems, such as the ability of some strains to utilize cellulosic materials directly.

Selection of substrates for microbial production of specialty chemicals, on the other hand, is often based upon considerations beyond simple, direct raw material cost. For products synthesized de novo from simpler substrates (e.g., during secondary metabolism), product value in general far exceeds substrate cost, thus minimizing economic constraints on substrate selection and enhancing the importance of other factors such as substrate purity or the ease of substrate/product separability from the reactor stream. In the case of products formed by bioconversions of structurally related substrates, the relatively narrow biochemistry of the reaction generally predetermines the substrate.

3. GENERAL PROCESS CONSIDERATIONS

The effects of elevated temperature on a variety of physical and chemical phenomena may have significant consequences on reactor design and operation. Many of these implications have been reviewed (13–15). At increased temperature, the viscosity and surface tension of water are reduced, and both the diffusion rate and the solubility of most nongaseous compounds are generally increased; these changes facilitate mass transfer and lower mixing costs. Solubility of most gases, on the other hand, decreases with increasing temperature (16). This facilitates attainment and maintenance of anaerobic conditions but causes limitations in the oxygen transfer rate in aerobic cultures. The general enhancement of chemical reaction rates at increasing temperature is also responsible for the observed elevation in the rate of corrosion and metal fatigue in fermentation equipment used for cultivation of thermophiles.

Despite earlier contentions that thermophilic systems are inherently less subject to contamination than are analogous mesophilic fermentations (17), it has since become clear that contamination is no less common or severe in thermophilic systems (13–15). In fact, the tremendous heat resistance of spores from some thermophilic species such as *Clostridium thermohydrosulfuricum* (18) and *Desulfotomaculum nigrificans* (19) poses special sterilization problems for thermophilic fermentations, with important consequences in energy costs and culture medium integrity.

4. PRODUCTION OF FUELS AND CHEMICALS

4.1. Ethanol

The ambitious ethanol program in Brazil provides a strong example of a rapidly developing, biomass-rich, fossil-fuel-poor nation building a massive transportation system based on fermentation ethanol. During the 1981–82 sugarcane season, fermentation ethanol production reached 1.4×10^9 gallons (20), enough to provide the bulk of domestic use and to support a healthy export market (21). Projected production will increase to 3.7×10^9 gallons by 1987 (22). The outlook for fuel ethanol in other nations, while not quite as sanguine, is nevertheless quite favorable. In the United States, use of ethanol as a fuel is almost wholly confined to its use as an octane enhancer in "super unleaded" gasoline (23). While this fuel blend represents less than 1 percent of total motor fuel consumption in the United States, ethanol use for this purpose translates to approximately 2×10^8 gal/yr. Currently, this fuel ethanol is obtained from both fermentation and chemical synthesis (ethylene hydration) with plant capacities (as of early 1983) of 5.3 and 2.7×10^8 gal/yr, respectively. Because the significant cost advantage of fermentation ethanol over its chemically synthesized competitor has continued to grow, projected new plant construction, at least in the United States, will be of only the fermentation type (24).

The attention lavished on the United States fuel alcohol program has obscured the fact that the most quantitatively important use of fermentation ethanol may ultimately be not as a fuel but as a source of ethylene. The latter compound is by far the largest volume organic chemical produced by the chemical process industry, with projected 1984 production of 14×10^9 kg in the United States alone (25), and has a wide variety of end uses in production of both polymers (e.g., polyethylene) and monomer derivatives (e.g., ethylene glycol, vinyl chloride, vinyl acetate, and styrene as well as ethanol). Because fermentation ethanol has become considerably cheaper than ethanol derived from ethylene, significant improvements in dehydration technology and/or further escalations in petroleum prices are likely to make ethylene obtained from fermentation ethanol a viable business opportunity. It is thus ironic that industrial ethanol fermentation, which was almost completely supplanted by the ethylene hydration process during the period from the 1950s through the 1970s, may be resurrected to become a source of ethylene.

All of the fermentation ethanol produced in the United States is made via yeast (*Saccharomyces*) fermentation, although the process mode (batch, semicontinuous, or continuous) varies from plant to plant (26). Utilization of thermophilic bacteria for ethanol production has been frequently proposed and has been the basis of several intensive research programs (13,15,28–43). The theoretical advantages of using thermophilic anaerobes for ethanol production are threefold (13–15,27,28). First, the elevated incubation temperatures should facilitate recovery of the volatile (b.p. = 78°C) ethanol/water azeotrope. Second, with proper engineering designs, the heating costs of thermophilic fermentations will probably be lower than the cooling costs of mesophilic fermentations, since optimum ethanol productivity in the *Saccharomyces* and *Zymomonas* systems occur at 30 and 25°C, respectively (15). Finally, and most significantly, thermophilic anaerobic bacteria are available which carry out a direct fermentation of polysaccharides (including the relatively recalcitrant cellulose) and a wide variety of monosaccharides to produce ethanol in moderate yield (13,15,27). This may be accomplished by the bacterium *Clostridium thermocellum*, particularly in defined coculture with other saccharolytic anaerobes (13,15,28). The discussion which follows includes some detail on the general growth features and catabolism of thermophilic anaerobic saccharolytic bacteria, followed by a summary of the technology of biomass fermented by these organisms.

C. thermocellum is a thermophilic, strictly anaerobic, spore-forming bacterium which ferments cellulose and cellodextrins ($\beta 1 \rightarrow 4$ linked oligomers of D-glucose) to produce a mixture of ethanol, acetic acid, lactic acid, H_2, and CO_2. The physiology of this organism (along with some previously described species now considered to be strains of *C. thermocellum*) has been recently reviewed (48). Cellulose breakdown is mediated by an extracellular cellulase complex composed of several distinct proteins which vary in substrate (chain length) specificity and in hydrolytic mechanism (49–51). Prominent among these is a very active endoglucanase ($\beta 1 \rightarrow 4$ glucan glucanohydrolase) which performs a random cleavage of interior glycosidic linkages, and which also hydrolyzes short chain oligomers (degree of polymerization (DP) ≥ 4) at moderate rates (52,53). In addition, the complex contains a rather weak exoglucanase ($\beta 1 \rightarrow 4$ cellobiohydrolase) activity which performs an endwise cleavage of cellobiosyl units from the reducing end of the cellulose polymer (46). The exoglucanase component has been proposed as the likely site of an observed O_2-mediated inactivation of the C. *thermocellum* cellulase complex (54). Despite the very low exoglucanase/endoglucanase activity ratios in culture supernatants of C. *thermocellum* relative to those of *Trichoderma reesei* (a mesophilic fungus known to produce a highly active cellulase complex), the growth rate of the bacterium on crystalline cellulose has been reported to be about twice that of the fungus (55). The cellulase components of C. *thermocellum* are remarkable in their relative insensitivity to end product inhibition. For example, reported inhibition of cellulose-solubilizing activity by 100 mM glucose was 22 percent for *T. reesei* QM9414 cellulase but only 2 percent for C. *thermocellum* cellulase (49). Similarly, 90 mM cellobiose inhibited *T. reesei* cellulase by 40 percent, while C. *thermocellum* cellulase was not affected (49).

An investigation (56) into the effects of fine structure of native, type I celluloses on cellulose degradability (initial hydrolytic rate) by the extracellular C. *thermocellum* cellulase complex revealed that: a) the enzyme system of C. *thermocellum* is more sensitive to variations in substrate structure than is the cellulase complex of the fungus *T. reesei*; b) degradability of substrate by both cellulase complexes correlates very strongly with the fiber saturation point (the total volume of the pores in the cellulose matrix); c) correlation of crystallinity parameters with degradability is considerably weaker in the *Clostridium* system than in the *Trichoderma* system; and d) the rate-limiting components of both enzyme complexes are of approximately similar size (43 Å on one axis when the enzyme is oriented properly in the catalytic site).

Although a cell-bound β-glucosidase activity has reportedly been purified from cell extracts of C. *thermocellum* (57), the function of this enzyme is unclear. The purified enzyme displayed an extremely poor affinity for cellobiose ($K_m = 84$ mM), and cellobiose caused only a weak competitive inhibition ($K_i = 183$ mM) of *p*-nitrophenyl-β-D-glucoside hydrolysis. By contrast, laminaribiose ($\beta 1 \rightarrow 3$ glucosyl glucopyranoside) and sophorose ($\beta 1 \rightarrow 2$ glucosyl glucopyranoside) exhibited K_i values in the same reaction system of 1.2 and 3.7 mM, respectively. In addition, V_{max} values for laminaribiose, laminaritriose, and sophorose were 16-, 6.9-, and 3.3-fold greater than for cellobiose, whose V_{max} was in turn greater than those of other cellodextrins. It is possible that this enzyme functions primarily in the hydrolysis of $\beta 1 \rightarrow 3$ or $\beta 1 \rightarrow 2$ glucan linkages present in noncellulosic plant oligosaccharides, rather than in hydrolysis of cellodextrins.

While the role of extracellular enzymes in cellulose depolymerization by C. *thermocellum* has been firmly established, it appears that the hydrolytic process is further aided by a physical association between growing cells and the solid substrate (46,58). Recently, Bayer et al. (58) and Lamed et al. (59) have attributed this interaction to cellulose-binding proteins localized in a discrete cell-surface organelle. Analysis of proteins within this structure has revealed the presence of both exo- and endoglucanase activities which, these workers have proposed, are arranged in a particular supramolecular orientation designed for optimum cellulose degradation (59). The relationships between the organelle proteins and the extracellular cellulase components with respect to structure, function, and possible mutual origin have not been determined.

The cellodextrin products of cellulose hydrolysis are transported into the cell via two unusual enzymes, cellobiose phosphorylase (60) and cellodextrin phosphorylase (61), whose phosphorylation reactions are pulled (thermodynamically displaced) by extracellular hydrolytic release of glucose during ATP-independent transport of the remaining portion of the carbohydrate molecule (62) as shown in Figure 1. The glucose-1-phosphate produced by cellobiose phosphorylase is converted via phosphoglucomutase to glucose-6-phosphate, whereupon it may enter the Embden-Meyerhof pathway of hexose catabolism (63,64). The low levels of exoglucanase activity in *C. thermocellum* supernatants is reasonable in light of the highly active cellodextrin phosphorylase, which is responsible for the rapid uptake of cellodextrins produced by the random hydrolytic action of the endoglucanase (65).

At least some strains of *C. thermocellum* are also capable of glucose utilization, but only under rather specialized growth conditions (33,62,66,67). Initial adaptation of cultures unable to utilize glucose to glucose utilization is characterized by a long lag period and by a requirement for moderate concentrations of factor(s) found in yeast extract. Recent reports that cultures adapted to glucose utilization retain their cellulolytic properties after prolonged growth in the absence of cellulose have eliminated earlier concerns that glucose fermentation by *C. thermocellum* cultures is due to non-cellulolytic contaminating organisms (44,45). Glucose uptake is apparently mediated by an inducible (or derepressible) ATP-linked permease which differs from classical sugar phosphotransferase systems (62). Significant hexokinase activity in cell extracts of glucose-grown (but not cellobiose-grown) cells suggests that this enzyme is responsible for metabolism of glucose taken up by the permease activity (62).

The biochemistry of product formation in *C. thermocellum* has been investigated in some detail (63,64,68). Glucose and phosphorylated hexose intermediates are catabolized via the Embden-Meyerhof pathway to yield pyruvate, which is then converted either via a fructose-1,6-diphosphate (FDP)-activated NAD^+-linked lactate dehydrogenase (to yield L-lactate) or via a pyruvate dehydrogenase to yield acetyl-CoA (AcCoA), CO_2, and reducing equivalents (as reduced ferredoxin). AcCoA is metabolized either to ethanol via pyridine nucleotide-linked AcCoA and acetaldehyde reductases, or to acetate via phosphotransacetylase and acetate kinase. Two points of particular interest in the above pathway merit attention. First, the mode of pyruvate catabolism differs greatly from the pyruvate decarboxylase system of yeast and of *Zymomonas*, both of which form acetaldehyde directly from pyruvate (Figure 2). Second, the observation that FDP is responsible for

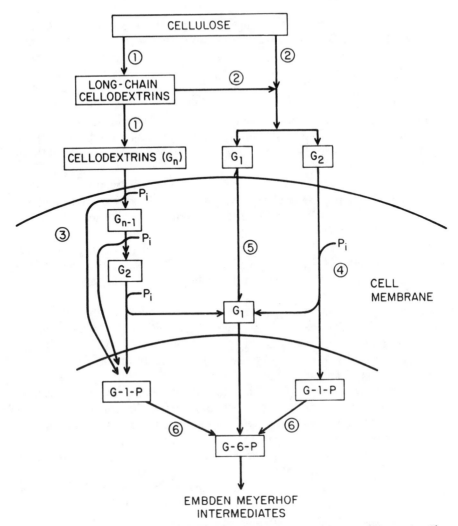

Figure 1. Proposed metabolism of cellulose, cellodextrins, cellobiose, and glucose by *Clostridium thermocellum*. Cellulose is hydrolyzed by endoglucanase (1) and exoglucanase (2) to yield predominantly cellodextrins (G_n) and cellobiose (G_2), respectively. Uptake of these products results in direct (ATP-independent) phosphorylation via cellodextrin phosphorylase (3) and cellobiose phosphorylase (4), to yield glucose-1-phosphate (G-1-P) units, which are converted intracellularly, via an active phosphoglucomutase (6), to yield the Embden-Meyerhof pathway intermediate glucose-6-phosphate (G-6-P). Free glucose (G_1) remaining from the phosphorylase reactions (3) and (4) is exported, resulting in its extracellular accumulation. Under specialized culture conditions, glucose is taken up via an unidentified ATP-linked permease system (5). The carriers involved in uptake and export of glucose have not been identified.

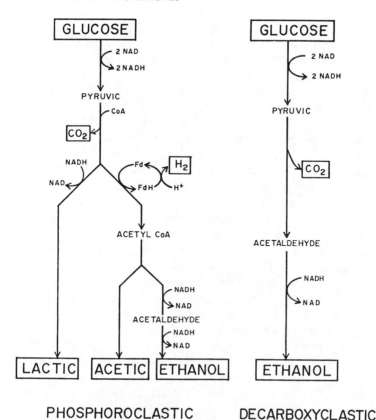

PHOSPHOROCLASTIC   DECARBOXYCLASTIC

Figure 2. Schematic comparison of glucose fermentation pathways. The decarboxyclastic fermentation carried out by *S. cerevisiae* and by *Zymomonas* species proceeds without formation of an acetyl-coenzyme A intermediate and results in production of ethanol and CO_2 as sole products. The phosphoroclastic pathway carried out by many anaerobic bacteria, including thermophilic saccharolytic anaerobes, results in formation of an acetyl-CoA intermediate, with subsequent production of both ethanol and acetic acid. Excess reducing equivalents are disposed of as H_2 or as lactic acid. From Zeikus (27); reproduced with permission from Annual Reviews of Microbiology, Vol. 34, © 1980 by Annual Reviews, Inc.

activation of lactate dehydrogenase suggests that carbohydrate fermentation in *C. thermocellum* is subject to catabolite repression and that control of product flux can be modulated by factors which control intracellular FDP concentration.

The utilization of cellulose by *C. thermocellum* currently has severe limitations from the standpoint of fermentation technology. These problems include: a) low yield of ethanol due to incomplete carbohydrate fermentation (accumulation of reducing sugars) (69) and to formation of acetate, lactate, and H_2 as coproducts (15,27-29); b) low volumetric productivity resulting from the slow rate of cellulose fermentation and from low cell yield and resulting low cellulase titers (31); and c) considerable toxicity of both ethanol and organic acid end products (45,70,71). Several attempts have been made to circumvent yield problems, most of which have employed the addition of another organism, viz. a thermophilic, anaerobic, noncellulolytic bacterium which ferments free monosaccharides left behind by *C. thermocellum* during growth on cellulosic substrates (for review, see 15 and 47). Organisms which have been tested for enhancement of ethanol yield in cocultural systems include *Thermoanaerobacter ethanolicus* (72), *Clostridium thermohydrosulfuricum* (37,73), and *Clostridium thermosaccharolyticum* (34–42). The species designation of this last organism has been questioned (47). Nutritional and physiological characteristics of these and other organisms potentially utilizable in coculture systems (72–81) are summarized in Table 1. All of these species have temperature optima similar to that of *C. thermocellum*, and some can ferment both the glucose and the xylose which would be produced by *C. thermocellum* enzymes from cellulose and hemicellulose, respectively. This capacity to ferment several carbohydrate substrates is a strong incentive to utilize these thermophilic anaerobes in conversion of cellulosic biomass to ethanol, as the organisms used in competing mesophilic processes, *Saccharomyces* and *Zymomonas*, both display narrow carbohydrate utilization ranges (15).

An additional benefit of combining an accessory organism with *C. thermocellum* lies in the inherently higher ethanol yields of some noncellulolytic species, due to a more favorable flux of substrate-derived electrons in the direction of AcCoA reduction to ethanol and away from the formation of acetate, lactate, and H_2 (47,73). These differences in product yield can be traced to differences in the levels of catabolic enzyme activities measurable in cell extracts (68,82,83). For example, the V_{max} of hydrogenase in *C. thermocellum* has been reported to be three times higher than in *Thermoanaerobium brockii* and this is reflected in a 14-fold greater production of H_2 in the former species relative to the latter. At low yeast extract concentrations, *T. brockii* produces higher ethanol yields than does *C. thermocellum* due to the presence of both NAD- and NADP- linked ethanol dehydrogenase activities (*C. thermocellum*'s enzyme is only NAD-linked) and a sixfold greater level of ferredoxin/NADP oxidoreductase. These data suggest only general trends, with actual product formation determined not only by V_{max} values of enzymes, but also by their K_m values and by the pool sizes of the intermediary metabolites (83).

The dependence of end product formation in some thermophilic saccharolytic anaerobes on the availability of intermediary metabolites can best be illustrated by the work of Ben-Bassat et al. (82). These workers demonstrated that growth of *T. brockii* on starch (relative to growth on glucose) was associated with a twofold increase in ethanol yield and a twofold decrease in lactate yield, resulting from a dramatic (25-fold) decrease in intracellular FDP concentration. The metabolic versatility of *T. brockii* is reflected in its relationship to exogenous H_2. In the absence of exogenous

TABLE 1. Culture characteristics of thermophilic, strictly anaerobic, ethanol-producing bacteria

Species	Growth temperature		pH for growth		Substrates utilized						References
	Range	Optimum	Range	Optimum	Cellulose	Hemicellulose	Starch	Glucose	Cellobiose	Xylose	
Clostridium thermocellum	45–65	62	6–8.5	~7.4	+	–	–	+	+	–	44, 46
C. thermosaccharolyticum	45–65	55–60	NR	NR	–	–	+	+	+	+	74
C. thermohydrosulfuricum	45–78	62	4.8–9.7	6.9–7.5	–	–	+	+	+	+	75,76
C. thermosulfurogenes	35–75	60	>4.0–<7.6	5.5–6.5	–	NR	+	+	+	+	77
Thermoanaerobium brockii	40–75	65–70	5.5–9.5	~7.5	–	–	NR	+	+	–	78
Thermoanaerobacter ethanolicus	37–78	70	4.4–9.9	5.8–8.5	–	–	+	+	+	+	79
Thermobacteroides acetoethylicus	>40–<80	65	5.5–8.5	~7	–	–	w	+	w	–	80
Thermoanaerobacter sp.	37–72	60	3.5–7.5	5–6	–	+	+	+	+	+	81

NR, not reported
w, weakly utilized

electron acceptors, H_2 completely inhibits growth at 1 atm partial pressure, due to inhibition of reoxidation of NADPH and/or reduced ferredoxin. Addition of electron acceptors such as acetone results in consumption of exogenous H_2 and concomitant reduction of the added electron acceptor. This reversibility in the in vivo direction of catabolically important enzymes finds its greatest expression in the organism's ability to grow on ethanol as its primary energy source if H_2 evolved during growth is simultaneously removed, for example, by coculture with *Methanobacterium thermoauto-trophicum*. The capacity of H_2-evolving, sugar-fermenting bacteria to grow on compounds such as ethanol when in coculture with H_2-consuming anaerobes may have ecological significance (84).

Superimposed upon the metabolic characteristics of individual species are the interactive effects among coculture organisms. For example, consumption of reducing sugars by the noncellulolytic accessory organism probably enhances cellulose degradation by the cellulolytic member, due to removal of end product inhibition on cellulase (although this effect may be minor, due to the relative insensitivity of *C. thermocellum* cellulase to such inhibition (49)). Competition between cellulolytic and noncellulolytic members for reducing sugars may also affect the end product ratios of individual organisms, although in general organisms such as *C. thermohydrosulfuricum* have been shown to outcompete *C. thermocellum* for both glucose and cellobiose (62).

Some strains of noncellulolytic, thermophilic anaerobes also appear capable of utilizing pentose oligomers or even polymeric hemicelluloses. Wiegel et al. (85) have reported that *T. ethanolicus* JW200Fe(4) can utilize xylose and xylooligomers through xylotetraose (X_1 through X_4) at approximately similar growth rates. At low substrate concentrations (5 g/l), a 1 percent (v/v) inoculum converted 95 to 98 percent of substrate in 45 hours to yield 1.3 to 1.5 mol ethanol/mol xylose monomer equivalent. In general, growth rates (and, presumably, volumetric productivities) on xylose are 2- to 3-fold lower than on glucose. In at least one case (that of *T. ethanolicus*), xylose fermentation occurs over a narrower pH range than does glucose fermentation. By using adapted strains of certain thermophilic anaerobes, some encouraging ethanol production data have been obtained from xylose fermentations. For instance, an ethanol-tolerant derivative of *C. thermosaccharolyticum* (designated HG-6-610) has been reported to ferment xylose to ethanol with a yield of 41 to 43 percent (weight basis, equivalent to 1.4 mol ethanol/mol xylose). In a 4 l fed-batch reactor with pH controlled at 6.6, a total of 66 g xylose/l was fermented by this strain to 27 g ethanol, 2.6 g acetate, and <2 g lactate (all per l) during a 60 hour run (37). It should be noted that this fermentation was conducted in a very rich medium containing, among other things, 10 g yeast extract and 0.2 g Na palmitate/l.

Some of the thermophilic anaerobic saccharolytic species described above have been shown to be capable of a limited hemicellulose fermentation. Although 93 to 100 percent utilization of the hemicellulose fraction of steam-treated birch (concentration = 0.8 percent w/v) was observed after 72 hours incubation (with periodic pH adjustment), ethanol production was poor (maximum concentration of 24 mM, corresponding to ~15 percent yield by weight) (78). Resting cells of *T. ethanolicus* JW200 wt, concentrated 40-fold to an unspecified absolute cell density, produced about 70 mM ethanol from 4 percent (w/v) of the above hemicellulosic substrate. As had been observed in the case of other carbohydrates, ethanol/acetate ratios were enhanced when lower substrate concentrations were used. More recently, thermophilic anaerobes which grow via an active hemicellulose fermentation have been isolated from geothermal sources (81,87,88). At least one of these strains can produce up to 1 percent (w/v) ethanol from purified hemicellulose (89).

Certain noncellulolytic saccharolytic species display enigmatic behavior during growth on mixed substrates. For example, Carreira et al. (90) have reported that while *T. ethanolicus* grown in batch culture on 1 to 2 percent carbohydrate plus 0.2 percent yeast extract produces high yields of ethanol from individual monosaccharides (1.9 mol ethanol/mol hexose, or 1.5 mol ethanol/mol xylose), growth on mixed substrates (e.g., glucose/xylose, or glucose/xylose/starch) results in considerable reductions in ethanol yield and concomitant formation of acetate and lactate. The regulatory mechanism controlling end product yields on mixed substrates has not been elucidated in *T. ethanolicus*, and the conclusion that glucose and xylose are catabolized by different pathways appears premature on the basis of the data presented. Eliminating the unfavorable product mixture obtained from mixed substrates is a challenging problem which limits the exploitation of an otherwise promising microorganism. A mutant strain of this species (strain JW200 Fe(7)) has produced 15.5 and 21.7 g ethanol/l from fermentation of 31 and 62 g starch, respectively, and can reportedly grow in the presence of 10 percent (v/v) ethanol (91).

Ethanolic fermentation of biomass by thermophilic anaerobes presents an engineering situation markedly different from the alcoholic fermentation of corn syrup by mesophilic yeast. Although the thermophilic processes operate in the 60 to 65°C range, much of the energy used in heating the fermentors may be derived from downstream unit operations (e.g., product distillation) via efficient use of heat exchangers (13,14). Formation of significant amounts of organic acids as side products of carbohydrate fermentation results in a detrimental acidification of the culture medium and a resultant requirement for pH control. Because the process is anaerobic, mixing requirements are minimal when compared to aerobic fermentations, where O_2 mass transfer is critical (14).

For fermentation of cellulosic biomass, considerable engineering complexity and costs are associated with processes involved in physical or chemical pretreatment (and subsequent cleanup) of substrate to obtain a fermentable material. The insoluble nature of biomass substrates also limits the concentrations of substrate which can be used in the reactor and encourages the use of fed-batch process modes. Organism considerations place other constraints on reactor design: for example, the facilitation of cellulose degradation by contact of *C. thermocellum* with its solid substrate limits the use of certain immobilization techniques designed to increase cell density, and prodigious gas formation may limit the use of large-scale packed-bed reactors.

C. thermocellum has been shown to grow in a defined medium containing mineral salts, a fermentable carbohydrate, and trace quantities of biotin, pyridoxamine, *p*-aminobenzoic acid, and vitamin B_{12} (92). However, both growth rate and growth yield were considerably less than in complex media and no data are available on product formation by cultures after prolonged cultivation (over many months) in this medium. Certain additives to salts/ carbohydrate/yeast extract medium have stimulatory effects on product formation (38), but none are sufficiently dramatic to justify these additions, at their observed effective concentrations, at an industrial scale. Virtually all of the accessory organisms described above have complex nutritional requirements (74–80). On a commercial scale, cheaper alternatives to yeast extract would almost certainly be needed, regardless of the organisms chosen for the fermentation.

Much of the fermentation technology of ethanol production by thermophilic anaerobes has been performed by Wang and his colleagues (29–43,93). The bulk of this work has involved the use of pure or mixed cultures of *C. thermocellum* in batch culture at relatively modest (≤5 l) scale, and has focused upon improving ethanol production via: a) empirical alteration of environmental conditions; b) development of bacterial strains which display improvements in cellulolytic activity, ethanol tolerance, or ethanol yield; and c) coculture with *C. thermosaccharolyticum* strains selected for enhanced yield of, or tolerance to, ethanol. Most of these experiments on cellulose fermentation have utilized Solka Floc, a purified noncrystalline holocellulose which contains small amounts of xylan. Wang (93) has reported that typical batch fermentations result in consumption of 7 to 26 g substrate/l, with production of only 1.1 to 4.0 g ethanol/l.

The relatively low concentration of ethanol accumulated by *C. thermocellum* has been widely attributed to the sensitivity of this species to ethanol and (to a lesser extent) to acetate and protons (30,45,70,71). The inhibitory effects of ethanol and acetate have been shown to be additive. For example, μ_{max} values in the presence of 10 g of exogenously added ethanol or acetate per liter have been reported to be 52 and 53 percent, respectively, of μ_{max} values observed in control cultures without the exogenous addition; when both products were added at 10 g/l, μ_{max} dropped to 21 percent of the control value (30).

Both ethanol and acetate have been shown to exert inhibitory effects on a number of fermentation parameters. Herrero and Gomez (70) have reported that, for strain ATCC 27405 grown on cellobiose, increasing ethanol concentrations caused a decrease in growth rate and optimum growth temperature, and caused an increase in the activation energy required for cellular growth. Kundu et al. (71) noted similar effects for the same strain grown on Solka Floc; in addition, these workers determined the kinetic parameters of ethanol and acetate inhibition, and demonstrated that high concentrations of both ethanol and acetate caused a decrease in the in vitro activities of the endoglucanase and exoglucanase components of cellulase.

Development of ethanol-tolerant mutant strains has been reasonably successful. *C. thermocellum* strain S-4, derived from ATCC strain 27405 by batch transfer of stock cultures through media containing successively higher ethanol concentrations, reportedly exhibits only a 35 to 50 percent inhibition of growth in the presence of 5 percent ethanol (v/v, equivalent to 40 g/l), while the parent strain exhibited no growth at 3 percent (v/v) ethanol (31). In addition, these ethanol-tolerant strains reportedly displayed both enhanced cellulolytic activity and higher ethanol yield (i.e., greater ethanol production and reduced formation of acetate, lactate, or H_2) (31). Polysaccharide fermentations employing these strains have exhibited moderate improvements in final ethanol accumulation, although they have not approached the concentrations to which the derived strains display ethanol tolerance with respect to growth.

The biochemical basis of the ethanol sensitivity of *C. thermocellum* has been elegantly investigated using ^{31}P-NMR techniques employing both wild-type cells and cells adapted to rapid growth in the presence of moderate concentrations (10 g/l) of ethanol (38). Current evidence—obtained from measurements of substrate uptake rates, intracellular metabolite concentrations, and transmembrane pH gradients—suggests that ethanol inhibition of growth is largely the result of a proportional reduction in the rate of intracellular ATP production, and further suggests that ethanol tolerance is due to an acquired ability to produce ATP at a rate independent of ethanol concentration. No mechanisms for ethanol tolerance in ethanol-adapted noncellulolytic thermophilic anaerobes (e.g., *T. ethanolicus* JW200 Fe(7) or *C. thermosaccharolyticum* HG-6-62) have been reported (38,91).

Although the fermentability of different substrates varies widely, it is clear that crude biomass polysaccharide is almost always fermented more poorly than are purified polysaccharides (73,93). *C. thermocellum* in general displays a slow fermentation of agricultural residues and an even slower fermentation of wood materials. A survey of the fermentation of a number of crude biomass substrates by Wang (93) bears this out. After 350 hours of fermentation at 25 g biomass/l, maximum ethanol concentrations varied from 0.2 to 1.2 g/l. A common feature of *C. thermocellum* biomass fermentations is the relatively high yield of acetic acid, which may reach ten times the yield of ethanol, and reducing sugars, whose concentrations generally exceed that of all other nongaseous fermentation products combined (73,93). Cocultures of *C. thermocellum* with other thermophilic saccharolytic anaerobes also produce lower ethanol/acetate ratios on crude biomass substrates than on purified celluloses (37,73). Appropriate chemical or physical pretreatment of the biomass generally results in considerable improvement in both the yield and the concentration of ethanol, thus underscoring the importance of developing effective pretreatment processes to render the fermentable components of biomass accessible to enzymatic attack and fermentation by thermophilic microflora. Nevertheless, ethanol concentrations in fermentations employing pretreated cellulosic biomass almost always remain below 4 g/l.

The desire to enhance final product concentration has stimulated examination of more easily degraded plant-derived materials, some of which appear to show promise. Both honey locust pods and mesquite beans contain large amounts of fermentable sugars (particularly fructose) along with lesser amounts of cellulose and pentosans (38). Wang et al. (38) have reported that mixed cultures of *C. thermocellum* S-7 and *C. thermosaccharolyticum* HG-6-62 (both of which are derived strains selected for ethanol tolerance) fermented ground beans from hybrid mesquite to produce ethanol at a final concentration of 30 g/l and a maximum productivity of 0.44 g/l·h. This fermentation was unusual in both substrate composition (44 percent sucrose, 10 percent pentosan, and 11 percent cellulose) and fermenter operation. The fermentation was conducted in a fed-batch mode (total mesquite addition: 100 g/l) with *C. thermosaccharolyticum* inoculated at time zero and *C. thermocellum* not inoculated until hour 42 of the 80 hour fermentation. The high ethanol/acetate ratio observed, along with the fact that *C. thermocellum* was not inoculated until most of the carbohydrate had been consumed, indicated that the fermentation was primarily a *C. thermosaccharolyticum* monosaccharide fermentation. Nevertheless, the reported consumption of 11 g cellulose/l during the last half of the fermentation suggests that *C. thermocellum* was metabolically active at high ethanol concentrations, even if its contribution to total ethanol production was quantitatively minor. While mesquite beans are unlikely to serve as an important source of fermentation ethanol, the data are rather encouraging and indicate that the general concept of mixed-culture fermentation of monosaccharide/ polysaccharide mixed substrates deserves further research efforts.

The inherently low volumetric productivity of anaerobic fermentations can in general be increased by proportional increase in biomass density (e.g., by cell holdback or cell immobilization) (94). This strategy has been shown to be successful for *C. thermocellum* in continuous culture on soluble sugars (39,40). Using glucose as substrate and reasonably rapid dilution rates (up to D=0.415/h), volumetric productivities for ethanol, when corrected for product loss due to gas sparging, attained values as high as 1.45 g/l·h. However, low dilution rates were necessary to reduce breakup of the bead matrix, and at all dilution rates, the concentration of ethanol in the effluent was relatively low (<4 g/l). As pointed out above, cell immobilization on metabolically inert materials is of questionable value when insoluble substrates are employed.

Although the economic viability of using starch as a substrate for thermophilic ethanol fermentations remains an open question, a recently reported (95) novel interaction between *C. thermohydrosulfuricum* 39E and *C. thermosulfurogenes* 4B growing as a coculture on starch is worthy of note. Both organisms are reasonably efficient starch fermenters, although the specific enzymes of starch hydrolysis and saccharide uptake, as well as the reversibility of certain key intracellular oxidoreductases, differ in two strains (96). Starch fermentation is mediated via either glucoamylase and pullulanase (strain 39E) or glucoamylase and β-amylase (strain 4B). Relative to monocultures of these strains, cocultures employing both strains exhibited more complete starch utilization and higher ethanol concentrations, due to the simultaneous expression of all three starch-hydrolyzing enzymes and the rapid removal of low-molecular-weight glucans, which cause catabolite repression of enzyme synthesis (95). Cocultures grown in rich media at 60°C for 70 hours accumulated a total of 284 mM ethanol ($\sim$1.3 percent, w/v), with a calculated product yield of 1.7 mol ethanol/mol anhydroglucose. This case represents an unusual example of mutalism in two organisms at the same trophic level, and suggests that a search for similar relationships among other polysaccharide-degrading organisms may prove rewarding.

Certain facultatively anaerobic thermophiles produce ethanol as a catabolic product of carbohydrate fermentation. Although wild-type *Bacillus stearothermophilus* has been shown to produce only small amounts of ethanol, mutants lacking lactate dehydrogenase (selected by use of the "suicide" substrate fluoropyruvate) displayed enhanced ethanol yield ($\sim$1 mol ethanol/mol glucose fermented) (97). Subsequent optimization of fermentation parameters led to a reported productivity of 1.3 g ethanol/g cells·h. As pointed out by Payton (97), *B. stearothermophilus* may represent a promising organism for generating ethanol hyperproducing mutants because of its inherently high ethanol tolerance; in addition, the comparatively well-developed genetic technology for *Bacillus* should aid in the introduction of additional catabolic capabilities (e.g., cellulase enzymes) which would enhance the utility of such organisms in industrial processes. Thus, future work on improving this organism (e.g., by eliminating pyruvate-formate lyase activity) may be worthwhile.

4.2. Organic Acids

As mentioned in Section 4.1 above, ethanol-producing thermophilic, anaerobic, saccharolytic bacteria all produce various amounts of organic acids (acetic and, usually, lactic acids) as coproducts of carbohydrate fermentation. Interest in biotechnological production of organic acids, however, has been focused upon the potential exploitation of other bacteria which produce an organic acid as virtually their sole fermentation product.

Acetic Acid We must at the outset distinguish between the common acetic acid of commerce, glacial acetic acid, which is produced chemically by catalytic carbonylation of methanol (the so-called Monsanto process), and dilute, food-grade acetic acid, or vinegar, which in the United States is produced solely by microbial action because of governmental regulatory requirements (98). Production of acetic acid from ethanol in the latter process is accomplished via aerobic, mesophilic bacteria of the genus *Acetobacter* (98). The resulting fermentation broth (containing 12–13 percent (w/v) acetic acid and having a pH of about 2) is diluted to 5 percent acetic acid prior to marketing.

Adaptation of the vinegar fermentation to produce commodity acetic acid is considered to be economically impractical for two reasons. First, the combined theoretical yield of acetic acid from glucose is low:

$$C_6H_{12}O_6 \rightarrow 2CH_3CH_2OH + 2CO_2 \text{ (yeast)} \tag{1}$$

$$2CH_3CH_2OH + 2O_2 \rightarrow 2CH_3COOH + 2H_2O \text{ (\textit{Acetobacter})} \tag{2}$$

(i.e., $1C_6H_{12}O_6$ yields $2CH_3COOH$, a 67 percent weight yield). Second, the costs associated with recovery of pure acetic acid from the 10 to 13 percent (w/v) acetic acid concentration in the *Acetobacter* broth are high. Because the boiling point of acetic acid (117°C) exceeds that of water, recovery by distillation is extremely expensive. Alternative recovery schemes, such as extraction by CO_2/solvent combinations (99,100) or by supercritical fluids (101) are characterized by low partition coefficients which render the process economically unfeasible unless new extractants are identified.

The impetus for investigating other microbial systems for their acetogenic potential lies in the fact that certain bacteria, the homoacetate fermenters, produce considerably higher yields of acetic acid from carbohydrates than does the above two-stage yeast/*Acetobacter* process (102). The homoacetate-producing (acetogenic) bacteria include both mesophilic and thermophilic species which are taxonomically divided into several genera. The physiological characteristics of the three described species of thermophilic acetogens are summarized in Table 2. Most, if not all, acetogens can grow both as chemolithotropic autotrophs on H_2/CO_2 or on CO, or as chemoorganotrophic heterotrophs on various organic substrates, particularly certain monosaccharides (103-108). The presence of hydrogenase in most species (109,110), along with the ability of at least some species (e.g., the mesophilic *Acetobacterium woodii*) to grow on carbohydrates in coculture with H_2-oxidizing methanogens (111), suggests that H_2 formation by acetogens not only occurs, but may be important in the physiology of these organisms in mixed bacterial populations.

Our understanding of the biochemistry of homoacetate fermentation has undergone considerable advancement over the past several years and has been the subject of an excellent recent review by Ljungdahl (102). It now appears that the pathway of acetate synthesis, first elucidated in detail in *C. thermoaceticum*, is utilized with only minor modifications by the other described acetogenic species.

TABLE 2. Physiological characteristics of thermophilic homoacetate-fermenting bacteria

Characteristic	*Clostridium thermoaceticum*	*Clostridium thermoautotrophicum*	*Acetogenium kivui*
Temperature			
optimum (°C)	55–60	56–60	66
range (°C)	45–65	36–70	50–72
pH optimum	7–8	5.7	6.4
range	6–8.5	4.5–7.6	5.3–7.3
Growth substrates	Xylose, fructose, glucose, H_2/CO_2, CO	Glucose, fructose, galactose, glycerate, formate, methanol (or methylamine)/CO_2, H_2/CO_2, CO	Glucose, fructose, mannose, pyruvate, formate, H_2/CO_2
Growth factors[a]	Biotin or nicotinic acid required. YE or tryptone enhances growth rate and yield	YE enhances growth yield	Trypticase or YE enhances growth yield
Spore formation	+	+	−
References	103,104	105	106

[a]YE = yeast extract

Inspection of Table 2 reveals that thermophilic homoacetogens (like their mesophilic counterparts) lack the acidophilic characteristics of the aerobic mesophiles used for vinegar production. Consequently, fermentations employing the homoacetogens must generally be carried out at pH values at or above the pK_a of acetic acid (which is 4.76 at 25°C and 4.80 at 60°C) (112). Under these conditions, the product exists in solution primarily in the form of the unprotonated acetate anion. Economical recovery of acetic acid or the acetate anion from this dilute aqueous acetate solution is beyond current technological capabilities. Consequently, fermentation technologists have attempted to optimize homoacetogenic fermentations for low pH and high total acetate (acetic acid plus acetate anion) concentration (29–42,113,114).

Attempts to develop an economical fermentation based on homoacetogenic bacteria have focused on *C. thermoaceticum*. This species alone among the acetogens can utilize both glucose and xylose, which are the chief monosaccharide products of enzymatic or chemical hydrolysis of the cellulose and hemicellulose components of plant biomass (102). The stoichiometries of carbohydrate catabolism

$$C_6H_{12}O_6 \rightarrow 3CH_3COOH \tag{3}$$

$$C_5H_{10}O_5 \rightarrow 2.5CH_3COOH \tag{4}$$

indicate a theoretical acetic acid yield of 100 percent on a weight basis. In practice, maximum observed yields in the 85 to 90 percent range are obtained, due to use of substrate in synthesis of cellular components and loss of substrate in chemical reactions (e.g., carmelization) at elevated incubation temperatures.

   C. thermoaceticum exhibits a differential monosaccharide fermentation with the preferred order of substrate utilization being first xylose, then glucose, then fructose. The implications of these preference patterns in the fermentation of biomass hydrolyzates in either batch or continuous culture are as yet uncertain and may demand modifications of either the bacterial strains or the process engineering if mixed carbohydrates are to be used efficiently. It is not surprising, therefore, that most studies of the homo-acetate fermentation in small-scale fermenters have used glucose (readily available on an industrial scale as corn syrup) as substrate (29–42,113,114).

   Wang et al. (29–42) have reported attempts to develop a viable fermentation process for production of acetic acid from this organism. Initial experiments (29) with the Ljungdahl strain of *C. thermoaceticum* (ATCC 39073 and DSM 521) in complex medium revealed that $\mu_{max}(0.13/h)$ and maximum specific productivity (0.5 g acetate/g cells·h) occurred at pH 7.1 to 7.2. At successively lower pH values, acetate production became uncoupled from growth; no growth was observed at pH ≤ 6.5, although small amounts of acetate were formed at pH values as low as 6.0. Control of pH during fermentation permitted prolonged growth of the organism and maximal acetate accumulation. For example, experiments with acetate-tolerant strain S3 under pH control at 7.0 resulted in conversion of 70 g glucose/l to 56 g acetate/l over a 125 hour incubation period (32). However, at pH 7.0, this concentration corresponds to only 0.35 g free acetic acid/l.

   Wang et al. developed a defined culture medium (mineral salts, glucose, and either biotin or nicotinic acid) for growth of this organism (32). Comparison of defined and complex media (using strain S3) revealed that in both media μ_{max} decreased linearly with increasing acetate concentration; the μ_{max} in defined medium was 0.07/h less than that in complex medium at all sodium acetate concentrations tested up to 40 g/l, the maximum acetate concentration permitting growth in defined medium (36). In complex medium, measurable growth rates were obtained at sodium acetate concentrations up to 70 g/l.

   A 130 hour batch fermentation in defined medium without pH control resulted in production of 11.6 g acetate/l plus 1.27 g cell mass/l from the consumption of 14.1 g glucose/l (38). During the fermentation μ_{max} was 0.07/h, and final culture pH a relatively high 6.1. Parallel batch-culture experiments in complex media resulted in more rapid growth ($\mu_{max} = 0.12/h$) and in conversion of 20 g glucose/l into 14 to 16 g acetate/l plus 4 to 4.6 g cell mass/l. *C. thermoaceticum* was also grown in continuous culture in complex medium (38). At dilution rates of 0.04 to 0.156/h, acetate concentrations of 17.4 to 6.0 g/l were obtained and yields (based on glucose removed from the medium) were in the 75 to 77 percent range.

Schwartz and Keller (113,114) have reported their attempts to improve the fermentation technology of this process, using strains of *C. thermoaceticum* selected for tolerance to acetate and to low pH. Beginning with the wild-type Ljungdahl strain, these workers isolated a variant capable of growth in 2 percent sodium acetate. This variant (designated strain 1745) was then successively adapted through a series of pH-controlled, fed-batch fermenters containing a rich medium and increasing concentrations of acetate. A culture (ATCC 31490) which grew at an initial pH of 4.5 and an initial acetate concentration of 2.2 g/l was ultimately obtained (113). During the course of the adaptation, it was discovered that decreasing the culture pH below 6.0 and increasing the initial acetate concentration above 1.5 g/l resulted in a considerable decrease of both growth rate and final acetate concentration, without significantly altering acetate yield ($\sim$90 percent, based on substrate consumed). The authors concluded that most of the fermentation goals required for an economically viable process either had been attained (growth at pH 4.5, maintenance of 90 percent yield) or could be attained by further selection (μ of 0.1/h); however, the goal of 50 g acetate/l at pH 4.5 (equivalent to 33.3 g free acetic acid/l) was considered unattainable (103).

It should be pointed out that the above experiments were performed in media with considerable amounts of peptone and yeast extract (2.5–8.5 g each/l) in addition to various vitamins and salts. These components add to total raw material costs and complicate recovery of the acetic acid product. Adaptation of cultures to a defined or minimal medium would circumvent these problems, but as indicated above, would result in a significant reduction of growth rate, acetate yield, and acetate tolerance.

Attempts to improve the productivity of *C. thermoaceticum* fermentation by cell immobilization have been only moderately successful. Rate of production of acetate in cells immobilized on agarose or *k*-carrageenan (0.13–0.16 g acetate/g cells·h) was similar to that obtained in free cells (0.18 g acetate/g cells·h). Because cell densities achieved within the gel matrix were considerably greater than those obtained with free nonimmobilized cells, volumetric productivities were improved 2.8-fold (to 2.2 g acetate/l·h). The carrageenan beads were easily harvested, and when operated in a fed-batch mode continued to produce acetate at the same rate for 300 hours upon transfer to fresh medium (37).

The fermentation of xylose by *C. thermoaceticum* has also been investigated by Wang et al. (26–39). Growth rate on xylose in complex media was 0.13/h, similar to that on glucose. Observed yields of acetate and cell mass were 0.76 and 0.21 g/g xylose, respectively. The latter value is quite high for fermentative growth, particularly in light of the high maintenance energy (0.26 g glucose/g cell·h) reported for this organism (37).

Considerable progress has recently been made in elucidating the biochemical mechanism of acetic acid toxicity in this organism. Baronofsky et al. (115) have measured transmembrane gradients of pH (ΔpH) and electrical charge ($\Delta\psi$) in *C. thermoaceticum* grown or maintained under various concentrations of H$^+$ and acetate. These workers have shown that an internal pH of $\geq$5.5 to 5.7 is required for growth of the organism. Reduction of external pH to values of $\leq$5.5 to 5.7 in the absence of acetate permits maintenance of a sufficient ΔpH to permit growth; however, if acetic acid (or other organic acids which passively traverse the membrane) is present, both ΔpH and $\Delta\psi$ are almost entirely abolished. These workers concluded that acetate at low pH ($\leq$5.0) causes an inability of the organism to generate sufficient ATP to export H$^+$ (via the H$^+$-ATPase) at a rate necessary to maintain the minimum intracellular pH permitting growth. Because acetate inhibition is the result of such a basic biochemical mechanism, these workers further concluded that development of multiple-site mutants tolerant to acetate at low pH are unlikely to be obtained.

The apparent inability of *C. thermoaceticum* to adapt to pH values of $\leq$4.5 suggests that acetate production by other, more acid-tolerant thermophilic anaerobes may be worth further investigation. For example, the wild-type strain of the recently described *Clostridium thermoautotrophicum* has a reported pH optimum of 5.7 and retains the ability to grow at pH 4.5 (105). Further attempts to adapt this organism to lower pH values and to improve final acetate concentration may prove fruitful.

The mixed-culture approach described in Section 4.1 for enhancing ethanol production from cellulose fermentation has also been attempted for conversion of cellulose to acetic acid. Crude mixed cultures of undefined composition have long been known to ferment cellulosic materials to mixtures of organic acids, particularly acetic acid (116,117). More recently, Le Ruyet et al. (118) have shown that cocultures of *C. thermocellum* and *Acetogenium kivui* grown on crystalline cellulose (Whatman CC31) at pH 6.8 performed a virtual homoacetate fermentation, according to the stoichiometry

$$1 \text{ anhydroglucose} \rightarrow 2.70 \text{ acetic acid} + 0.11 \text{ lactic acid} + 0.07 \text{ ethanol} + 0.05 \text{ formic acid} + 0.13 \text{ H}_2 + 0.11 \text{ CO}_2 \tag{5}$$

Due to the weight gain of substrate during cellulose fermentation (i.e., hydration of anhydroglucose), this molar yield of 2.7 translates to essentially a 100 percent weight yield, and a final acetate concentration of 10 g/l (or 0.1 g free acetic acid/l). The rate of cellulose degradation in the coculture (63 mg/l·h) was about 30 percent above that in *C. thermocellum* monocultures tested at the same pH, due to enhanced utilization of carbon in production of acetate and therefore enhanced ATP synthesis. However, the rate of substrate utilization was far below the 470 mg/l·h observed in *C. thermocellum* monocultures grown near their pH optimum of 7.4. These data again point to the necessity of developing more acid-tolerant cellulolytic strains if the full potential of thermophilic anaerobic cocultures for bulk chemical production from cellulose is to be realized.

The tendency of *C. thermocellum* to produce higher amounts of acetate and lower amounts of ethanol when grown on crude lignocellulosic substrates than when grown on purified carbohydrates (one of the limitations of the use of this organism in ethanol production) suggests that *C. thermocellum/A. kivui* cocultures may produce even higher acetate yields on crude substrates than on the purified cellulose used in the above experiments.

The discovery that acetogenic bacteria can utilize CO and H_2/CO_2 as grown substrates has prompted suggestions that these organisms may be useful in production of acetic acid from readily available C_1-containing feedstocks (e.g., synthesis gas, a mixture of gases—predominantly H_2 and CO—produced from coal gasification or from natural gas conversion) (107,108). The observed stoichiometries of these fermentations:

$$4CO \rightarrow 1.1CH_3COOH + 2CO_2 \qquad (6)$$

and

$$4H_2 + 2.1CO_2 \rightarrow 0.9CH_3COOH \qquad (7)$$

suggest that a nearly 100 percent weight yield may be obtained from the H_2 and CO components of synthesis gas. CO toxicity does not appear to be a severe problem, since strains capable of growth under a pure CO gas phase have already been isolated (104). The two chief factors which would probably limit the success of these fermentations are the slow growth rates of the organisms (doubling times during the above reactions have been reported as 18 and 16 hours, respectively) (104) and their poor tolerance to low pH and to acetic acid. Finally, it makes little sense to attempt development of a fossil-fuel-based fermentation process when the analogous chemical synthesis routes are well established and reasonably efficient.

The outlook for commercial utilization of homoacetogens for production of commodity acetic acid appears bleak. Of the various possibilities employing homoacetogens, thermophilic cocultures utilizing inexpensive lignocellulosic substrates would be most worth further investigation, since the cost advantage of the substrate would help offset some of the weaknesses of the system. Further research in this area must be directed toward: a) enhanced acid tolerance of the cellulolytic, hemicellulolytic, and homoacetogenic organisms (perhaps via isolation of novel species), along with narrowing the differences between the temperature or pH optima of the coculture organisms; b) increased effective biomass density (e.g., through cell immobilization) to overcome the inherently low volumetric productivities resulting from the low cell densities common to anaerobic systems; and c) development of a durable and efficient recovery process converting a dilute aqueous acetate broth to a more concentrated acetic acid solution. Unless these requirements are met, it would seem that any microbial production of commodity acetic acid would come from the *Acetobacter* (vinegar) process.

Lactic Acid   This organic acid (CH₃CHOHCOOH) is produced commercially by fermentation of glucose, using homofermentative lactobacilli, particularly strains of *Lactobacillus delbruckii* (119). The theoretical reaction stoichiometry (1 glucose→2 lactic acid) translates to a 100 percent weight yield of product. Although commercial *Lactobacillus* fermentations are run in the 45 to 50°C range, the organisms may be regarded as thermotolerant or facultatively (rather than truly) thermophilic, and they will not be discussed further here.

L-Lactic acid is produced by numerous thermophilic bacteria (e.g., *C. thermocellum, C. thermohydrosulfuricum, Thermoanaerobium brockii*, and *Thermoanaerobacter ethanolicus*) as one of several coproducts of carbohydrate fermentation (28). In the case of *T. brockii*, lactate yields of up to 0.94 mol lactate/mol glucose (47 percent weight yield) have been reported (83). As noted above, the biochemical basis for enhanced flux of carbon and electrons toward lactate during rapid growth may lie in the fact that the lactate dehydrogenase (at least in the cases of *C. thermocellum* and *T. brockii*) requires activation of fructose-1,6-diphosphate (FDP), a key Embden-Meyerhof pathway intermediate whose concentration is presumably elevated at rapid growth rates.

The facultative thermophilic, facultative anaerobe *Bacillus coagulans* may have some potential for lactic acid production. This bacterium is a homofermentative lactic acid producer when grown on hexoses, but produces moderate amounts of acetate and ethanol coproducts when grown on pentoses (32). Wang et al. (32) have carried out studies using an isolate with minimum, optimum, and maximum temperatures of 30, 45 to 55, and 63°C, respectively. They reported that a 9 hour fermentation of a glucose/cellobiose mixture (approximately equal weights of each, totaling 19.6 g reducing sugar/l) was characterized by consumption of glucose prior to cellobiose utilization and by net production of lactate (17.7 g/l) as sole catabolic product (32). Fermentation of 20 g xylose/l yielded 16.4 g lactate and 0.9 g each acetate and ethanol per liter after 38 hours incubation. The relatively low concentrations of lactate (compared to concentrations of 120 g/l in commercial *Lactobacillus* fermentations) (119) suggests that the *B. coagulans* system would not compete with the latter process in most applications.

While *B. coagulans* is noncellulolytic, its ability to catabolize mono- and disaccharides suggests that it may serve as a useful accessory organism with *C. thermocellum* growing on cellulose, wherein the *Bacillus* could both utilize free reducing sugars and stimulate lactate formation. Wang et al. (38) have quantitated product formation from cellulose by *C. thermocellum/B. coagulans* cocultures. Inoculation of the latter organism into a 12 hour culture of *C. thermocellum* growing on Solka Floc under pH control at 6.8 resulted in an eventual lactate yield of about 35 percent on a weight basis. After 200 hours incubation, 65 g of substrate was fermented to 23 g lactic acid, 6 g acetic acid, and 3.5 g ethanol. Lactate concentration in the broth, however, was only 7 g/l. Thus, the chief limitations of the coculture system appear to be low product concentration and the relatively low yield of lactate, which would probably decrease even further if crude, pentose-containing cellulosic materials were used as substrates.

Because the purification of the nonvolatile lactic acid from dilute aqueous solution is rather bothersome and expensive, the product is normally concentrated to a 60 percent (w/v) syrup for shipment. Since fermentation lactic acid is used primarily as a food acidulent, this syrup must be acidic. Consequently, industrial production of fermentation lactic acid employs moderately acidophilic *Lactobacillus* strains to minimize acidification costs during processing of the product. Fermentations requiring extensive pH control to obtain reasonable rates of lactate production (e.g., the *B. coagulans* fermentation) increase these acidification costs.

For most applications, the isomeric composition of the lactic acid is of little importance. However, specific end uses for the individual purified isomers do exist. Both D- and L-isomers are used for research purposes, particularly the L-isomer as a substrate for diagnostic testing of mammalian lactate dehydrogenases. L-Lactate may also be particularly useful in the synthesis of polylactates used for self-dissolving surgical sutures, or lactide/caprolactam copolymers (120). Commercial lactate fermentations employing *Lactobacillus* strains produce the L-isomer almost exclusively.

It thus appears that the use of thermophiles for lactic acid production is not advantageous (from the standpoints of product concentration, degree of acidity, or isomeric form) compared with existing *Lactobacillus* fermentations.

4.3. Hydrocarbon Modification

Thermophiles may be useful in the pretreatment of natural hydrocarbon materials to improve their subsequent utility as sources of chemicals and fuels. One potential example (121) is the possible use of the thermophilic, acidophilic, aerobic bacteria such as *Thiobacillus* or *Sulfolobus* to remove pyrite from coal, resulting in a low-sulfur coal which would release less sulfur dioxide during combustion in electric power plants.

A second potential example is the use of thermophilic fungi in the decomposition or "retting" of natural rubber from the guayule plant (*Parthenium argentatum*) (4). Such a process enhances the quality of the rubber by reducing its resin content and increasing its tensile strength. Natural, self-heating rets contain a great diversity of thermophilic microflora, including several thermophilic fungi which can produce good rets when inoculated as a single pure culture onto the crude guayule rubber. Several species (*Mucor pusillus, Penicillium duponti, Humicola lanuginosa,* and *Malbranchea pulchella var. sulfurea*) after 7 days' incubation at 45°C reduced the resin content from an initial value of about 18 percent to a final value of 8.6 to 10.8 percent (4). Recent proposals (122) to develop guayule as an arid land biomass crop may give impetus to a reinvestigation of the thermophilic retting process.

Finally, both *Bacillus stearothermophilus* (123) and *B. coagulans* (124) can perform monoterminal (ω) hydroxylation of alkanes. This property is also widely distributed among mesophilic bacteria and yeast. In general, there is no specific advantage of using thermophiles for these reactions, except perhaps for conversion of materials (e.g., fats, waxes, or higher alkanes) whose melting points are in the growth temperature range of thermophiles.

4.4. Specialty Chemicals

Although thermophilic fermentations to produce bulk chemicals and fuels have received considerable recent research effort, there has been little work in the area of thermophilic microbial production of specialty chemicals (i.e., compounds whose use is based primarily on function rather than on structure, and whose markets are characterized by low volume but high value-in-use). It is likely that thermophilic microorganisms have a greater economic potential for production of specialty chemicals than bulk chemicals. This opinion is based upon the facts that: a) isolation of novel thermophilic species (and thus genetic capability for novel metabolic reactions) is proceeding at an unprecedented rate; and b) many of the thermophilic bacteria now available for biotransformation studies are taxonomically related to members of mesophilic genera which have been shown to be capable of synthesizing a wide variety of useful specialty chemicals. Potential examples include *Thermoactinomyces* and *Thermomonospora* (related to *Streptomyces* and *Micromonospora*) (125) and *Thermoleophilum album* (related, at least nutritionally, to *Pseudomonas* and *Acinetobacter*) (126).

Vitamins Cartenoids have considerable commercial utility as food coloring agents and also find minor use as vitamin A precursors. Both α- and β-carotene possess provitamin A activity, that is, they are convertible in mammalian intestinal mucosa to vitamin A (127). The stoichiometries of the conversion (moles vitamin A formed per mole carotene consumed) are 1:1 and 2:1 for the α- and β-forms, respectively. Microbial production of either isomer may represent a viable route to replace the existing chemical synthetic routes to commercial vitamin A. The best (though still not commercially operated) mesophilic microbial process for carotene production employed the fungus *Blakeslea trispora* in a highly complex medium which included not only starch, glucose, and various nutritive meals but also various inducers and permeability-modifying agents (viz., β-ionone, isoazid, and kerosene) (128). This process, tested at the 320 l scale, produced mycelium-associated β-carotene at an effective concentration of 3000 mg/l broth and a yield of 30 mg/g added carbohydrate. An alternative microbial route might employ the thermophilic aerobe *Thermus aquaticus*. Up to 80 percent of the total cell lipid in this organism (or 16 percent of cell dry weight) has been reported to be α-carotene (129). Despite the poor growth yield (0.1–0.2 g cells/g substrate) and incomplete substrate utilization reported (130), significant yields of α-carotene are obtainable. In view of the ability of this organism to grow on a variety of substrates in a relatively simple medium, attempts at optimizing cell growth and α-carotene production might prove fruitful.

Cinquina (131) has isolated a tocopherol similar to or identical with α-tocopherol (vitamin E) from whole cells or cell membranes of *Desulfotomaculum nigrificans* or from its spent culture medium. This finding is of some interest, as α-tocopherol is normally a eucaryotic product whose presence in procaryotes is not widespread (132).

Krzycki and Zeikus (133) have reported that *Methanobacterium thermoautotrophicum* contained 0.66 nmol corrinoid/mg cell dry weight), a value somewhat lower than the corrinoid content of most mesophilic methanogens. Because mesophilic anaerobes which catabolize methanol (e.g., *Methanosarcina barkeri* or *Butyribacterium methylotrophicum*) contain four- to fivefold higher levels of corrinoids than do methanogens or acetogens which do not catabolize methanol, measurement of corrinoid levels in some recently isolated methanol-fermenting thermophiles (e.g., *Methanosarcina* strain TM-1) (134) would appear to be warranted. However, the high concentrations of corrinoids produced by thermophilic anaerobes may not be sufficient to warrant their commercial use since they form low biomass densities. A more viable approach may involve recovery of corrinoids from thermophilic aerobic methylotrophs, which should produce considerably higher biomass densities. Nonmethylotrophic thermophilic aerobes in general produce only small amounts of corrinoids. For example, Desai and Dhala (135) reported that 67 of 80 thermophilic actinomomycete strains tested produced vitamin B_{12}, but the maximum amount found at 45 to 50°C and pH 7.4 to 7.6 was only 0.25 mg/l.

Amino Acids and Carbohydrates Matsumoto et al. (136) have reported that *Bacillus coagulans* strain B_9-17 grown aerobically in a suitable medium produced moderate amounts of DL-alanine. Optimization of fermentation conditions led to the accumulation of 16 g DL-alanine/l from 43 g glucose/l after 56 hours' incubation at 56°C. Both the yield (37 percent by weight) and productivity (0.29 g/l·h) are well below those of commercial fermentation of other amino acids (137), and the racemic mixture of the product is of relatively low utility. The above data are nevertheless encouraging when one considers that the highly successful commercial production of amino acids by mesophilic fermentations developed only slowly as a result of painstaking effort in strain selection and process development (138). Similar efforts in thermophilic systems may be potentially very rewarding.

Certain thermophilic cyanobacteria, like their mesophilic counterparts, can accumulate soluble carbohydrates intracellularly upon exposure to NaCl-mediated osmotic stress. Both *Mastigocladus laminosus* and *Phormidium inundatum* have been reported to accumulate the disaccharide trehalose (1-O-α-D-glucopyranosyl-α-D-fructofuranoside), although the amounts accumulated were not reported (139).

Antibiotics   Many members of the thermophilic actinomycete group, like their mesophilic counterparts, are capable of synthesizing compounds which display antimicrobial activity (140–152). In general, thermophilic production of these compounds has been limited to *Thermoactinomyces* strains, thermophilic or thermotolerant *Streptomyces* strains, or related organisms of uncertain taxonomic status. Although few of the products have been characterized in detail (Table 3), most of the products appear to be unique to the producing strain (i.e., have not been reported to be produced by mesophilic organisms). Several of the antibiotics display considerable potency, particularly against Gram-positive bacteria and exhibit low toxicity in mammalian test systems (e.g., mice).

It is likely that many more thermophile-derived antibiotics await discovery and characterization. The few reported surveys of antibiotic production by natural isolates of thermophilic bacteria have shown that up to one half of the strains tested displayed some sort of antimicrobial activity, at least in vitro (Table 4).

Other Bioactive Products   Microbial production of compounds with biological activity toward eucaryotic systems represents another potential approach to biotechnological exploitation of nonrecombinant microorganisms. The few (and unfortunately rather incomplete) reports of bioactive products of thermophilic metabolism are thus of some interest.

Lesage (156) described the properties of unidentified factors produced by a *Phormidium* isolated from the hot springs of Daux and Bagnères de Bigorre (T = 50°C) in southern France. When grown at suboptimal temperatures (30–32°C) for 15 days, this strain produced one or more heat-labile extracellular products which significantly aided the healing of infected surface wounds in clinical tests on human subjects. The product(s) also stimulated growth of a variety of cell types, including various animal and plant tissue cultures and unicellular algae. Interestingly, the bioactive material was not produced during growth at the optimum growth temperature (50°C) or if it was produced was inactivated by the heat.

TABLE 3. Antibiotics reportedly produced by thermophilic actinomycetes

Antibiotic	Producing organism	Biological activity	Remarks	References
Thermomycin	*Streptomyces thermophilus* (S form)	Bacteriostatic against *Corynebacterium diptheriae*	Optimal production in 4 days at 60°C; partially purified product nontoxic to mice	140
A25	Unclassified actinomycete	Bactericidal in vitro against *Staphylococcus aureus*	Optimal production at 45–50°C; resembles litmocidin	141
A26	Unclassified actinomycete	Not detailed	Produced at 50–60°C; butanol-soluble yellow pigment	141
Thermoviridin	*Thermoactinomyces vulgaris*	Most active against Gram + bacteria	Produced at 45°C; excreted into medium; partially purified	142
T-12	*Thermoactinomyces* strain	Active against Gram + bacteria; purportedly antineoplastic	Production conditions at 51°C and 70-l scale described; chromatographically distinct products; localized in mycelium	143
M19A	*Streptomyces* MA-568	Active in vitro against *E. coli* and in *Salmonella*-infected mice	Mixture of antibiotics produced at both 28 and 50°C; purified antibiotic partially characterized	144, 145
Granaticin	*Streptomyces* strain TA/30		Also produced by some mesophiles	146
Thermothiocin	*Thermoactinopolyspora* strain	Bactericidal in vitro and in vivo	Sulfur-containing polypeptide; low toxicity toward mice	147
Thermorubin	*Thermoactinomyces antibioticus*	Bactericidal in vitro and in vivo	Production conditions at 51°C and 10-l scale described; excreted into medium; minimum inhibitory concentration, toxicity, and chemical structure determined	148–151

TABLE 4. Summary of reported testing of thermophilic actinomycetes for antibiotic activity

Source of isolates (and ref.)	No. of isolates tested	No. showing antibiotic activity (%)	Remarks
Pamir Mtns., USSR (153)	289	137 (48)	Isolates grown at 55°C; 66% of mesophiles from same source displayed antibiotic activity
Southern China (154)	1100	354 (32)	Four strains active against both Gram + and Gram − bacteria
Worldwide (5 continents) (155)	500	250 (50)	

Daskalyuk and Godzinskii (157) have reported that another thermophilic cyanobacterium, *Plectonema boryanum*, excreted a compound which activated growth of isolated tomato roots and which partially protected roots from γ-irradiation as well as stimulated recovery of roots when added after γ-irradiation.

Omura et al. (158) have reported that a *Thermoactinomyces* strain when grown in complex media produced an alkaloid-like secondary metabolite. Approximately 300 mg of purified product was recovered from 20 l of culture broth after 48 to 72 hours incubation at 45°C. Although neither the structure nor the utility of the compound was described, the data suggest that at least some thermophilic bacteria are capable of producing alkaloids, a class of compounds which include many analgesics, anesthetics, and stimulants.

Structural Components   Thermophiles possess a variety of structural molecules which may be worth exploring for possible end uses. Aerobic acidophilic thermophiles, such as *Sulfolobus* (159), *Thermoplasma* (160), and *Bacillus acidocaldarius* (161,162), contain large amounts of glycolipids. Detailed studies by Langworthy et al. (161) have resulted in the isolation and characterization of one of the chief glycolipids from *B. acidocaldarius* (see Chapter 5, this volume). This compound, a fully saturated pentacyclic hydrocarbon with a six-carbon carbohydrate side chain, is one of a class of pentacyclic trietherpenoids known as hopanoids. Although many bacteria (including certain mesophiles) produce hopanoids, they are found in relatively large amounts in thermophiles, where they have been proposed to confer thermostability to membranes. A report (163) that hopanoid content in *B. acidocaldarius* increased with increasing growth temperature (up to 16 percent of total lipid at pH 3 and 65°C, the highest temperature tested)

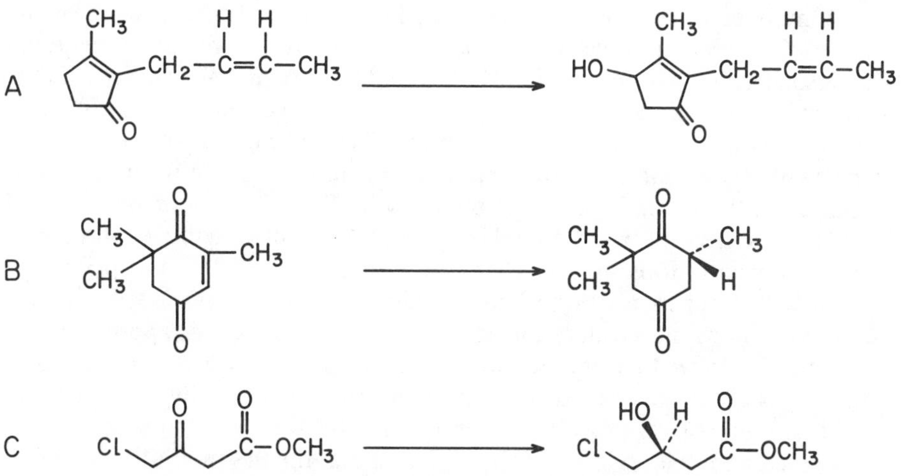

Figure 3. Examples of stereoselective bioconversions performed by thermophilic microorganisms. See text for details of the transformations. (a) Hydroxlylation of cinerone to cinerolone by "thermophilic streptomycete" strains NRRL 3232 and NRRL 3233. Ratio of enantiomeric forms of product varies with strain (158). (b) Reduction of 4-oxoisophorone to 3-R-dihydro-4-oxoisophorone by *Thermoactinomyces* or *Thermomonospora* strains (159). (c) Reduction of 4-chloro-3-oxobutanoate methyl ester to 4-chloro-3S-hydroxybutanoate methyl ester by *Thermoanaerobium brockii*, an example of a ketone or ketoester reduction (162).

is especially interesting in light of a report (164) that a closely related organism, *B. caldolyticus* (where hopanoid content is unknown but presumably significant), has been reported to be capable of growth at 105°C after proper adaptation. Hopanoids are useful in condensing and stabilizing lipid bilayers (e.g., in artificial membranes) (165,166). Other lipids from thermophiles, including novel sulfonolipids (162) and n-1,2- or sn-2,3-diphytanylglycerol dither lipids (167) may also be useful in preparation of membranes with specifically desired properties (see Chapter 5).

Other nonlipid structural components, such as cell wall macromolecules (168) or constituents of outer membranes or S layers (169–171), appear to have unique structures in certain thermophiles and might be explored for novel end uses.

Biotransformation Products Several thermophilic streptomycete strains (viz., NRRL strains 3232 and 3233) have been shown (172) to convert cinerone (2-['-cis-butenyl]-3-methyl-2-cyclopenten-1-one) to cinerolone (the corresponding 4-hydroxy compound), an important precursor in the chemical synthesis of insecticidal pyrethrins and their analogs (Figure 3a). Despite the relatively low yield (~9 percent) obtained with strain NRRL 3233, the greater tolerance of this strain to substrate (2 g/l, versus 0.5 g/l for other strains) resulted in accumulation of 0.18 g cinerolone/l in the 50°C reaction mixture. This concentration is comparable to the highest values obtained for the best fungal or bacterial transformations carried out at 28°C.

A number of thermophilic bacteria, including several members of the genera *Thermoactinomyces* and *Thermomonospora*, as well as unidentified soil isolates, have been reported (173) to reduce 4-oxoisophorone (6R-2,2,6-trimethyl-1,4-cyclohexadione) to the optically active 3R-dihydro-4-oxoisophorone (Figure 3b). The latter compound is an important intermediate in the synthesis of natural chiral xanthophylls and has some utility as a fragrance. Optimal conversion was obtained by *Thermonospora curvata* IFO 12384 in a glycerol-based complex medium. Complete conversion of substrate concentrations up to 1.25 g/l were obtained at 50°C and pH 6.5 to 7.0 in a 17 hour incubation period. Under these conditions, a specific productivity of 86 mg product formed/g cells·h was obtained, approximately four times greater than that reported for analogous mesophilic conversions. Product stability decreased upon prolonged incubation due to further reductive metabolism to yield a racemic mixture of reduced analogues.

Lamed and Zeikus (174,175) have shown that the thermophilic anaerobic non-spore-forming bacterium *Thermoanaerobium brockii* contains a heat-stable, stereoselective secondary alcohol dehydrogenase with a very broad substrate specificity. Experiments with whole cultures of this organism grown at 60°C in complex medium with glucose as the energy source have resulted in reduction of 0.25 percent (w/v) of exogenously added acetone, butanone, or 2-methyl cyclohexanone to the corresponding secondary alcohols. Recent work by Seebach et al. (176) has demonstrated that whole cultures of this organism can also perform stereoselective reduction of other carbonyl compounds (for example, Figure 3c). Cultures grown in complex media at 72°C reduced exogenously added 3-ketoesters, as well as the ethyl ester of 2-formylpropionate, with varying degrees of stereoselectivity. Single charges of 20 to 100 mmol of carbonyl substrate in the 3 1 continuous reactor (dilution rate, 0.3–0.5/h) were converted in 50 to 80 percent yield to the corresponding hydroxyester with enantiomeric excesses of up to 89 percent. Of considerable significance was the observation that some of the reactions proceeded in a stereochemical sense opposite that observed in *Saccharomyces cerevesiae*, the organism most commonly used for stereoselective reduction of carbonyl compounds.

Whole cells of some *Bacillus* species have been shown to transform certain carbohydrates in one step to desirable products. For example, *B. coagulans* performs a conversion of D-glucose to D-fructose (177), and *B. stearothermophilus* strains used for commercial production of β-galactosidase carry out lactose hydrolysis, though not as efficiently as that of the isolated enzyme (178).

The existence of potentially useful thermophilic biochemical reaction types may also be inferred from biodegradation studies in natural systems. For example, Kaplan and Kaplan (179) have developed an aerobic biotransformation pathway for 2,4,6-trinitrotoluene (TNT) in composting environments incubated at 55°C, based on isolation and identification of pathway intermediates produced from ^{14}C-labeled TNT substrate. Among the proposed reaction types were reduction of nitroaryl compounds via aryl hydroxylamines to arylamines and condensation of aryl hydroxylamines to form azoxy compounds. Although it is not clear whether these transformations were being carried out by any of the several organisms isolated from the composting mixture (*Bacillus coagulans, B. stearothermophilus,* or *B. subtilis,* among others) or by uncultured organisms, the mere existence of some of these reaction types suggests that they might be exploited for production of specialty chemicals.

5. OUTLOOK AND RESEARCH NEEDS

Commercial exploitation of thermophilic bioprocesses will require that they demonstrate clear economic advantages over both chemical (nonbiological) synthetic routes and analogous mesophilic bioprocesses. On this basis, it is unlikely that thermophilic fermentations to acetic or lactic acid are worth intensive development, unless novel organisms are discovered and employed.

In the case of ethanol fermentation, thermophilic bioprocesses show some promise, but considerable improvement in several areas must be obtained: among these are enhanced cellulolytic rate (e.g., via higher cellulase titers), improved product yield (e.g., via naturally isolated or genetically engineered organisms employing a decarboxyclastic pyruvate metabolism), and improved product concentration. As noted above, some improvements in these areas have already been attained; further improvements are doubtless obtainable, perhaps by use of genetic engineering techniques which are now being applied to thermophiles (see Chapters 6, 7, and 8). Genetic engineering may also be used to improve competing mesophilic processes (for example, the development of cellulolytic or pentose-fermenting yeast). Product concentration constraints may be mitigated somewhat if novel recovery procedures are discovered (e.g., those in which recovery efficiency is a linear rather than exponential function of product concentration). This is especially critical for thermophilic fermentations, since the highest reported ethanol concentrations (30 g/l) are well below the 100 to 200 g/l observed in conventional yeast fermentations (7).

Utilization of cellulosic biomass for fermentative production of any chemical will require development of an efficient pretreatment process which permits recycling of pretreatment reactant and generates a fermentable product devoid of reactant residue. Taken together, the above problems represent a challenging research area in which true advancements will require multidisciplinary research approaches and will (unfortunately) be strongly stimulated only by another consumer-level shortage in petroleum supplies.

Among the specialty chemicals, commercial exploitation of thermophilic bioprocesses may show promise if novel thermophile-derived materials can establish unique functional niches. While many thermophile-derived compounds with novel structures certainly await discovery, they must demonstrate marketplace superiority to mesophile-derived bioproducts. The most likely candidates in this regard are compounds which confer or support the thermophilic lifestyle, such as novel structural components. A search for novel end-uses for such compounds may prove very rewarding.

Identification of potentially useful reactions performed by thermophiles would undoubtedly be aided by the use of rapid screening methods for specific metabolic products. Such techniques, which permit testing of a large number of organisms and enrichment cultures, have been employed for identifying overproducers of anabolic (180) and catabolic (181) products. The development of microtiter dish systems in particular has enormously simplified the handling and analysis of samples. Adaptation of such screening methods for thermophiles or the development of new methods, represents a relatively unexplored field for future research.

The many recent reports of newly isolated thermophilic bacteria suggest that a large number of novel thermophiles await future discovery and isolation. Among these would be organisms which perform reactions known to be mediated by mesophilic organisms but which currently have no known counterparts among the thermophiles, as well as novel reactions not known to occur biologically at any temperature. The former case presents certain incentives for further investigation owing to particular advantages of thermophilic processes in some unit operations (e.g., recovery of volatile products). For instance, acetone-producing thermophiles might be of particular use owing to the low boiling point (56°C) of the acetone/water azeotrope. As for novel reactions not known to occur biologically, detailed examination of microorganisms inhabiting certain unique thermal environments (e.g., the "black smoker" environment (182–184) or even certain surface geothermal features (3) may reveal the existence of such organisms, since these areas may have fostered the development of microflora with novel, unanticipated energy metabolisms. As scientists' interest in thermophily continues to undergo a renaissance, the discovery and exploitation of microbial reactions for products and processes useful to human society can be expected to flourish.

ACKNOWLEDGMENTS

I wish to thank J. G. Zeikus for the use of Figure 2 and for providing preprints of several papers. I also thank Lorraine Savage and Varetta Manlove for typing the manuscript.

REFERENCES

1. M. Shilo, Ed., *Strategies of Life in Extreme Environments*, Dahlem Konferenzen, Verlag Chemie, Weinheim, 1979.

2. D. J. Kushner, Ed., *Microbial Life in Extreme Environments*, Academic Press, New York, 1978.

3. T. D. Brock, *Thermophilic Microorganisms and Life at High Temperatures*, Springer-Verlag, New York, 1978.

4. D. G. Cooney and R. Emerson, *Thermophilic Fungi*, Freeman, San Francisco, 1964.

5. J. L. Stokes, in A. H. Rose and J. S. Harrison, Eds., *The Yeasts*, Vol. 2, Academic Press, London, 1971, p. 120.

6. Agricultural Statistics-1980, U. S. Dept. Agriculture, 1981, p. 419.

7. T. K. Ng, R. M. Busche, C. C. McDonald, and R. W. F. Hardy, *Science* **219**, 733 (1983).

8. R. M. Busche, in *Scientific Conf. Corn Refiners Assoc.*, Vol. 4, St. Charles, IL, 1984.

9. C. E. Dunlap, J. Thomson, and L. C. Chiang, *Amer. Inst. Chem. Eng. Symp. No. 158* **72**, 56 (1976).

10. *1982 Corn Annual*, Corn Refiners Association, Washington, 1982, p. 15.

11. P. L. Farris, in R. L. Whistler, J. N. BeMiller, and E. F. Paschall, Eds., *Starch: Chemistry and Technology*, 2nd Ed., Academic Press, New York, 1984, p. 11.

12. G. Moulin and P. Galzy, *Biotechnol. Genet. Eng. Rev.* **1**, 347 (1984).

13. J. G. Zeikus, *Enz. Microb. Technol.* **1**, 243 (1979).

14. B. Sonnleitner, *Adv. Biochem. Eng. Biotechnol.* **28**, 69 (1984).

15. L. H. Carreira and L. G. Ljungdahl, in D. L. Wise, Ed., *Liquid Fuel Developments*, CRC Press, Boca Raton, FL, 1983, p. 1.

16. W. Gerrard, *The Solubility of Gases and Liquids*, Plenum, New York, 1976.

17. C. L. Cooney and D. W. Wise, *Biotechnol. Bioeng.* **17**, 1119 (1975).

18. H. H. Hyun, J. G. Zeikus, R. Longin, J. Millet, and A. Ryter, *J. Bacteriol.* **156**, 1332 (1983).

19. L. S. Donnelly and F. F. Busta, *Appl. Environ. Microbiol.* **40**, 721 (1980).

20. T. P. Lyons, *Dev. Ind. Microbiol.* **25**, 231 (1984).

21. E. Anderson, *Chem. Eng. News*, Dec. 3, 1984, p. 27.

22. S. C. Trinidade, *Abstr. Int. Conf. on Energy from Biomass*, Brighton, U.K., 1980, p. 1, No. 2.

23. J. Haggin and J. H. Krieger, *Chem. Eng. News*, March 14, 1983, p. 28.

24. *Chem. Eng. News*, May 2, 1983, p. 18.

25. *Chem. Eng. News*, Oct. 24, 1983, p. 13.

26. G. Reed, in G. Reed, Ed., *Prescott and Dunn's Industrial Microbiology*, 4th ed. AVI, Westport, CT, 1982, p. 835.

27. J. G. Zeikus, *Ann. Rev. Microbiol.* **34**, 423 (1980).

28. J. G. Zeikus, A. Ben-Bassat, T. K. Ng, and R. Lamed, in A. Hollander, Ed., *Trends in the Biology of Fermentations for Fuels and Chemicals*, Plenum, New York, 1981.

29. D. I. C. Wang, C. L. Cooney, A. L. Demain, R. F. Gomez, and A. J. Sinskey, U. S. Dept. of Energy Report COO-4198-4 (1977).

30. ———, Report COO-4198-5 (1978).

31. ———, Report COO-4198-6 (1978).

32. ———, Report COO-4198-7 (1978).

33. ———, Report COO-4198-8 (1978).

34. ———, Report COO-4198-9 (1979).

35. ———, Report COO-4198-10 (1979).

36. ———, Report COO-4198-11 (1980).

37. ———, Report COO-4198-12 (1980).

38. ———, Report COO-4198-13 (1981).

39. ———, Report COO-4198-14 (1981).

40. C. L. Cooney, A. L. Demain, A. J. Sinskey, and D. I. C. Wang, U. S. Dept. of Energy Report COO-4198-15 (1982).

41. ———, Report COO-4198-16 (1982).

42. ———, Report COO-4198-17 (1983).

43. G. Avgerinos and D. I. C. Wang, *Ann. Report Ferm. Processes* **4**, 165 (1981).

44. R. H. McBee, *Bacteriol. Rev.* **14**, 51 (1950).
45. P. J. Weimer and J. G. Zeikus, *Appl. Environ. Microbiol.* **33**, 289 (1977).
46. T. K. Ng, P. J. Weimer, and J. G. Zeikus, *Arch. Microbiol.* **114**, 1 (1977).
47. J. G. Zeikus and T. K. Ng, *Ann. Report Ferm. Processes* **5**, 263 (1982).
48. T. V. C. Duong, D. A. Johnson, and A. L. Demain, in A. Wiseman, Ed., *Topics in Enzyme and Fermentation Biotechnology*, Vol. 7, Ellis-Horwood, Chichester, U.K., 1983, p. 156.
49. T. K. Ng and J. G. Zeikus, *Appl. Environ. Microbiol.* **42**, 231 (1979).
50. N. Ait, N. Creuzet, and P. Forget, *J. Gen. Microbiol.* **113**, 399 (1979).
51. A. Shinmyo, D.V. Garcia-Martinez, and A. L. Demain, *J. Appl. Biochem.* **1**, 202 (1979).
52. T. K. Ng and J. G. Zeikus, *Biochem. J.* **199**, 341 (1981).
53. J. Petre, R. Longin, and J. Millet, *Biochimie* **63**, 629 (1981).
54. E. A. Johnson and A. L. Demain, *Arch. Microbiol.* **137**, 135 (1984).
55. T. K. Ng and J. G. Zeikus, *Anal. Biochem.* **103**, 42 (1980).
56. P. J. Weimer and W. M. Weston, *Biotechnol. Bioeng.* **27**, 1540 (1985).
57. N. Ait, N. Creuzet, and J. Cattaneo, *J. Gen. Microbiol.* **128**, 569 (1982).
58. E. A. Bayer, R. Kenig, and R. Lamed, *J. Bacteriol.* **156**, 818 (1983).
59. R. Lamed, E. Setter, and E. A. Bayer, *J. Bacteriol.* **156**, 828 (1983).
60. J. K. Alexander, *J. Biol. Chem.* **243**, 2899 (1968).
61. K. Sheth and J. K. Alexander, *Biochem. Biophys. Acta* **148**, 808 (1967).
62. T. K. Ng and J. G. Zeikus, *J. Bacteriol.* **150**, 1391 (1982).
63. N. J. Patni and J. K. Alexander, *J. Bacteriol.* **105**, 220 (1971).
64. N. J. Patni and J. K. Alexander, *J. Bacteriol.* **105**, 226 (1971).
65. T. K. Ng, personal communication.
66. J. G. Vidrine and L. Y. Quinn, *Bacteriol. Proc.* **69**, 135 (1969).
67. W. S. Park and D. D. Y. Ryu, *J. Ferm. Technol.* **61**, 563 (1983).
68. R. Lamed and J. G. Zeikus, *J. Bacteriol.* **144**, 569 (1980).
69. J. Gordon, M. Jiminez, C. L. Cooney, and D. I. C. Wang, *Amer. Inst. Chem. Eng. Symp. No. 181* **74**, 91 (1978).
70. A. A. Herrero and R. F. Gomez, *Appl. Environ. Microbiol.* **40**, 571 (1980).
71. S. Kundu, T. K. Ghose, and S. N. Mukhopadhyay, *Biotechnol. Bioeng.* **25**, 1109 (1983).
72. L. G. Ljungdahl, L. H. Carreira, and J. Wiegel, *Int. Symp. Wood and Pulp Chem.*, Stockholm **4**, 23 (1981).
73. T. K. Ng, A. Ben-Bassat, and J. G. Zeikus, *Appl. Environ. Microbiol.* **41**, 1337 (1981).
74. L. S. McClung, *J. Bacteriol.* **29**, 189 (1935).
75. H. Klaushofer and E. Parkkien, *Zeitschrift for Zuckerindustrie* **15**, 445 (1965).
76. J. Wiegel, L. G. Ljungdahl, and J. R. Rawson, *J. Bacteriol.* **139**, 800 (1979).
77. B. Schink and J. G. Zeikus, *J. Gen. Microbiol.* **129**, 1149 (1983).
78. J. G. Zeikus, P. W. Hegge, and M. A. Anderson, *Arch. Microbiol.* **122**, 41 (1979).
79. J. Wiegel and L. G. Ljungdahl, *Arch. Microbiol.* **128**, 343 (1981).
80. A. Ben-Bassat and J. G. Zeikus, *Arch. Microbiol.* **128**, 365 (1981).
81. P. J. Weimer, L. W. Wagner, S. Knowlton, and T. K. Ng, *Arch. Microbiol.* **138**, 31 (1984).
82. A. Ben-Bassat, R. Lamed, and J. G. Zeikus, *J. Bacteriol.* **146**, 192 (1981).
83. R. Lamed and J. G. Zeikus, *J. Bacteriol.* **141**, 1251 (1980).
84. J. G. Zeikus, A. Ben-Bassat, and P. W. Hegge, *J. Bacteriol.* **143**, 432 (1980).
85. J. Wiegel, L. H. Carreira, C. P. Mothershed, and J. Puls, *Biotechnol. Bioeng. Symp.* **13**, 193 (1984).
86. J. Wiegel and J. Puls, in A. Strub, P. Cartier, and G. Schleser, Eds., *Energy from Biomass*, 2nd E. C. Conference, Berlin, 1982. Applied Science Publishers, London, 1983, p. 994.
87. L. Leighton, M. Himmel, N. Burris, and T. Ng, Absract, Ann. Mtg. Amer. Soc. Microbiol. I18 (1982).
88. M. Himmel, L. Leighton, J. Janssens, R. Askeland, and K. Grohmann, *Abstr. 183rd Mtg. Amer. Chem. Soc.*, Las Vegas, 1982.
89. P. J. Weimer, *Arch. Microbiol.* **143**, 130 (1985).

90. L. H. Carreira, J. Wiegel, and L. G. Ljungdahl, *Biotechnol. Bioeng. Symp. 13*, 183 (1984).
91. L. G. Ljungdahl and L. H. Carreira, U. S. Patent 4,385,117 (1983).
92. E. A. Johnson, A. Madia, and A. L. Demain, *Appl. Environ. Microbiol.* **41**, 1060 (1981).
93. S. D. Wang, M. S. Thesis, Massachusetts Institute of Technology (1979).
94. W. R. Vieth and K. Venkatsubramanian, in K. Venkatsubramanian, Ed., *Immobilized Microbial Cells*, Am. Chem. Soc. Symp. Ser. 106, 1979, p. 1.
95. H. H. Hyun and J. G. Zeikus, *Appl. Environ. Microbiol.* **49**, 1174 (1985).
96. H. H. Hyun, G.-J. Shen, and J. G. Zeikus, *J. Bacteriol.* **164**, 1153 (1985).
97. M. A. Payton, *Trends Biotechnol.* **2**, 153 (1984).
98. G. B. Nickol, in H. J. Peppler and D. Perlman, Eds., *Microbial Technology*, Vol. II, 2nd ed., Academic Press, New York, 1979, p. 155.
99. R. M. Busche, E. J. Shimshick, and R. A. Yates, *Biotech. Bioeng. Symp. 12*, 249 (1982).
100. R. A. Yates, U. S. Patent 4,282,323 (1981).
101. E. J. Shimshick, *Chem. Tech.* **13**, 374 (1983).
102. L. J. Ljungdahl, in D. L. Wise, Ed., *Organic Chemicals from Biomass*, Benjamin-Cummings, Menlo Park, 1983, p. 219.
103. F. E. Fontaine, W. H. Peterson, E. McCoy, M. J. Johnson, and G. J. Ritter, *J. Bacteriol.* **43**, 701 (1971).
104. R. Kerby and J. G. Zeikus, *Curr. Microbiol.* **8**, 27 (1983).
105. J. Wiegel, M. Braun, and G. Gottschalk, *Curr. Microbiol.* **5**, 255 (1981).
106. J. A. Leigh, F. Meyer, and R. S. Wolfe, *Arch. Microbiol.* **129**, 275 (1981).
107. P. F. Levy, G. W. Barnard, D. V. Garcia-Martinez, J. E. Sanderson, and D. L. Wise, *Biotechnol. Bioeng.* **23**, 2293 (1981).
108. J. G. Zeikus, in D. L. Wise, Ed., *Organic Chemicals from Biomass*, Benjamin-Cummings, Menlo Park, CA, 1983, p. 359.
109. H. L. Drake, *J. Bacteriol.* **150**, 702 (1982).
110. R. Kellum and H. L. Drake, *J. Bacteriol.* **160**, 466 (1984).
111. J. U. Winter and R. S. Wolfe, *Arch. Microbiol.* **124**, 73 (1980).
112. R. C. Weast, *Handbook of Chemistry and Physics*, 50th ed., Chemical Rubber Co., Cleveland, 1969, p. D-120.
113. R. D. Schwartz and F. A. Keller, Jr., *Appl. Environ. Microbiol.* **43**, 117 (1982).
114. R. D. Schwartz and F. A. Keller, Jr., U. S. Patent 4,371,619 (1983).
115. J. J. Baronofsky, W. J. A. Schreurs, and E. R. Kashket, *Appl. Environ. Microbiol.* **44**, 1134 (1984).
116. F. R. Olson, W. H. Peterson, and E. C. Sherrard, *Ind. Eng. Chem.* **29**, 1026 (1937).
117. C. J. Hajny, C. H. Gardner, and G. J. Ritter, *Ind. Eng. Chem.* **43**, 1384 (1951).
118. P. LeRuyet, H. C. Dubourgier, and G. Albagnnc, *Appl. Environ. Microbiol.* **48**, 893 (1984).
119. L. M. Miall, in A. H. Rose, Ed., *Economic Microbiology*, Vol. 2: *Primary Products of Metabolism*, Academic Press, London, 1978.
120. E. S. Lipinsky, *Science* **212**, 1465 (1981).
121. F. Kargi, *Adv. Biotechnol. Processes* **3**, 245 (1984).
122. C. W. Hinman, *Science* **225**, 1445 (1984).
123. T. Kachholz and H.-J. Rehm, *Eur. J. Appl. Microbiol.* **4**, 10 (1977).
124. T. Kachholz and H.-J. Rehm, *Eur. J. Appl. Microbiol.* **6**, 39 (1978).
125. R. E. Buchanan and N. E. Gibbons, Eds., *Bergey's Manual of Determinative Bacteriology*, Williams and Wilkins, Baltimore, 1974.
126. K. A. Zarilla and J. J. Perry, *Arch. Microbiol.* **137**, 286 (1984).
127. W. Crueger and A. Crueger, *Biotechnology: A Textbook of Industrial Microbiology*, Engl. Ed. Transl. by T. D. Brock, Sinauer Associates, Sunderland, MA, 1984, p. 192-194.
128. L. Ninet and J. Renaut, in H. J. Peppler and D. Perlman, Eds., *Microbial Technology*, Vol. I, 2nd ed., Academic Press, New York, 1979, p. 529.
129. U. J. Heinen, G. Klein, H. P. Klein, and W. Heinen, *Arch. Mikrobiol.* **76**, 18 (1971).
130. S. Cometta, B. Sonnleitner, and A. Fiechter, *Eur. J. Appl. Microbiol. Biotechnol.* **15**, 69 (1982).

131. C. L. Cinquina, *J. Bacteriol.* **95**, 2436 (1968).
132. J. Green, S. A. Price, and L. Gare, *Nature* **184**, 1339 (1959).
133. J. Krzycki and J. G. Zeikus, *Curr. Microbiol.* **3**, 243 (1980).
134. S. H. Zinder and R. A. Mah, *Appl. Environ. Microbiol.* **38**, 996 (1981).
135. A. J. Desai and S. A. Dhala, *Indian J. Microbiol.* **8**, 41 (1968).
136. T. Matsumoto, T. Yano, and K. Yamada, *Agr. Biol. Chem.* **31**, 1381 (1967).
137. J. C. Johnson, *Amino Acids Technology: Recent Developments*, Noyes Data Corp., Park Ridge, NJ, 1978.
138. K. Yamada, S. Kinoshita, T. Tsunoda, and K. Aida, *The Microbial Production of Amino Acids*, Wiley, New York, 1972.
139. R. H. Reed, D. L. Richardson, S. R. C. Warr, and W. D. P. Stewart, *J. Gen. Microbiol.* **130**, 1 (1984).
140. R. Schone, *Antibiotics and Chemotherapy* **1**, 176 (1951).
141. K. Aiso, T. Arai, T. Hashimoto, I. Shidara, and K. Ogi, *Rept. Inst. Putrefaction Res. Chiba Univ.*, Japan, **5**, 100 (1952).
142. D. M. Schuurmans, B. H. Olson, and C. L. San Clemente, *Appl. Microbiol.* **4**, 61 (1956).
143. A. E. Kosmachev, *Microbiology* (USSR) Eng. Transl. **31**, 52 (1961).
144. B. M. Miller, A. Baretto, Jr., and H. B. Woodruff, in M. Finland and G. M. Savage, Eds., *Proc. Amer. Soc. Microbiol. Conf. Antimicrobial Agents Chemother.*, Detroit, MI, 1961, p. 445.
145. I. Putter and F. J. Wolf, in M. Finland and G. M. Savage, Eds., *Proc. Amer. Soc. Microbiol. Conf. Antimicrob. Agents Chemother.*, Detroit, MI, 1961, p. 453.
146. R. Craveri and H. Pagani, *Ann. Microbiol. Enzymol.* **12**, 131 (1962).
147. C. Coronelli, R. Craveri, H. Pagani, P. Sensi, and G. Tamoni, *Ann. Microbiol.* **13**, 125 (1963).
148. R. Craveri, C. Coronelli, H. Pagani, and P. Sensi, *Clin. Med.* **71**, 511 (1964).
149. R. Craveri, et al., Ger. Patent 1,180,891 (1964).
150. G. Pirali, S. Somma, G. C. Lancini, and F. Sala, *Biochem. Biophys. Acta* **366**, 310 (1974).
151. C. E. Moppett, D. T. Dix, F. Johnson, and C. Coronelli, *J. Amer. Chem. Soc.* **94**, 3269 (1972).
152. A. B. Silaev, N. L. Sokolova, and N. S. Agre, *Biol. Nauk.* (USSR) **10**, 115 (1967).
153. N. S. Agre and V. K. Orelanskii, *Microbiology* (USSR) Eng. Transl. **31**, 75 (1962).
154. E. S. Kudrina and T. S. Maksimova, *Microbiology* USSR Eng. Transl. **32**, 532 (1963).
155. R. Craveri, *Abst. 8th Int. Cong. Microbiol.*, Montreal, 1962, No. B.15.8, p. 60; through *Biol. Abstr.* **41**, 15956 (1963).
156. A. Lesage, *Bull. Acad. Vet.* **18**, 272 (1945); through M. Lefevre, in D. F. Jackson, Ed., *Algae and Man*, Plenum, New York, 1964, p. 337.
157. A. P. Daskalyuk and D. M. Godzinskii, *Fiziol. Biokhim. Kul't. Rast.* **4**, 619 (1972), through *Chem. Abstr.* **56**, 57164 (1973).
158. S. Omura, Y. Suzuki, C. Kitao, Y. Takahashi, and Y. Konda, *J. Antibiotics* (Japan) **28**, 609 (1975).
159. T. A. Langworthy, P. F. Smith, and W. R. Mayberry, *J. Bacteriol.* **112**, 1193 (1972).
160. T. A. Langworthy, W. R. Mayberry, and P. F. Smith, *J. Bacteriol.* **119**, 106 (1974).
161. T. A. Langworthy and W. R. Mayberry, *Biochem. Biophys. Acta* **431**, 570 (1976).
162. T. A. Langworthy and W. R. Mayberry, *Biochem. Biophys. Acta* **431**, 550 (1976).
163. K. Poralla, T. Hartner, and E. Kannenberg, *FEMS Microbiol. Lett.* **23**, 253 (1984).
164. W. Heinen and A. M. Lauwers, *Arch. Microbiol.* **129**, 127 (1981).
165. K. Poralla, E. Kannenberg, and A. Blume, *FEBS Lett.* **113**, 107 (1980).
166. P. Bisseret, G. Wolff, A-M. Albrecht, T. Tanaka, Y. Nakatani, and G. Ourisson, *Biochem. Biophys. Res. Commun.* **110**, 320 (1983).
167. T. A. Langworthy, G. Holzer, J. G. Zeikus, and T. G. Tornabene, *System. Appl. Microbiol.* **4**, 1 (1983).
168. I. T. Forrester and A. J. Wicken, *Biochem. Biophys. Res. Commun.* **25**, 23 (1966).
169. U. B. Sleytr and P. Messner, *Ann. Rev. Microbiol.* **37**, 311 (1983).

170. Z. Kupcu, L. Marz, P. Messner, and U. B. Sleytr, *FEBS Lett.* **173**, 185 (1984).
171. M. Rasch, W. O. Saxton, and W. Baumeister, *FEMS Microbiol. Lett.* **24**, 285 (1984).
172. B. Tabenkin, R. A. LaMahieu, J. Berger, and R. W. Kierstead, *Appl. Microbiol.* **17**, 714 (1969).
173. N. Hori, T. Hieda, and Y. Makami, *Agr. Biol. Chem.* **48**, 123 (1984).
174. R. Lamed and J. G. Zeikus, *Biochem. J.* **195**, 183 (1981).
175. R. Lamed and J. G. Zeikus, U. S. Patent 4,352,885 (1982).
176. D. Seebach, M. F. Zueger, F. Giovannini, B. Sonnleitner, and A. Fiechter, *Angew. Chem. Int. Ed. Engl.* **23**, 151 (1984).
177. J. F. T. Spencer and P. A. J. Gorin, *Prog. Ind. Microbiol.* **7**, 177 (1965).
178. M. W. Griffiths, D. D. Muir, and J. D. Phillips, U. K. Patent Applic. 2,022,595 (1979).
179. D. L. Kaplan and A. M. Kaplan, *Appl. Environ. Microbiol.* **44**, 757 (1982).
180. F. Kavanaugh, Ed., *Analytical Microbiol.*, Vol. 2, Academic Press, New York (1972).
181. B. Stieglitz and P. J. Weimer, *Appl. Environ. Microbiol.* **49**, 593 (1985).
182. East Pacific RISE Group, *Science* **207**, 1421 (1980).
183. H. W. Jannasch and D. C. Nelson, in M. J. Klug and C. A. Reddy, Eds., *Current Perspectives in Microbial Ecology*, Amer. Soc. Microbiol., Washington, D.C., 1984, p. 170.
184. J. A. Baross, J. W. Deming, and R. P. Becker, in M. J. Klug and C. A. Reddy, Eds., *Current Perspectives in Microbial Ecology*, Amer. Soc. Microbiol., Washington, D.C., 1984, p. 186.

11

THERMOPHILIC WASTE TREATMENT SYSTEMS

STEPHEN H. ZINDER

Department of Microbiology, Stocking Hall, Cornell University, Ithaca, New York 14853 USA

1. INTRODUCTION

A wide variety of organic wastes, including municipal, agricultural, and industrial wastes, can be treated using microbiological processes. Among

257

the goals of biological waste treatment processes are the stabilization of that waste in a cost- and energy-efficient manner, and, if possible, the production of a useful product such as methane, or a sludge which can be used as fertilizer. Typically, biological waste treatment occurs in the mesophilic range, 20–40°C. However, there are waste treatment processes available which operate in the thermophilic range, generally 45–65°C. The advantages of thermophilic processes include potentially greater reaction rates, and the destruction of pathogenic organisms. In this chapter, three waste treatment processes involving thermophiles are discussed: thermophilic anaerobic digestion, composting, and thermophilic activated sludge. The basic properties of these processes and the microorganisms involved will be discussed, as will be the advantages and disadvantages relative to mesophilic systems. The most attention will be paid to thermophilic anaerobic digestion, since there has been the greatest amount of recent research done on the microbiology of that process.

2. THERMOPHILIC ANAEROBIC DIGESTION

2.1. General Description of Anaerobic Digestion

The goals of anaerobic digestion of organic wastes include the conversion of as much of that waste as possible to methane and the production of a solid residue which is stabilized such that it no longer contains significant amounts of readily degradable organic material. Compared with aerobic processes, the major advantages of anaerobic digestion are: 1) most of the free energy present in the organic matter is conserved as methane, which can be collected and used for energy production; 2) there is no requirement for aeration, an energy-intensive process; 3) because anaerobes conserve less energy per unit substrate consumed than do aerobes, less cellular biomass is produced meaning that less sludge needs to be disposed of. Anaerobic digestion is used on the greatest scale to treat the particulate fraction of sewage (called sludge), and nearly every sewage treatment plant in the United States has an anaerobic bioreactor. Other particulate wastes such as animal manures, crop wastes, and the organic fraction of municipal refuse, are also being treated to some extent by anaerobic digestion. It was once considered that anaerobes were rather limited in the diversity of organic substrates that they were capable of degrading. It is now appreciated that many aromatic compounds, halogenated compounds, and other xenobiotics are biodegradable under anaerobic conditions (1). This new knowledge, coupled with new bioreactor configurations which facilitate the treatment of dilute soluble wastes, now makes the treatment of a variety of industrial wastes feasible.

Because the bacteria which produce methane are capable of catabolizing only a limited number of one-carbon substrates and acetate (1), the bioconversion of complex organic matter to methane requires cooperative in-

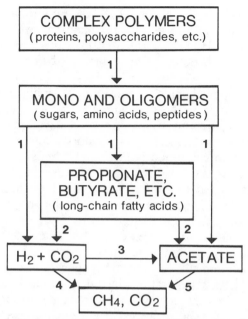

Figure 1. Microbial groups involved in the flow of carbon from polymeric materials to methane in an anaerobic bioreactor. Group 1, fermentative bacteria; group 2, hydrogen-producing acetogenic bacteria; group 3, hydrogen-consuming acetogenic bacteria; group 4, CO_2-reducing methanogens; group 5, aceticlastic methanogens. From (1), reprinted with permission.

teractions among a variety of nonmethanogenic and methanogenic bacteria. Figure 1 shows a typical model for the interactions of five major microbial groups involved in methanogenesis from a polymeric substrate. The model and others like it are described in greater detail elsewhere (1–3). The fermentative bacteria (Group 1) are a diverse group responsible for the production of hydrolytic enzymes, such as cellulases, amylases, and proteases, which break down polymers to soluble products. These monomers and oligomers are then fermented to volatile organic acids, H_2 and CO_2. Fatty acids longer than acetate (including long chain acids derived from cell lipids), and certain aromatic rings such as benzoic acid (4), are metabolized to acetate, H_2, and CO_2 by the hydrogen-producing acetogenic bacteria (Group 2). These Group 2 reactions are generally only thermodynamically favorable if H_2 partial pressures are maintained below 10^{-3} atm by hydrogen-consuming bacteria such as methanogens or sulfate reducers (2,5). Presently, cultures of fatty acid-oxidizing acetogens have been obtained only in mixed culture with hydrogen-consuming anaerobes. The hydrogen-consuming acetogenic bacteria carry out the reaction: $4H_2 + 2HCO_3^- + H^+ \rightarrow CH_3COO^- + 2H_2O$, and can compete for H_2 with methanogens, but have not been found to play a major role in the anaerobic bioreactors (6). It is generally accepted that the CO_2-consuming methanogens, which carry out the reaction

$4H_2 + HCO_3^- + H^+ \rightarrow CH_4 + 3H_2O$, account for about one-third of the methane produced, while aceticlastic methanogens, which carry out the reaction $CH_3COO^- + H_2O \rightarrow CH_4 + HCO_3^-$, produce the remaining two-thirds (1). Some methane can also be produced from formate, methanol, and methylamines.

The bacteria responsible for the consumption of higher organic acids, the aceticlastic methanogens, and the hydrogen-producing acetogens, generally grow slowly, with doubling times of several days typical for mesophilic cultures (1). Thus, most anaerobic bioreactors, which are tank reactors fed continuously or semicontinuously (e.g., daily), must be operated with retention times greater than ten days to prevent washout of these slow growing organisms. It is critical to bioreactor stability that these acid-consuming organisms are able to keep pace with acid production in order to prevent pH drop and reactor failure.

New reactor configurations are designed to uncouple biomass retention from the hydraulic retention times, thereby preventing washout of slow-growing organisms and allowing for higher concentrations of active biomass in the reactor. These configurations include the anaerobic contact process, in which sludge solids are recycled back into the reactor in a manner similar to the aerobic activated sludge process. There are now several reactor designs which employ thin films of microorganisms attached to solid substrates. These include the anaerobic filter, in which organisms are attached to rocks or stationary plastic matrices, and upflow fluidized or expanded bed reactors, in which the microbial films are attached to small particles such as alumina or diatomaceous earth. The liquid flow by these particles, provided by a recycle pump, is rapid enough to keep them suspended, thereby preventing clogging. Upflow sludge blanket reactors involve the formation of macro-scopic "granules" of microorganisms which are retained by the system. These attached-film systems allow hydraulic retention times of less than a day, greater bioreactor stability, and are suitable for the treatment of dilute soluble wastes. Tank reactors are really only suitable for treatment of high-strength particulate wastes such as sewage sludges or manures. A more complete description of these reactor configurations and their properties can be found in recent reviews (7,8).

2.2. Thermophilic Anaerobic Systems

Generally, conventional anaerobic bioreactors are operated in the meso-philic range of 35–40°C, but it has been known for over 50 years (9) that bioreactors can also be operated in the thermophilic range, 45–65°C. Full-scale application of the thermophilic process has been sporadic and until recently confined to sewage sludge or animal manures as substrates (Table 1). However, the successful treatment of pulp mill wastes containing chlor-inated aromatic compounds by a pilot-scale thermophilic reactor has been reported (12), indicating that it is likely that a much wider range of substrates

TABLE 1. Selected full-scale and pilot-scale thermophilic anaerobic bioreactors, and some of their characteristics

Location	Substrate	Reactor size (m³)	Temperature range (°C)	Comments	Reference
Los Angeles, CA	Sewage sludge	9460	46–51	Temperatures greater than 51°C caused rapid increases in volatile acids	10
Chicago, IL	Sewage sludge	8900	52–55	Retention time could be decreased from 14 (mesophilic) to 7 days, doubling capacity of reactor	11
Helsinki, Finland	Pulp mill waste	"Pilot scale"	50–60	Fluidized bed reactor, wastes were 40–60°C, improved BOD and COD reduction over mesophilic	12
Urbana, IL	Beef cattle manure	0.775	58–60	Significantly improved gas production rate and yield versus mesophilic	13
Urbana, IL	Municipal refuse	0.4	57–60	Temperatures greater than 60°C adversely affected process	14
Haifa, Israel	Dairy cattle manure	1–10	55	Reactor at 55°C performed better than 35°C reactor at low retention times (6 days)	15

All reactors use stirred tanks unless otherwise stated.

can be treated using a thermophilic process. The Hyperion sewage treatment plant in the Los Angeles area (10) is the only full-scale sewage treatment plant in the United States routinely using thermophilic digestion at this time. It should also be mentioned that landfills, another full-scale waste treatment process, are frequently anoxic, producing large amounts of methane, and can reach temperatures around 50°C as a result of self heating (16).

Methanogenesis in laboratory-scale reactors has been studied for a much wider variety of substrates and reactor configurations. Table 2 provides a partial list of laboratory-scale thermophilic systems, showing the diversity of substrates, temperatures, and reactor configurations studied. It has only recently been demonstrated that thermophilic sludge blanket reactors (17), which require formation of macroscopic granular microbial consortia, and attached film systems such as the expanded bed (18,29), are feasible.

From these and other studies, some generalizations about the advantages and disadvantages of thermophilic anaerobic digestion relative to the mesophilic process can be made. Table 3 presents a list of some of these, adapted from a 1977 review of thermophilic anaerobic digestion by Buhr and Andrews (9). The essential advantages are killing of pathogens and decreased retention times, while the primary disadvantage is the increased energy input into a thermophilic system.

Decreased retention time is possible in thermophilic processes because chemical and biochemical reactions can proceed more rapidly at higher temperature. This does not necessarily mean that thermophiles grow more rapidly than corresponding mesophiles. However, the growth rate of many anaerobes can often be limited by the rate of catabolism of energy-yielding substrates, rather than the speed of other cellular processes such as DNA synthesis. Thus, thermophilic anaerobes may be capable of carrying out rate-limiting metabolic reactions more rapidly than comparable mesophiles, and thereby grow faster. For example, *Methanobacterium thermoautotrophicum* has a minimum doubling time near 1 hour at 65°C (31), while that for the mesophilic *Methanobacterium formicicum* is near 8 hours at 35°C (32). Mesophilic *Methanosarcina* cultures have a minimum doubling time on acetate of 24 to 36 hours at 35°C (33), while the thermophilic *Methanosarcina* sp. strain TM-1 can double in 12 hours at 50°C (34).

The greatest advantage of thermophilic digestion is that the retention times can be decreased (11,13,20,21). It is likely that the more rapid growth of thermophilic acid-consuming anaerobes prevents washout of these organisms at shorter retention times. Stable digestion in stirred reactors with retention times as short as 3 days at 60°C have been reported (20), while about 10 days is considered to be the lower limit for the mesophilic process. Indeed, if digestion efficiency is compared between mesophilic and thermophilic reactors with retention times greater than 10 days, often little difference is seen in their overall digestion efficiency (6). The advantages of being able to operate a bioreactor at shorter retention times is illustrated

TABLE 2. Selected laboratory-scale thermophilic bioreactors studied

Substrate	Temperature range (°C)	Reactor configuration	Reactor size (liters)	Comments	Reference
Municipal refuse	35–60	Stirred tank	15	Extensive study on temperature effects on digestion parameters	19
Beef cattle manure	55–65	Stirred bottle	3	Greatly lowered efficiency at 65°C, stable digestion with 3-day retention time at 60°C	20
Beef cattle waste	60	Stirred fermentor	3	Extensive comparison between 40 and 60°C reactors of turnover of ^{14}C-labeled substrates	6
Beef cattle waste	55	Stirred bottle	3	Reactor at 55°C could be operated at three times the loading rate of a reactor at 37°C	21
Municipal refuse	58	Stirred fermentor	3	Methanogenesis from acetate was inhibited at 65°C when *Methanosarcina* was predominant, but not with *Methanothrix*	22,23
Sewage sludge	52, 60	Stirred bottle	NS[a]	Higher volatile acids, decreased efficiency versus mesophilic	24
Chicken manure	40–60	Filter flask	1	Optimal methanogenesis at 50°C	25
Municipal refuse	65	Stirred carboy	50	Greater methanogenesis rates than at 37°C	26
Palm oil mill effluent	55	Stirred flasks	2	Efficient BOD reduction at retention times as short as 5 days	27
Paper mill condensate	60	Stirred fermentor	10	Soluble waste containing acetic acid, furfural and sulfite, *Methanosarcina* dominant	28
Sucrose	55	Expanded bed	2	Hydraulic retention times as short as 1 hour possible at low loading rates	18
Cellulose	55	Expanded bed	2	Could accept much higher loads of cellulose without physical overload than corresponding mesophilic unit	29
Sucrose, fatty acids	NS[a]	Sludge blanket	NS[a]	Stable methanogenesis at high loading rates	17
Sewage sludge	58	Stirred bottle	3	200 mg/l Ni or 300 mg/l Cd or Cu completely inhibited methanogenesis	30

[a]NS, not stated

263

TABLE 3. **Advantages and disadvantages of thermophilic anaerobic digestion compared to the mesophilic process**

Advantages:
   1) Increased reaction rates and decreased retention time
   2) Increased destruction of pathogenic microorganisms
   3) Lower viscosity: less mixing energy and easier sludge dewatering

Disadvantages:
   1) High energy requirement for heating
   2) Reports of poor supernatant quality
   3) Reports of poor process stability

After Buhr and Andrews (9)

by the study by Rimkus et al. (11) of a full-scale sewage sludge digestor operating in the Chicago area. They found that they could decrease the retention time from 14 days to 7 days with no loss in efficiency when they increased the operating temperature from 37 to 52–55°C. Since they were able to feed sludge of equivalent strength at twice the rate, the capacity of the bioreactor had been doubled, representing a major savings in the capital costs of constructing a new reactor.

Another major advantage of thermophilic digestion is that pathogenic organisms including bacteria, viruses, and parasites are killed. The resultant sludge is essentially pasteurized by the exposure to heat. For example, Berg and Berman (35), in their studies of the Hyperion sewage treatment plant, found that fecal coliform counts dropped 10- to 100-fold in mesophilic sludge, when compared to the raw sludge influent, but dropped 10^4 to 10^6-fold in thermophilic (50°C) sludge. Virus counts dropped about 10-fold in mesophilic sludge, but over 100-fold, often to undetectable levels, in the thermophilic sludge. They suggested that killing would be even greater in thermophilic digestors if the sewage plant operators, during the daily feeding, drew the digestor down for discharge and then fed raw sludge. The current practice is to feed before drawdown, thereby allowing some pathogens to reach the digestor effluent without extensive exposure to heat.

The lower viscosity of water at higher temperatures means that less energy has to be expended for mixing of stirred bioreactors. More importantly, solid-liquid separation of the sludge effluent by filtration or centrifugation can be more efficient (10). This is an important consideration in the treatment of high-strength particulate wastes, because a final sludge with low water content is desirable for disposal.

The primary disadvantage of thermophilic anaerobic digestion is the greater energy input needed in the form of heat. This is not always a disadvantage because some industrial wastes are already heated, such as pulp mill (12), distillery, and certain food-processing wastes. Also, many industries have waste heat which can be used to heat the reactor. High-strength wastes, such as sewage sludge and manures, can produce more methane than is needed for heating the reactor. Rimkus et al. (11) calculated that their full-scale thermophilic bioreactor produced more methane than was required to heat it, and that the energy balance would be even more favorable if heat exchangers were used to partially heat the influent using the effluent. Thus, the cost of increased energy input must be weighed against the potential advantages of the thermophilic process to determine whether that process is to be economically feasible.

Apparently, thermophilic sludges, such as those at the Hyperion sewage treatment plant, can have higher levels of organic acids and other soluble materials than corresponding mesophilic sludges (10,24), meaning that their effluents may need further treatment before discharge into the environment. It is not clear whether this phenomenon is widespread. Mackie and Bryant (6), for example, found that a 60°C laboratory-scale reactor operating with a 10 day retention time had lower fatty acid levels than did a corresponding 40°C reactor.

Reports of greater instability for the thermophilic process process are generally related to temperature shifts. As would be expected, downward temperature shifts are generally considered less harmful than upward ones. Our experience with a 58° laboratory-scale reactor using a cellulosic municipal refuse as a substrate (22) is that occasional 1 to 2 day interruptions in heating, during which the reactor temperature fell to room temperature, caused methanogenesis to completely stop; after restoration of heating, the reactor commenced methanogenesis immediately as though nothing had occurred. It is possible that reactors which receive a high inoculum of mesophilic fermentative anaerobes with their feed, such as in manures and sewage, would commence an unbalanced fermentation at low temperatures which could cause pH drop and failure.

Upward temperature shifts can be fatal to microbial populations in the reactor. This can be especially troublesome if the populations killed include acid-consuming anaerobes, since they are slow to recover, and if acid production outstrips its consumption, reactor failure can occur. Garber et al. (10) found rapid increases in organic acid concentrations if the temperature of a full-scale reactor at the Hyperion sewage treatment plant exceeded 52°C, although many other reactors are successfully operated at higher temperatures. In general, they considered thermophilic reactor stability to be comparable to that in mesophilic reactors. We have found that increasing the temperature of our 58°C laboratory-scale reactor to 64° for one day caused a rapid increase in acetic acid production and a corresponding pH drop (23). About 10 days were required for acetate to decrease to normal levels. The dominant aceticlastic population in the reactor at that time consisted of *Methanosarcina* which was sensitive to the temperature increase. We also concluded that it would be prudent to operate thermophilic bioreactors near 55°C, rather than near 60°C, so that they would be more tolerant of slight upward temperature shifts. In general, it is advisable to operate thermophilic reactors with back-up temperature control systems which prevent drastic upward shifts.

2.3. Microbial Populations in Thermophilic Bioreactors

In general, our understanding of the microbial populations in anaerobic bioreactors is rudimentary. The development of easier techniques for the culture of stringent anaerobes (36) and the relatively recent culturing of two important microbial groups, the aceticlastic methanogens (37–39) and fatty acid-oxidizing acetogens (40,41), have significantly improved our understanding of the microbiology of bioconversion of organic wastes to methane and has led to the model of digestion presented in Figure 1. In general, it has been found that thermophilic anaerobic digestion follows essentially the same pattern (6,22), in that about two-thirds or more of the methane is derived from acetate and the remainder is from the CO_2 reduction. Some thermophilic members of the microbial groups outlined in Figure 1 have been isolated from thermophilic bioreactors and other environments and characterized. A partial listing of isolated thermophilic anaerobes is presented in Table 4.

Fermentative bacteria usually give the highest counts of the different microbial groups in thermophilic anaerobic bioreactors. For example, Mackie and Bryant (6), using roll tubes containing a trypticase-digestor fluid medium, found 1.3×10^{10} colony forming units (CFU)/ml of sludge from a 60°C cattle waste digestor. Varel (42), using roll tubes with a sugar-trypticase-digestor fluid medium, found 3.5×10^9 CFU/ml in a 55°C cattle waste reactor, representing about 10% of the total microscopic counts. We have determined most probable number (MPN) values in a 58°C municipal refuse reactor (22) using a sugar-yeast extract-digestor fluid medium and have found 2×10^9/ml which represented about 40% of the total microscopic counts.

The only published systematic study on the fermentative bacteria in a thermophilic reactor is that of Varel (42), who examined microbial populations in a 55°C beef cattle waste reactor. He picked 104 isolated colonies found in roll tubes of nonselective medium. Of these, 64 grew well enough to partially characterize. He divided the 64 isolates into five groups on the basis of fermentation products, hydrogen sulfide production, and morphology. All of the isolates were Gram-negative pleomorphic rods, although one group formed spores. Common fermentation products of these groups were acetate, ethanol, lactate, and formate. None of the isolates used cellulose or xylan. Not enough information was available to classify the isolates, and it is likely that some of them represented new taxonomic groups. Essentially nothing is known about thermophilic anaerobes responsible for proteolysis and amino acid fermentation.

Cellulose is an important constituent of many agricultural and municipal wastes. Varel (42) found that MPNs of cellulolytic bacteria were about two

TABLE 4. Characteristics of some thermophiles possibly useful in methanogenesis of organic wastes

Organism name	Growth temp. (°C)	Growth substrates	Major products	Habitats found	Reference
Clostridium thermocellum	40–68	Cellulose, hexoses	Acetate, ethanol, lactate, H_2, CO_2	Soils, sewage	43
Clostridium stercorarium	45–70	Cellulose, hexoses, pentoses	Acetate, ethanol, lactate, H_2, CO_2	Compost pile	44
Clostridium thermohydrosulfuricum	40–78	Sugars	Ethanol, lactate, acetate, H_2, CO_2	Hot springs, soils, sewage	45
Thermoanaerobium brockii	40–80	Sugars	Ethanol, acetate, lactate, H_2, CO_2	Hot springs	46
Thermoanaerobacter ethanolicus	40–75	Sugars	Ethanol, acetate, H_2, CO_2	Hot springs	47
Clostridium thermoaceticum	45–65	Sugars, H_2-CO_2, CO	Acetate	Soils	48
Acetogenium kivui	50–70	Sugars, pyruvate, H_2-CO_2	Acetate	Lake Kivu sediments	49
Methanobacterium thermoautotrophicum	45–75	H_2-CO_2	CH_4	Hot springs, sewage	50,51
Methanococcus thermolithotrophicus	30–70	H_2-CO_2, formate	CH_4 (CO_2)	Marine hot springs	52
Methanococcus jannaschii	50–86	H_2-CO_2	CH_4	Deep sea vents	53
Methanogenium thermophilicum	37–65	H_2-CO_2, formate	CH_4 (CO_2)	Marine	54
Methanogenium frittonii	26–62	H_2-CO_2, formate	CH_4 (CO_2)	Freshwater	55
Methanothermus fervidus	65–97	H_2-CO_2	CH_4	Hot springs	56
Methanosarcina sp. strain TM-1	<37–57	Acetate, methanol, methylamines	CH_4, CO_2	Thermophilic sludge	34
Methanothrix sp. strain CALS-1	40–65	Acetate	CH_4, CO_2	Thermophilic sludge	22,57

orders of magnitude lower than total fermentative bacteria, an observation similar to results for a municipal refuse reactor (22). This is surprising considering that both wastes were high in cellulose and other fibrous polymers. At least a partial explanation is that numerous bacteria, apparently cellulolytic, have been found attached to fibrous particles in the reactor (22). In thin-section electron micrographs (S. Zinder, unpublished) the cells were attached by a fibrous matrix, had a Gram-positive type cell wall, and were similar in appearance to *Clostridium thermocellum* (43), the most commonly isolated thermophilic cellulolytic anaerobe. Another thermophilic cellulolytic clostridial species has been described, *C. stercorarium* (44), which was reported to use a greater number of sugars than *C. thermocellum*, including xylose and lactose, and to be capable of growth at temperatures up to 70°C.

There has been little work done on H_2-producing acetogens responsible for fatty acid oxidation in anaerobic thermophilic reactors. Mackie and Bryant (6) demonstrated that ^{14}C-labeled butyrate and propionate were broken down in a 60°C cattle waste reactor. Zinder et al. (25) found that 5 mM butyrate or isobutyrate added to sludge from a 58°C munipal refuse reactor was metabolized within 24 hours, with the concomitant accumulation of acetate, while added valeric acid increased propionate accumulation. Added propionate was metabolized much more slowly, with 96 hours required for complete degradation of an addition of 5 mM.

The only report of isolation of a thermophilic fatty acid-oxidizing acetogen was that of Henson and Smith (57a), who gave a description of a thermophilic butyrate-oxidizing acetogen in coculture with *Methanobacterium thermoautotrophicum*. The bacterium was a nonmotile curved rod which stained Gram-negative and grew at 55°C. Methanogenic enrichment cultures of butyrate-, isobutyrate-, and propionate-oxidizing acetogens have been established which have doubling times for methanogenesis near 24 and 120 hours, respectively, at 60°C (22). The doubling time for the butyrate-oxidizing enrichment was much less than that for the mesophilic *Syntrophomonas wolfii* methanogenic coculture (40), while that for the propionate enrichment is similar to that in the mesophilic *Syntrophobacter wolinii* (41) coculture. Another group of thermophilic H_2-producing acetogens which may be interesting to culture would be those capable of metabolizing aromatic ring compounds, since a thermophile may grow faster than the mesophilic aromatic-oxidizing *Syntrophus buswellii*, which had a doubling time of 130 hours (4).

The only determination of the extent of acetate synthesis from carbon dioxide in a thermophilic reactor was that of Mackie and Bryant (6), who, on the basis of [14]CO_2 incorporation into acetate, calculated that 3 to 4% of the acetate formed arose via reduction of carbon dioxide. The isolation of a thermophilic homoacetogen, *Clostridium thermoaceticum*, was described in 1942 (48), and it was recently demonstrated (58) that this culture was capable of growth using H_2-CO_2 or carbon monoxide as sole energy source.

The thermophilic homoacetogen, *Acetogenium kivui*, a Gram-positive non-sporeforming rod capable of growth on H_2-CO_2, has also been described (49).

Counts of CO_2-reducing methanogens in thermophilic reactors are typically 10^8–10^9/ml (6,22). The first thermophilic CO_2-reducing methanogen isolated was *Methanobacterium thermoautotrophicum* (50), which was enriched from a mesophilic reactor. The culture, a Gram-positive crooked rod to filament, grew on H_2-CO_2 in a completely mineral growth medium. Growth was optimal at temperatures near 65°C, and the upper limit was around 75°C. Several other strains of thermophilic *Methanobacterium* have been isolated with properties similar to the original strain, although some can use formate in addition to H_2-CO_2 (59). We have found that the *Methanobacterium* morphotype was the only one detected in colonies in H_2-CO_2 roll tubes derived from a 58°C reactor (22). Most methanogens have high concentrations of the fluorescent cofactor F_{420} and autofluoresce when viewed with an epifluorescence microscope. When sludge from a thermophilic reactor was viewed using an epifluorescence microscope, the only morphotype consistently seen was that of *Methanobacterium* (22), and numbers of these fluorescing bacteria were comparable with viable counts of CO_2-reducing methanogens (S. Zinder, unpublished).

Other thermophilic CO_2-reducing methanogenic genera described include marine (54) and freshwater (55) strains of *Methanogenium* sp. A coccoid methanogen isolated from a thermophilic kelp digestor by Ferguson and Mah (60) was not classified, but was likely to be a *Methanogenium* sp. The marine methanogens *Methanococcus thermolithotrophicus* (52) and *Methanococcus jannaschii* (53), are capable of growing in mineral media with doubling times of 55 and 26 minutes, respectively. *Methanothermus fervidus* (56), a member of the Methanobacteriales, is capable of growth at temperatures approaching the boiling point of water (see Chapter 3).

Experiments involving the addition of $^{14}CH_3COO^-$ or $CH_3^{14}COO^-$ demonstrated that acetate can be metabolized by the aceticlastic reaction in a thermophilic reactor (22), and acetate can account for 60 to 86% of the methane produced (6,22). The first thermophilic aceticlastic methanogen isolated was *Methanosarcina* sp. strain TM-1 (34), which forms large irregular packets, and grows optimally on acetate at 50°C with a doubling time of 12 hours, more than twice as fast as the fastest-growing mesophilic strain (33). It is capable of growth on methanol, methylamines, and H_2-CO_2. It can grow in a mineral acetate medium supplemented with the vitamin para-aminobenzoic acid (61). Recently, Sowers et al. (62), using nucleic acid hybridization, demonstrated that strain TM-1 represents a new species, and the species epithet *Methanosarcina thermophila* (63) has been proposed. We have also isolated *Methanosarcina* sp. strain CALS-1 which has similar properties to strain TM-1 except that it grows optimally at 55°C, and can grow at 60°C (23). Sludge from various thermophilic reactors, including the full-scale Hyperion sewage treatment plant (34), which have been examined using epifluorescence microscopy often show numerous autofluorescent *Methanosarcina*-like aggregates (12,22,28,34). It is of interest that the Hyperion reactors, the source of *Methanosarcina* sp. strain TM-1, which grows optimally at 50°C, were reported to accumulate acetate if the temperature was increased past 52°C (10).

We have recently isolated *Methanothrix* sp. strain CALS-1 from our 58°C municipal refuse reactor (22,57). *Methanothrix* is believed to play an important role in methanogenesis in many mesophilic anaerobic reactors (8). Like its mesophilic counterpart (39), strain CALS-1 forms sheathed, septate filaments and grows only on acetate. It grows optimally at 60°C with a doubling time of 24 hours, considerably more rapid growth than that reported for the mesophilic culture. It can grow at temperatures up to 70°C, and contains gas vacuoles. Initially, *Methanosarcina* sp. was the dominant aceticlastic methanogen, but was eventually displaced by *Methanothrix* (22). Startup conditions thus may have favored the *Methanosarcina*, which was capable of rapid growth and had the relatively high apparent K_m value for acetate near 5 mM (28,34). The *Methanothrix* sp., which had an apparent K_m below 1 mM (22,39), presumably was eventually able to outcompete the *Methanosarcina* for acetate. Little is known about the relative importance of *Methanothrix* in thermophilic reactors.

We have also isolated a thermophilic two-membered culture in which one member oxidizes acetate to CO_2 and H_2, which serve as methanogenic substrates for the other member, a CO_2-reducing *Methanobacterium* (59).

In summary, representatives of each of the five microbial groups presented in Figure 1 have been isolated from thermophilic anaerobic reactors. While much remains to be learned about representatives from all the groups, we know little about the diversity of fermentative bacteria in thermophilic anaerobic reactors, and there has been only slight characterization of thermophilic fatty acid-oxidizing acetogens.

2.4. Startup of Thermophilic Anaerobic Bioreactors

Anaerobic bioreactors are artificial ecosystems which obviously must obtain organisms from other habitats. This problem is less acute for mesophilic systems, since diverse mesophilic anaerobes are abundant in environments such as muds, soils, and rumens. It is not clear that thermophilic representatives of all the microorganisms needed for a balanced conversion of wastes to methane are abundant in temperate environments. Many thermophiles capable of growth at 50–60°C can be found in temperate habitats such as soil (64). Evidence indicates that at least certain thermophilic anaerobes, such as *Clostridium thermocellum* (43) and *Methanobacterium thermoautotrophicum* (50), are quite cosmopolitan. Varel (42) reported counts in cattle manure of fermentative bacteria capable of growth at 55°C of about 10^6 CFU/ml. Chen (65) counted various groups of fermentative and methanogenic bacteria capable of growth at 35, 50, and 60°C in mesophilic anaerobic sewage sludge and concluded that there were only low numbers of cellulolytic and aceticlastic bacteria capable of growth at the highest temperature. Fatty acid-oxidizing acetogens were not enumerated.

There are two basic methods of starting up a thermophilic reactor. One is to simply start it at the desired temperature and then take precautions to maintain neutral pH values until a balanced fermentation is achieved. The other is to slowly increase the temperature, usually at 1°C per day or less, from mesophilic conditions until the desired temperature is reached. An example of the former method is the study of Varel et al. (20), who found that 55°C cattle waste reactors stabilized within 8 days when started up with an inoculum from a 55°C municipal refuse reactor, while reactors started up using cattle manure, mesophilic sewage sludge, or rumen fluid all stabilized within 12 days. The reactors with the mesophilic inocula required periodic additions of alkali to maintain neutral pH until stabilization occurred. Engineers tend to favor the method for startup which uses a slow temperature increase (9–11). This conservatism is understandable when one considers the possibility of a 10,000 liter reactor going into complete failure.

2.5. An Upper Temperature Limit for Thermophilic Anaerobic Digestion?

The upper temperature limit for thermophilic anaerobic digestion is probably of more theoretical interest than of practical significance except for the possible treatment of certain high-temperature industrial wastes. Several studies have indicated that there is a significant decrease in methanogenesis and an increase in fatty acid levels when the temperature of an anaerobic bioreactor is increased above 60°C (19,20,66). This indicates that the aceticlastic methanogens and the fatty acid-oxidizing acetogens, the predominant microbial populations present in the bioreactor responsible for acid consumption, were not well adapted to growth at temperatures greater than 60°C. We have not been able to isolate cultures of *Methanosarcina* capable of growth at temperatures greater than 60°C. However, we found that *Methanothrix* sp. strain CALS-1 can grow well at 65°C. When this organism was apparently predominant in the bioreactor, sludge incubated at 65°C for 24 hours in serum vials did not accumulate acetate, and these vials actually produced more methane than those incubated at 60°C (23). In fact, although methanogenesis was greatly inhibited in vials incubated at 70°C, there was no accumulation of H_2 or acetate, suggesting that the fermentative anaerobes were the group most inhibited by this temperature. Such short-term experiments indicate that efficient conversion to methane is feasible at 65°C if *Methanothrix* is predominant, but this can only be proven by long-term trials and may not be true for wastes from which larger amounts of fatty acids longer than acetate are produced.

The conversion of a waste to methane in a reactor operating at 65°C or higher would require at least three organisms: a fermentative bacterium capable of using the primary substrate, a CO_2-reducing methanogen, and an aceticlastic methanogen. Fatty acid-oxidizing acetogens (and possibly acetogens capable of oxidizing ethanol, lactate, etc.) may also be needed for efficient conversion to methane. If cellulose is a primary constituent of the waste, the upper temperature limit for a cellulose-digesting bacterium presently known is 70°C for *Cl. stercorium* (44). However, a soluble sugar waste could be metabolized at temperatures up to 80°C by *Thermoanaerobium brockii* (46). There are several CO_2-reducing methanogens, including *Methanobacterium thermoautotrophicum*, which are capable of growth at 70°C or higher. In nonmarine systems, *Methanothermus fervidus* can reduce CO_2 at temperatures up to 97°C, while in marine systems, *Methanococcus jannaschii* can grow up to about 90°C. Presently, the upper temperature limit for rapid methanogenesis from acetate is about 65°C for *Methanothrix* sp. strain CALS-1. It is possible that aceticlastic methanogens capable of growth at higher temperatures could be isolated from hot springs, but so far attempts at isolation of any aceticlastic methanogens from these habitats have failed (51; S. Zinder, unpublished). Nothing is known about the temperature ranges for thermophilic fatty acid-oxidizing acetogens. Thus, the present upper limit for methanogenesis from a fermentable substrate is near 65°C until aceticlastic methanogens capable of growth at higher temperatures are isolated.

3. COMPOSTING

3.1. General Description of Composting

Composting is an aerobic process for the treatment of moist solid organic wastes. During composting, self-heating of the waste occurs due to exothermic oxidative biochemical and chemical reactions. Composting has traditionally been used to treat a variety of agricultural wastes, including crop wastes and manures (68). Composting more recently has been proposed as a method for the treatment of sewage sludges (69). Among the described advantages of composting are: 1) destruction of pathogens by heat; 2) mass and bulk reduction, which can be 50 to 75%, making the waste more suitable for landfilling; 3) the production of a stabilized humuslike product with a low C/N ratio suitable for use as a fertilizer (68).

Composting can be done as a batch process, in which the waste is formed into individual large piles which go through a heating and cooling cycle over several days. Temperatures generally reach 60–70°C, although temperatures greater than 80°C have been reported (68). Generally, these piles are turned over occasionally to maintain aerobiosis, or aeration is applied through aeration holes under the waste. The optimal surface area-volume ratio for the pile, moisture content of the waste, extent of aeration, and time for completion of the process are determined empirically for the given waste and climate condition. There are also semicontinuous plug flow processes in which the incoming waste is essentially instantaneously heated by the waste already present in the reactor (68). Retention times are typically five days. Reviews covering the process requirements and microbiological aspects of composting can be consulted (68,69).

3.2. Microbiological Aspects of Composting

It is relatively clear that most of the self heating is due to microbial activity, although some chemical reactions may become important, especially at higher temperatures. As the self heating occurs, one would expect a microbial succession from mesophilic to thermophilic populations. In general, evidence has supported this. For example, calorimetric measurements during the course of microbial self heating of wool have revealed two temperature optima for heat production, one near 40°C, and a second broader one near 60°C (68). Suler and Finstein (70) in a laboratory-scale study of a continuous system using food scraps and newspaper as a substrate, found optimal CO_2evolution and dry weight reduction at 56–60°C, indicating that a thermophilic population had established itself.

Recently, Nakasaki et al. (71,72) studied changes of microbial populations in a laboratory-scale sewage sludge composting unit. In compost which was seeded only with sewage sludge, the temperature increased to 60°C within 48 hours, and slowly declined to 30°C within 120 hours of starting. Plate counts of mesophilic bacteria peaked at 10^{10} CFU/gm dry wt and dropped to 10^9 CFU/gm dry wt during the heated phase. Numbers of thermophilic bacteria increased from 10^6 to 5×10^9 CFU/gm dry wt, and thermophilic actinomycetes increased from 10^4 to 5×10^7 CFU/gm dry wt. In compost samples which had been seeded with spent compost at a 24% seed/sludge ratio, heating was more rapid, with 60°C being reached within 24 hours. The relative increase in thermophilic populations was not as dramatic (from 10^7 to 10^9 CFU/gm dry wt), since greater numbers were present initially. The mesophilic bacteria isolated during the high-temperature phase were mainly Gram-positive endospore-forming rods which were only slowly killed by temperatures of 60°C, explaining the large numbers of mesophilic bacteria surviving the heated phase (72).

In contrast to these findings, a recent report by McKinley and Vestal (73) presented evidence that thermophilic microbial populations were not established in full-scale sewage sludge compost piles being operated in Ohio. They examined mineralization of ^{14}C-labeled glutamate to CO_2 and the incorporation of ^{14}C-acetate into cellular lipids. Their results indicated that the temperature optima for these reactions, even in areas of the pile that had self heated to 55–74°C, were always less than 55°C, and these activities were severely depressed at temperatures of 60°C and higher. It is possible that these methods were not applicable for determining the activity of thermophiles present in the compost piles.

Microorganisms cultured from various compost systems include fungi, actinomycetes, and other bacteria. Not surprisingly, the composition of the waste and the conditions of the process can greatly affect the nature of the microbiota found (69). Fungi have an upper temperature limit of 60°C (64) and apparently can play an important role in the earlier stages of composting, especially if the moisture content is low, since many fungi can grow well at low water potentials. Among the fungi cultured from compost heaps are *Aspergillus fumigatus*, *Cladosporium*, *Chaetomium thermophile*, and *Geotrichum candidum* (68). Generally counted among the actinomycetes are *Thermomonospora* and *Thermoactinomyces* (68), but recent phylogenetic evidence based on 16S ribosomal RNA catalogue sequencing (74) indicates that *Thermoactinomyces*, which forms a heat-stable dipicolinic acid-containing endospore, is actually more closely related to the members of the genus *Bacillus* than to the actinomycetes. Other bacteria isolated from compost piles at high temperature include *Bacillus stearothermophilus* and *Bacillus coagulans*, as well as other bacilli (68). In general, our understanding of the microorganisms involved in the high-temperature phases of composting clearly could use improvement.

4. THERMOPHILIC AEROBIC ACTIVATED SLUDGE

A thermophilic variation of the activated-sludge process has been operated at a pilot-plant scale (75) and at temperatures of 50–67°C. Efficiencies of organic matter decomposition were similar to the mesophilic process (75). The primary advantage of the thermophilic process is that it kills pathogens in the sludge, and it is claimed to do so in a cost-effective manner compared to other sanitation processes such as chlorination (75). There has been limited study of the microbiology of such processes. Sonnleiter and Fiechter (75) counted bacteria using plate-count agar and similar growth media, and found that during the operation of the plant at temperatures of 50–67°C, numbers of bacteria capable of growth at 50°C varied between 10^5 to 5×10^9 CFU/ml and those capable of growth at 80°C varied between 10^2 to 10^5 CFU/ml. When they examined bacteria isolated from the reactor, they found that 95% of them were in the genus *Bacillus* and most of those fit the description of *Bacillus stearothermophilus* (76). It is likely that a wider range of bacteria would have been cultured if a growth medium with a lower concentration of nutrients was used for counts and isolation (64).

5. SUMMARY AND CONCLUSIONS

Most thermophilic processes for waste treatment are in the development stage and need to prove themselves in the marketplace as practical and economic solutions for waste disposal. Thermophilic anaerobic digestion has been used in full- and pilot-scale applications to some extent for treatment of manures and sewage sludge, but it remains to be seen whether its advantages will outweigh the disadvantage of increased energy input, and whether it will develop the record for reliability that is required for a novel technology before it is widely accepted. Thermophilic anaerobic treatment of various industrial and food processing wastes has been demonstrated in the laboratory, and future research will indicate whether such processes will be viable on a full scale.

Composting differs from other thermophilic processes in that it generates its own heat via self heating and therefore does not require energy input in the form of heating. The process is limited to solid wastes such as agricultural wastes and sewage sludge. Thermophilic activated sludge treatment has the potential to rapidly disinfect sewage sludge while stabilizing its organic matter, but it is likely that such problems as increased odor and decreased solubility of oxygen at high temperature will limit its use.

An increased understanding of the microorganisms involved in thermophilic processes cannot help but increase our overall understanding of the processes themselves and their control. This understanding should help the processes become better controlled and can lead to their becoming more widely accepted. Further study of the microbiology of these processes also will undoubtedly lead to the isolation and characterization of scientifically interesting and industrially useful microorganisms.

ACKNOWLEDGMENTS

The author's research on thermophilic anaerobic digestion has been supported by contract no. DE-AC02-81ER10872 from the U. S. Department of Energy.

REFERENCES

1. S. H. Zinder, *ASM News* **50**, 294 (1984).
2. M. J. McInerney and M. P. Bryant, in D. L. Wise, Ed., *Fuel Gas Production from Biomass*, Chemical Rubber Co., West Palm Beach, Florida, 1981, p. 26.
3. D. B. Archer, *Enzyme Microb. Technol.* **5**, 162 (1983).
4. D. M. Mountfort and M. P. Bryant, *Arch. Microbiol.* **133**, 249 (1982).
5. M. J. Wolin and T. L. Miller, *ASM News* **48**, 561 (1982).
6. R. I. Mackie and M P. Bryant, *Appl. Environ. Microbiol.* **41**, 1363 (1981).
7. M. S. Switzenbaum, *ASM News* **49**, 532 (1983).
8. L. van den Berg, *Can. J. Microbiol.* **30**, 975 (1984).
9. H. O. Buhr and J. F. Andrews, *Water Res.* **11**, 129 (1977).
10. W. F. Garber, G. T. O'Hara, J. E. Colbaugh, and S. K. Ranskiti, *J. Wat. Poll. Cont. Fed.* **47**, 950 (1975).
11. R. R. Rimkus, J. M. Ryan, and E. J. Cook, *J. Wat. Poll. Cont. Fed.* **54**, 1447 (1982).

12. M. Salkinoja-Salonen, R. Valo, and J. Apajalahti, in *Third International Symposium on Anaerobic Digestion: Proceedings,* Third International Symposium on Anaerobic Digestion, Boston, 1983, p. 107.
13. J. T. Pfeffer and G. E. Quindry, Biological Conversion of Biomass to Methane: Beef Lot Manure Studies. US DOE Report UILU-ENG-78-2011 (1978).
14. J. W. Brown, J. T. Pfeffer, and J. C. Liebman, *Biological Conversion of Organic Refuse to Methane,* NSF/RANN Rept. UILU-ENG-76-2019 (1976).
15. G. Shelef, S. Kimchie, and J. Grynberg, *Biotechnol. Bioeng. Symp.* **10,** 341 (1980).
16. K. L. Jones, J. F. Rees, and J. M. Grainger, *Eur. J. Appl. Microbiol. Biotechnol.* **18,** 957 (1983).
17. G. Lettinga, in *Third International Symposium on Anaerobic Digestion: Proceedings,* Third International Symposium on Anaerobic Digestion, Boston, 1983, p. 139.
18. G. Schraa and W. J. Jewell, *J. Wat. Poll. Cont. Fed.* **56,** 226 (1984).
19. J. T. Pfeffer, *Biotechnol. Bioeng.* **16,** 771 (1974).
20. V. H. Varel, H. R. Isaacson, and M. P. Bryant, *Appl. Environ. Microbiol.* **33,** 298 (1977).
21. A. G. Hashimoto, *Biotechnol. Bioeng.* **24,** 2039 (1982).
22. S. H. Zinder, S. C. Cardwell, T. Anguish, M. Lee, and M. Koch, *Appl. Environ. Microbiol.* **47,** 796 (1984).
23. S. H. Zinder, T. Anguish, and S. C. Cardwell, *Appl. Environ. Microbiol.* **47,** 808 (1984).
24. F. G. Pohland and D. E. Bloodgood, *J. Wat. Poll. Cont. Fed.* **35,** 11 (1963).
25. J. J. H. Huang and J. C. H. Shih, *Biotechnol. Bioeng.* **23,** 2307 (1981).
26. C. L. Cooney and D. L. Wise, *Biotechnol. Bioeng.* **17,** 1119 (1975).
27. K. K. Chin and K. K. Wong, *Water Res.* **9,** 993 (1983).
28. G. Brune, S. M. Schoberth, and H. Sahm, *Process Biochem.* **17**(3), 10 (1982).
29. W. H. Clarkson, Ph.D. Thesis, Cornell University (1985).
30. B. K. Ahring and P. Westerman, *Eur. J. Appl. Microbiol. Biotechnol.* **17,** 365 (1983).
31. P. Schonheit, J. Moll, and R. K. Thauer, *Arch. Microbiol.* **127,** 59 (1980).
32. H. Schauer and J. G. Ferry, *J. Bacteriol.* **142,** 800 (1980).
33. M. R. Smith and R. A. Mah, *Appl. Environ. Microbiol.* **36,** 870 (1978).
34. S. H. Zinder and R. A. Mah, *Appl. Environ. Microbiol.* **38,** 996 (1979).
35. G. Berg and D. Berman, *Appl. Environ. Microbiol.* **39,** 361 (1980).
36. W. E. Balch, G. E. Fox, L. J. Magrum, C. R. Woese, and R. S. Wolfe, *Microbiol. Rev.* **43,** 260 (1979).
37. R. A. Mah, M. R. Smith, and L. Baresi, *Appl. Environ. Microbiol.* **35,** 1174 (1978).
38. P. J. Weimer and J. G. Zeikus, *Arch. Microbiol.* **119,** 175 (1978).
39. B. A. Huser, K. Wuhrmann, and A. J. B. Zehnder, *Arch. Microbiol.* **132,** 1 (1982).
40. M. J. McInerney, M. P. Bryant, R. B. Hespell, and J. Costerton, *Appl. Environ. Microbiol.* **41,** 1029 (1981).
41. D. R. Boone and M. P. Bryant, *Appl. Environ. Microbiol.* **40,** 626 (1980).
42. V. H. Varel, *Microbial Ecol.* **10,** 15 (1984).
43. T. K. Ng, P. J. Weimer, and J. G. Zeikus, *Arch. Microbiol.* **114,** 1 (1977).
44. R. H. Madden, *Int. J. Syst. Bacteriol.* **33,** 837 (1983).
45. J. Wiegel, L. G. Ljungdahl, and J. R. Rawson, *J. Bacteriol.* **139,** 800 (1979).
46. J. G. Zeikus, P. W. Hegge, M. A. Anderson, *Arch. Microbiol.* **122,** 41 (1979).
47. J. Wiegel and L. G. Ljungdahl, *Arch. Microbiol.* **128,** 343 (1981).
48. F. E. Fontaine, W. H. Peterson, E. McCoy, M. J. Johnson, and G. J. Ritter, *J. Bacteriol.* **43,** 701 (1942).
49. J. A. Leigh, F. Mayer, and R. S. Wolfe, *Arch. Microbiol.* **129,** 275 (1981).
50. J. G. Zeikus and R. S. Wolfe, *J. Bacteriol.* **109,** 707 (1972).
51. J. G. Zeikus, A. Ben-Bassat, and P. W. Hegge, *J. Bacteriol.* **143,** 432 (1980).
52. H. Huber, M. Tomm, H. Konig, G. Thies, and K. O. Stetter, *Arch. Microbiol.* **132,** 47 (1982).
53. W. J. Jones, J. A. Leigh, F. Mayer, C. R. Woese, and R. S. Wolfe, *Arch. Microbiol.* **136,** 254 (1983).

54. C. J. Rivard and P. H. Smith, *Int. J. Syst. Bacteriol.* **32**, 430 (1982).
55. J. E. Harris, P. A. Pinn, and R. P. Davis, *Appl. Environ. Microbiol.* **48**, 1123 (1984).
56. K. O. Stetter, M. Thomm, J. Winter, G. Wildgruber, H. Huber, W. Zillig, D. Janecovic, H. König, P. Palm, and S. Wunderl, *Bakt. Hyg. I. Abt. Orig. C* **2, 166 (1981).**
57a. J. M. Henson and P. H. Smith, *Appl. Environ. Microbiol.* **49**, 1461 (1985).
57. S. H. Zinder and T. Anguish, *Abst. Ann. Mtg. Amer. Soc. Microbiol.*, 126 (1984).
58. R. Kerby and J. G. Zeikus, *Curr. Microbiol.* **8**, 27 (1983).
59. S. H. Zinder and M. Koch, *Arch. Microbiol.* **138**, 263 (1984).
60. T. J. Ferguson and R. A. Mah, *Appl. Environ. Microbiol.* **45**, 265 (1983).
61. P. A. Murray and S. H. Zinder, *Appl. Environ. Microbiol.* **51**, 49 (1985).
62. K. R. Sowers, J. L. Johnson, and J. G. Ferry, *Int. J. Syst. Bacteriol.* **34**, 444 (1984).
63. S. H. Zinder, K. R. Sowers, and J. G. Ferry, *Int. J. Syst. Bacteriol.* **35**, 522 (1985).
64. T. D. Brock, *Thermophilic Microorganisms and Life at High Temperature*, Springer-Verlag, New York, 1978.
65. M. Chen, *Appl. Environ. Microbiol.* **45**, 1271 (1983).
66. V. H. Varel, A. G. Hashimoto, and Y. Chen, *Appl. Environ. Microbiol.* **40**, 217 (1980).
67. Deleted in proof.
68. M. S. Finstein and M. L. Morris, *Adv. Appl. Microbiol.* **19**, 113 (1975).
69. M. S. Finstein, J. Cirello, D. J. Suler, M. L. Morris, and P. F. Strom, *J. Wat. Poll. Cont. Fed.* **52**, 2675 (1980).
70. D. J. Suler and M. S. Finstein, *Appl. Environ. Microbiol.* **33**, 345 (1977).
71. K. Nakasaki, M. Sasaki, M. Shoda, and H. Kubota, *Appl. Environ. Microbiol.* **49**, 37 (1985).
72. K. Nakasaki, M. Sasaki, M. Shoda, and H. Kubota, *Appl. Environ. Microbiol.* **49**, 42 (1985).
73. V. L. McKinley and J. R. Vestal, *Appl. Environ. Microbiol.* **47**, 933 (1984).
74. G. E. Fox, E. Stackebrandt, R. B. Hespell, J. Gibson, J. Maniloff, T. W. Dyer, R. S. Wolfe, W. E. Bälch, R. S. Tanner, L. J. Magrum, L. B. Zablen, R. Blakemore, R. Gupta, L. Bonen, B. J. Lewis, D. A. Stahl, K. R. Luehrsen, K. N. Chen, and C. R. Woese, *Science* **209**, 457 (1980).
75. B. Sonnleitner and A. Fiecter, *Eur. J. Appl. Microbiol. Biotechnol.* **18**, 47 (1983).
76. B. Sonnleitner and A. Fiecter, *Eur. J. Appl. Microbiol. Biotechnol.* **18**, 174 (1983).

12

MICROBIAL MINING USING THERMOPHILIC MICROORGANISMS

JAMES A. BRIERLEY AND CORALE L. BRIERLEY

Advanced Mineral Technologies, Inc., 5920 McIntyre, Golden, Colorado 80403

1. INTRODUCTION

1.1. Metal Leaching

In this chapter thermophilic microorganisms are examined in terms of their application in metal recovery from ore and mineral wastes. This technology, called "bacterial leaching," is a hydrometallurgical process for solubilizing metals and separating them from their mineral matrices. Once solubilized, the metals are concentrated by nonbiological methods. Microbial leaching has been industrially applied to leach copper from mine waste and to extract uranium from ore.

The principal benefits of bacterial leaching are low operating costs and mitigation of air pollution. The lowering of operating costs allows the processing of lower grade ores and the scavenging of residual metal values from mine wastes. Because metals are extracted hydrometallurgically, sulfur dioxide, generated by conventional pyrometallurgical processing, is not produced.

Bacterial leaching technology has primarily been linked to the mesophilic *Thiobacillus* species; however, recent investigations have revealed the presence of thermophilic bacteria in commercial leaching operations. These thermophiles carry out reactions similar to those performed by the thiobacilli. Although the actual contributions of thermophiles to metal extraction in commercial leaching operations are not quantified, it is apparent from laboratory study that these microorganisms can substantially aid in metal extraction from mineral matrices. Because the thermophiles function at elevated temperatures, the kinetics of biological and chemical reactions, both of which are important in leaching operations, are enhanced. Faster reaction rates reduce the time that ores and wastes must be processed to extract metal values.

This chapter presents information about the types of thermophilic bacteria involved in leaching technology, relevant aspects of their physiology, and the role these organisms play in metal extraction from minerals.

1.2. Historical Perspectives

The process of dump and heap leaching was invented and commercially employed in the 1700s to extract copper in Spain. This technology involves the piling of crushed ore onto pads, the application of sulfuric acid to the top surface of the ore pile, and the collection of copper-laden solutions from the base of the ore pile. This technology was introduced into the United States in the 1900s and since that time has gained widespread acceptance worldwide for the treatment of low-grade copper ores.

At times, dump and heap leaching technology, implemented much like it was 200 years ago, has accounted for over 10% of the United States production of copper. This technology has been practiced on an enormous scale with dumps containing millions of tons of waste rock and rising to over 300 m in height. Similar technology has also been employed to leach uranium from pyritic ores, but its use for this metal is not as extensive as that for copper.

Research initiated in the 1950s served to elucidate chemical, physical, and biological activities and interactions in commercial copper and uranium leaching operations. A complex microflora was detected and the vital role of *Thiobacillus* in metal extraction from mineral matrices was confirmed.

Most research on microbial leaching and microbial activities in commercial operations deals with *Thiobacillus ferrooxidans*. These acidogenic, iron-oxidizing bacteria were first reported in 1947 following their discovery in acidic effluents of coal mining operations (1). Subsequent studies revealed that these microbes participate in the solubilization of metals from minerals (2). Recent reviews (3–7) discuss the involvement of *T. ferrooxidans* and other microbes (e.g., *Leptospirillum ferrooxidans, Sulfolobus* species, and facultative, thermophilic, iron-oxidizing bacteria) in the leaching process.

Leach dumps can become heated because of the oxidation of pyrite, which is an exothermic reaction. Because the immense size of these operations prevents dissipation of the heat, the internal temperatures rise; these elevated temperatures may prohibit the growth and activity of *T. ferrooxidans* and other mesophilic bacteria. Beck (8), who observed temperatures of 60 to 80°C in leach dump environments, concluded that one may have "to reconsider the role of microbial activity in leaching operations" where these temperatures occur. Harries and Ritchie (9) also reported that some regions in waste rock dumps at Rum Jungle in Australia's Northern Territory exceed 50°C. A large scale study (10), in which a 1.7×10^5 kg mass of low-grade copper ore was leached in an insulated tank, also demonstrated an increase in temperature; temperatures greater than 59°C were reported. These reports suggest that thermophilic microbes may either participate in generating heat or be used at elevated temperatures for metals solubilization.

2. CHEMISTRY OF THE LEACHING PROCESS

The role of *T. ferrooxidans* and thermophilic bacteria in leaching is complex and not precisely defined. Some research supports the concept that their function may be "indirect," whereby the microbes generate ferric iron, which oxidizes the mineral. Other investigations indicate "direct" leaching in which the microbes contact and adhere to the mineral surface, oxidizing the mineral without the use of the ferric iron oxidant.

The fundamental reaction for indirect leaching is the microbial oxidation of ferrous iron (equation 1) in acid conditions for the purpose of energy generation:

$$4FeSO_4 + O_2 + 2H_2SO_4 \rightarrow 2Fe_2(SO_4)_3 + 2H_2O \qquad (1)$$

The ferric sulfate which is generated serves to oxidize minerals such as chalcopyrite (equation 2), chalcocite (3), covellite (4), and uraninite (5):

$$CuFeS_2 + 2Fe_2(SO_4)_3 \rightarrow CuSO_4 + 5FeSO_4 + 2S \qquad (2)$$

$$Cu_2S + 2Fe_2(SO_4)_3 \rightarrow 2CuSO_4 + 4FeSO_4 + S \qquad (3)$$

$$CuS + Fe_2(SO_4)_3 \rightarrow CuSO_4 + 2FeSO_4 + S \qquad (4)$$

$$UO_2 + Fe_2(SO_4)_3 \rightarrow UO_2SO_4 + 2FeSO_4 \qquad (5)$$

The resulting soluble metal sulfates are recovered by solvent extraction, ion exchange, or other methods. The iron, now reduced to the ferrous state, is reoxidized by the microorganisms according to equation 1. The sulfur, which is often present as an end product of the metal solubilization (equations 2, 3, and 4), may also be oxidized to produce sulfuric acid.

The "direct" mechanism of metal leaching takes place without ferrous iron as an oxidant. Pyrite may be oxidized directly by the microbes, equation 6:

$$FeS_2 + 3\tfrac{1}{2}O_2 + H_2O \rightarrow FeSO_4 + H_2SO_4 \qquad (6)$$

This results in the solubilization of iron. The iron is subsequently oxidized according to equation 1 and the ferric iron then participates in the "indirect" leaching process. Copper-containing minerals can also be leached by the "direct" process (equations 7 and 8).

$$2CuFeS_2 + 8\tfrac{1}{2}O_2 \rightarrow 2CuSO_4 + Fe_2(SO_4)_3 + H_2O \qquad (7)$$

$$2Cu_2S + 2H_2SO_4 + 5O_2 \rightarrow 4CuSO_4 + 2H_2O \qquad (8)$$

The "direct" leaching mechanism has not been conclusively demonstrated for iron-containing minerals such as chalcopyrite (equation 7). Because solubilized iron facilitates the "indirect" mechanisms, even minerals without iron, such as chalcocite (equation 8) are oxidized in part by the "indirect" process. The presence of any iron as a contaminant will initiate an "indirect" leaching reaction.

TABLE 1. Sources of facultative, thermophilic, iron-oxidizing bacteria

Designation	Source description	Reference
TH-1	Thermal spring, Iceland	19
TH-2	Copper-leaching test system, Socorro, New Mexico, USA	22
TH-3	Copper-leaching dump, Chino Mine, Kennecott Copper Corp., Hurley, New Mexico, USA	21
TH-4	Copper-leaching dump, Bingham Mine, Kennecott Copper Corp., Bingham, Utah, USA	Unpubl. (Brierley)
TH-5	Gas vent, Kiluea Volcano, Hawaii, USA	Unpubl. (Brierley)
TH-6	Copper-leaching dump test, Chuquicamata Mine, CODELCO-Chile, Chuquicamata, Chile	Unpubl. (Brierley)
Alvecote	Coal spoil tip, Alvecote, Warwickshire, UK	15
Birch Coppice	Birch Coppice Colliery, Warwickshire, UK	15
Evenwood	Coal-rich spoil heap, Evenwood, County Durham, UK	15
Kingsbury	Coal-rich spoil heap, Kingsbury, Warwickshire, UK	15
Lake Myvam	Thermal spring, Lake Myvam, Iceland	15
Sulfobacillus thermosulfidooxidans	Nikolaev copper-zinc-pyrite ore, Eastern Kazakhstan, USSR	23

Cell suspensions of the TH-1 strain oxidizing ferrous iron show increased respiration over the pH range 1.3 to 3.2 (20). Iron oxidation by these organisms is favored as the pH approaches the level at which iron spontaneously oxidized. At pH 1.6, iron oxidation has a K_m of 7.3 mM $FeSO_4$ and it is competitively inhibited by its end product, ferric iron (24).

Studies of growth on pyrite (20) suggest that moderate thermophiles oxidize metal sulfides as well as ferrous iron. Whereas the TH-1 strain has an apparent maximum temperature for growth on ferrous iron of about 50°C, the maximum temperature for growth on pyrite was determined to be around 55°C. This suggests a difference in the heat stability of the enzyme systems that oxidize these substrates. Pyrite oxidation occurred over a pH range of 1.1 to 3.5 with an optimum for oxidation at 2.6. In contrast, the ferrous iron oxidation rate increased with increasing pH up to 3.5. Table 2 provides a listing of the metal sulfides which support the growth of moderate thermophiles.

Few of the iron-oxidizing, moderately thermophilic microbes oxidize sulfur. The initial investigation (20) of the TH-1 strain indicated that sulfur may be used as an energy source for growth. However, it was subsequently

TABLE 2. Metal sulfide minerals that serve as substrates for growth of the moderately thermophilic iron-oxidizing bacteria

Metal sulfide	Strains examined	Reference
FeS$_2$ (pyrite)	TH-1, TH-3	19,20,24
	Sulfobacillus thermosulfidooxidans	23,25,26,27
(NiFe)$_9$S$_8$ (pentlandite)	TH-1	19
CuFeS$_2$ (chalcopyrite)	TH-1, TH-3	19,24
	Sulfobacillus thermosulfidooxidans	23,27
CuS (amorphous copper sulfide)	TH-1	24
CuS (covellite)	*Sulfobacillus thermosulfidooxidans*	23
FeAsS (arsenopyrite)	*Sulfobacillus thermosulfidooxidans*	23,27
ZnS (sphalerite)	*Sulfobacillus thermosulfidooxidans*	23,27
Sb$_2$S$_2$ (antimonite)	*Sulfobacillus thermosulfidooxidans*	23,27
PbS (galenite)	*Sulfobacillus thermosulfidooxidans*	23,27

determined that serial subculture of the microorganism on sulfur resulted in less growth with each transfer and eventual growth failure (24). Respirometry showed little oxygen consumption with sulfur as the substrate, and no pH decline during growth was noted (20). The observed growth on sulfur by TH-1 (20) may have resulted from utilization by the organism of yeast extract, which was added to the culture media. One strain, isolated by Marsh and Norris (15) from a thermal spring in Lake Myvam, Iceland, is an autotrophic sulfur- and iron-oxidizing bacterium. This microbe oxidizes sulfur in the form of tetrathionate.

Because better growth was observed under mixotrophic conditions, early investigators (19,22) cultured the moderately thermophilic microorganisms using yeast extract (0.02% w/v) and an inorganic energy source (ferrous iron or pyrite). Yeast extract, however, is not required by these organisms for oxidation of either ferrous iron or pyrite; respirometry measurements even suggest a slight inhibition of pyrite oxidation by yeast extract. The TH-1 strain grows using yeast extract as a sole energy source, but respiration is not observed. It is apparent that when using yeast extract, the organisms revert to a fermentative mode of metabolism.

When preliminary studies (19,20) suggested a requirement for yeast extract, the potential for use of the moderate thermophiles in metal leaching systems was in doubt. From an operational standpoint it would be prohibitively expensive to supplement growth media or leach solutions with yeast extract. Subsequent studies revealed that yeast extract could be replaced by glutathione (24) or cysteine (21) when the moderate thermophile, TH-1, was growing on ferrous iron or pyrite. These data suggested that the organism required a sulfur-containing organic compound. Glucose and nonsulfur

TABLE 3. $^{14}CO_2$ uptake by TH-3 and *Thiobacillus ferrooxidans* when growing on ferrous iron

Culture age (h)	TH-3		*T. ferrooxidans*	
	CPM	CPM/mg protein ($\times 10^6$)	CPM	CPM/mg protein ($\times 10^6$)
0	102	—	19,009	—
24	79,026	—	269,602	—
48	112,494	—	278,296	—
72	108,208	12.9	266,023	11.9

CPM, counts per minute

From Schacklett (28)

amino acids (e.g., serine, glutamic acid, sodium acetate, pyruvic acid), coupled with a partially reduced sulfur compound (e.g., nickel sulfide, pyrite, tetrathionate, thiosulfate), supported growth of the microbe on ferrous iron (25) without addition of yeast extract. The microbe appeared to require a reduced form of carbon and sulfur for biosynthesis. However, recent studies (P. Norris, personal communication) indicate reduced sulfur sources are not required by all strains of the moderately thermophilic bacteria. Sulfate serves as a sulfur source in energy-rich conditions such as a glucose culture medium.

When growing on yeast extract and ferrous iron, the TH-1 strain incorporated only 1% of its carbon from CO_2; *T. ferrooxidans* derived 78% of its carbon from CO_2 under the same conditions (24). Although these data suggested that the moderate thermophiles do not fix CO_2, recent studies (28,29) have demonstrated that CO_2 can be used as a carbon source. This makes these microbes more attractive for use in metal leaching systems. A comparative study (28) of $^{14}CO_2$ fixation by TH-3 and *T. ferrooxidans* (Table 3) demonstrated $^{14}CO_2$ incorporation by TH-3 grown on ferrous iron. The TH-3 culture fixed nearly the same amount of $^{14}CO_2$ per unit of protein as *T. ferrooxidans*.

When early investigators attempted to grow the moderately thermophilic microbes under autotrophic conditions, the lack of growth may have been due to the requirement by these microbes for higher concentrations of CO_2. The growth of TH-3 was enhanced when the culture atmosphere was supplemented with 1 to 10% (v/v) CO_2 (Figure 2). Culture growth, as reflected by ferrous iron oxidation, is very slow under ambient atmospheric CO_2 concentration ($\sim 0.03\%$, v/v) (28).

Wood and Kelly (29,30) reported that the Alvecote strain of the thermophilic, iron-oxidizing bacteria possesses a preferentially mixotrophic mode of metabolism. Carbon assimilation by this organism was strictly dependent upon iron oxidation and could occur simultaneously by CO_2-fixation via

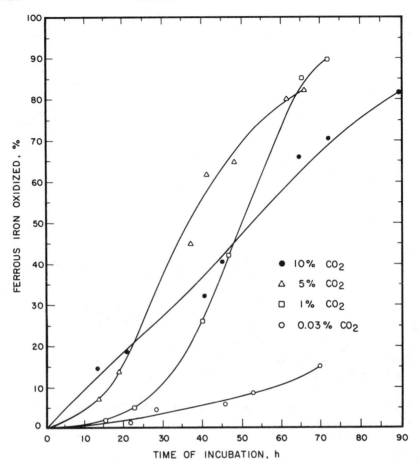

Figure 2. Effect of CO_2 concentration on growth, as indicated by iron oxidation, of the TH-3 strain of facultative, thermophilic microbes.

the Calvin cycle and direct assimilation of sugar carbon. Ribulose biphosphate carboxylase was demonstrated to be present. Glucose was oxidized by the oxidative pentose phosphate cycle, but little of this sugar was used for energy.

The taxonomy and nomenclature for the facultative, thermophilic, iron-oxidizing bacteria have not been resolved. First reports (19,20,24) described the microbes as "*Thiobacillus*-like" and suggested the genus, *Thiobacillus*, for classification (24). However, further studies indicate that the differences between the moderate thermophiles and *Thiobacillus* are too great to consider these thermophiles as part of the genus, *Thiobacillus*. Using 5S ribosomal RNA sequences of *Thiobacillus* and *Thiomicrospira* genera to establish phylogenetic relationships, two microbial isolates from the Chino

Mines (USA) copper leach dump were assessed (31). One strain was clearly *T. ferrooxidans*; the second was neither a *T. ferrooxidans* nor a *T. thiooxidans*, but rather a distinct microbial grouping. Although not confirmed, it could represent the facultative thermophile, which is present in the Chino mine dump (21) in relatively large concentrations (2.6×10^4 to $>7.0 \times 10^6$ cells per gram). Golovacheva and Karavaiko (23) proposed the name *Sulfobacillus thermosulfidooxidans* for their thermophilic isolate; this represented both a new genus and new species. Work is currently progressing on in-depth characterizations and comparison of the various moderately thermophilic isolates to establish a satisfactory classification and nomenclature (A. P. Harrison and P. R. Norris, personal communication).

4. METAL SULFIDE LEACHING

Leaching studies, using *Sulfolobus* and the moderate thermophiles, have explored metal extraction from metallic sulfide minerals and removal of pyrite from coal. This section discusses metallic sulfide leaching only and presents pertinent information on microbe-metal interactions.

Leaching studies using *Sulfolobus* have been confined to molybdenite, a molybdenum disulfide mineral, and several copper sulfide minerals. The first reported study of metal sulfide leaching using a thermophilic microorganism concerned the solubilization of molybdenum and copper from minerals by *Sulfolobus* (32). The published literature on leaching by the moderate thermophilic bacteria has principally dealt with copper leaching.

4.1. Molybdenite Leaching by *Sulfolobus*

The first successful leaching of molybdenum by a microorganism was accomplished using *Sulfolobus* (32). The mineral, molybdenite (MoS_2), is particularly refractory to leaching; in part this is due to the strength of the molybdenum disulfide bond. A second problem with leaching of molybdenite is toxicity of the resultant soluble molybdenum (molybdate) to the leaching organism. Although the mesophilic *T. ferrooxidans* is inhibited by 5 to 90 mg Mo/l (33), *Sulfolobus* has a remarkable resistance to this metal.

Molybdenite serves as an energy source for growth of *Sulfolobus* (32,34). The microbes attach to the surface of finely ground molybdenite concentrate (Figure 3). The addition of yeast extract and yeast extract plus ferrous iron to molybdenite increased the rate of molybdenum extraction. Presumably, the enhanced leaching by *Sulfolobus* after additions is in response to enhanced growth, which results from the culture additives.

Because molybdenum is solubilized during molybdenite leaching (Table 4), resistance by the leaching microorganisms to this product is of paramount importance in commercial applications in which soluble metal concentra-

Figure 3. *Sulfolobus* growing on a molybdenite concentrate (10,000× magnification). The cells are the irregular, spherical structures covering the platelike molybdenite mineral concentrate.

tions must be maximized. Respiration rate and growth for *Sulfolobus*, using sulfur, were determined to establish the limiting concentrations of molybdenum (Table 5). Respiration and growth were reduced as molybdenum concentration increased. Respiration, using sulfur as an energy source, occurred at 2000 mg Mo/l; cell growth was inhibited near a concentration of 750 mg Mo/l as indicated by decrease in formation of cell protein and lack of sulfuric acid produced from sulfur oxidation.

TABLE 4. Molybdenum solubilization during molybdenite leaching by *Sulfolobus*[a]

Culture conditions	Leach period (days)	Mo solubilized (mg/l·day)	Mo leached (%)
No additions	30	6.6	3.3
Plus 0.02% (w/v) yeast extract	30	16.6	8.3
Plus 0.02% (w/v) yeast extract and 1% (w/v) FeSO$_4$ · 7H$_2$O	30	26.5	13.3

[a]Leaching conditions: 60°C; 1 g MoS$_2$ mineral concentrate in 100 ml medium; initial pH 2.5
From Brierley (34)

TABLE 5. *Sulfolobus* **respiration and growth at increasing concentrations of molybdenum (MoO_4^{2-})**

Mo conc. (mg/l)	Respiration (μl O_2/mg protein·h)	Growth		
		Culture pH[a]	Microscopic comparison	Cell protein (mg/ml)
0	158	1.3	4+	18
500	118	2.3	3+	8.5
750	n.d.[b]	2.4	2+	6.5
1000	117	2.5	1+	0
1500	78	2.6	−	0
2000	83	n.d.	n.d.	n.d.

[a]Uninoculated control, pH 2.6.
[b]n.d., not determined.
From Brierley (34)

Leaching of mine waste and tailings for molybdenum recovery was not successful (34). The wastes tested contained acid-consuming materials, which caused the pH to stabilize above the optimum for biogenic leaching. Although molybdenite concentrates and ores can be leached by *Sulfolobus* in laboratory experimentation, there has been no commercial application of the process by industry to recover molybdenum.

4.2. Leaching of Copper Minerals by *Sulfolobus*

Sulfolobus was examined for possible use in extraction of copper from copper-sulfide minerals. A preliminary study (32) indicated that the microbe could oxidize a chalcopyrite ($CuFeS_2$) concentrate with copper solubilization occurring at a rate of 10 to 16 mg Cu/l·day over a 30-day period at 60°C. Another study of copper leaching (35) suggested that microbes resembling *Sulfolobus* were more effective than "sulfur bacteria" (presumably *T. ferrooxidans*) in leaching chalcopyrite, which is a refractory copper mineral.

In a more detailed study, *S. acidocaldarius* was used for copper leaching from porphyry copper ore with chalcopyrite as the primary mineral (36). The ore sample, obtained from Duval Sierrita Corporation, AZ, possessed a mineralization of primarily chalcopyrite, with some digenite (Cu_9S_5) and covellite (CuS). The ore assayed 0.31% Cu, 0.05% Mo, 0.02% Zn, 5.7% Fe, 0.01% Pb, and 0.01% Ni. The ore was leached in heated columns, 1.8 m length and 0.15 m diameter, containing 47.6 kg of −3+100 mesh (6.7–150 mm) sample. Two test columns were run for evaluating the efficacy of *S. acidocaldarius* for copper leaching. Column A was inoculated with 6.8 × 10^8 cells; column B served as a sterile control, using 50 mg/l "panacide" (2,2-methylenebis-4-chlorophenol) as an inhibitor. The column temperature was maintained at 60°C. Solution concentrations of copper during leaching

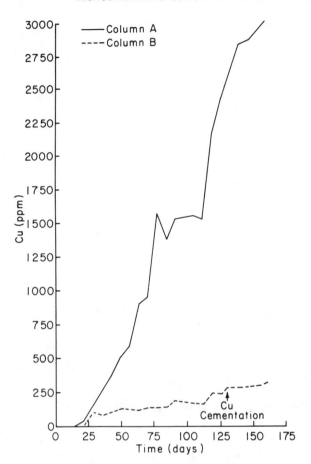

Figure 4. Percent copper leached from a *Sulfolobus*-inoculated column (A) and an uninoculated control column (B) containing chalcopyrite ore sized to −6.7 mm+150 mm. Reproduced with permission from *Developments in Industrial Microbiology* of the Society for Industrial Microbiology (36).

are shown in Figure 4. The copper concentration was decreased after day 130 by cementation, a process using scrap iron to precipitate copper by reducing cupric ion to metallic copper. This is practiced by some leaching operations for recovery of the solubilized copper. It is clear that *S. acidocaldarius* facilitates copper leaching from chalcopyrite at 60°C. The lack of leaching between days 90 and 112 was a response to shut-down of column operation with subsequent cooling; this provided further substantiation of the importance of *S. acidocaldarius* for leaching copper from the refractory chalcopyrite mineral. The amount of copper leached by *S. acidocaldarius* was 38% at an average rate of 21 mg/l·day; only 4% of the copper was leached from the control column at an average rate of 1.9 mg/l·day.

Leach results similar to those reported above were achieved using a chalcopyrite ore obtained from the Pinto Valley Mine, Cities Services Corporation, AZ (37). The leach experiment was similar to actual commercial practice in that leaching was cyclic; following an initial period of 92 days of leaching, a cycle of 4-day leach and 3-day rest was initiated. Periodically, the soluble copper was removed by cementation. Figure 5 illustrates the leaching results using *S. acidocaldarius* at 60°C. The copper extraction from column E was 7.5 mg/l·day; after initiation of the leach/rest cycle the rate was 12.5 mg/l·day. The control (column F) had a copper extraction rate of 1.0 mg/l·day prior to cycling with an increase to 1.9 mg/l·day after cycling. Following leaching, the approximate numbers of *S. acidocaldarius* were determined for samples from regions within the column (Table 6). The cell concentrations decreased from the column top, where solution input occurred, to the bottom. This may reflect a response to oxygen depletion. During leaching, the column effluent contained from 2.4 to 1.2 × 10⁴ cells/ml. Most of the *S. acidocaldarius* apparently remained attached to mineral particles during leaching.

The largest leaching test using *Sulfolobus* was conducted by the U.S. Bureau of Mines (38). A mass of chalcopyrite ore (5.8 metric tons) containing 0.77% copper was leached for a total period of 2,165 days. An initial leach of 25°C was run for 1,135 days with an extraction of only 5% of the total copper. This extraction was associated with *T. ferrooxidans* activity, which provided a leach rate of 0.12 kg Cu/metric ton·year. The temperature of the ore mass was then increased to 50°C, resulting in a decreased copper leach rate and destruction of the resident *T. ferrooxidans* population. The ore was then heated to 60°C. The system was inoculated with a total of 295 l of *Sulfolobus* culture between days 1,957 and 2,002. The copper extraction rate increased by a factor of 6.64 times over that observed at 25°C. However, after 120 days of active leaching, the production of copper decreased. Presumably the decreased leaching rate was caused by depletion of exposed chacopyrite minerals. This was the only reported study in which inoculation of a large ore mass with thermophilic bacteria was successfully accomplished.

Sulfolobus appears to be advantageous in extracting copper from the refractory mineral, chalcopyrite. *T. ferrooxidans* and *Sulfolobus* were compared for copper leaching from various copper-containing minerals, including chalcocite, covellite, and chalcopyrite (39). Increased leaching of copper from nonrefractory copper minerals at 60°C was found not to be an inherent microbial characteristic, but rather a response to temperature, since copper extraction in uninoculated controls was also increased by temperature. When the difference between the inoculated and uninoculated conditions was determined, the copper extracted by microbial activity at 25°C (*T. ferrooxidans*) and 60°C (*Sulfolobus*) was nearly the same except for a mineral concentrate sample consisting of chalcopyrite. Following 60 days' incubation, 78% of the copper in the calcopyrite concentrate was leached by *Sulfolobus*,

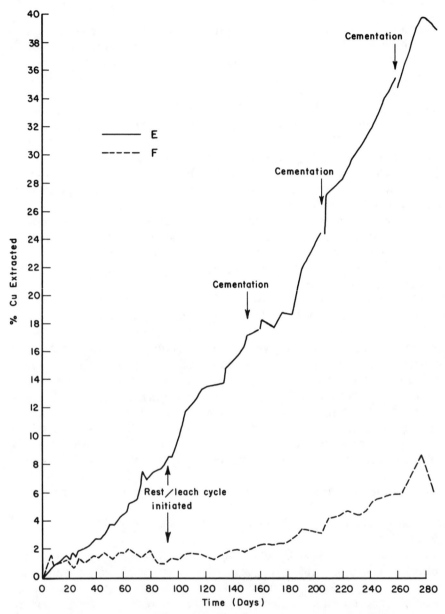

Figure 5. Percent copper leached from a *Sulfolobus*-inoculated column (E) and an uninoculated control column (F) containing chalcopyrite ore sized to -1.3 cm $+300$ μm. Reproduced with permission from *Developments in Industrial Microbiology* of the Society for Industrial Microbiology (37).

TABLE 6. Concentration of *Sulfolobus acidocaldarius* in column-leached ore

Ore sample location	Cell numbers (organisms/g ore)
Top	1.4×10^7
Middle	2.8×10^5
Bottom	$< 10^3$

From Brierley (34)

with only 8% leached by *T. ferrooxidans*. The basis for this difference is unknown since, presumably, the mechanism for copper leaching involves oxidation by ferric iron, which both microbes are capable of producing from ferrous iron.

Marsh et al. (40) made a comparative study of *Sulfolobus* species with regard to their ability to oxidize minerals. The rate and extent of mineral dissolution were found to be strain dependent. Using pyrite and chalcopyrite concentrates as substrates, some isolates were capable of almost complete mineral oxidation. The yield of copper from the chalcopyrite concentrate was greater with increasing temperature up to 70°C.

4.3. Leaching with Facultative, Thermophilic Bacteria

The facultative, thermophilic, iron-oxidizing microbes catalyze important reactions for leaching certain metals from low-grade ores and mine wastes. Strains TH-1, TH-2, and TH-3 have been examined for metal resistance, a required attribute of metals-leaching microbes (41). The ability of these organisms to oxidize ferrous iron at high metals concentration was evaluated by measuring respiration of resting cells (Table 7). High concentrations of the metal cations, Cu^{2+}, Ni^{2+}, and Zn^{2+}, inhibited iron oxidation by the three strains to similar degrees. Although inhibition of the moderate thermophiles by the oxy-cation, uranium, was not tested at high ion concentration, inhibition of the TH-1 strain, which was isolated from a thermal spring, was strong. The strains TH-2 and TH-3, both isolated from copper leaching systems, exhibited greater tolerance to uranyl ion. Growth of TH-1 on ferrous iron in the presence of uranyl ion was strongly inhibited at 4.2 mM UO_2^{2+} concentration; inhibition was 75 to 90% at 48 hours. Low concentrations of the oxy-anion, molybdate, inhibited iron oxidation by all three moderate thermophiles. These results are comparable to the sensitivity of *T. ferrooxidans* to molybdate. The two strains TH-2 and TH-3, isolated from the copper-leaching systems, were more resistant to the molybdate than TH-1. Culture growth, as determined by iron oxidation, occurred at 393 mM Cu^{2+} for all strains, with 7 to 40% inhibition after 48 hours. The overall response of the facultative, thermophilic, iron-oxidizing microbes

TABLE 7.   Respiration rates for iron oxidation by three strains of facultative, thermophilic, iron-oxidizing bacteria in the presence of metals

Metal species	Metal conc. (mM)	Strain TH–1		Strain TH–2		Strain TH–3	
		Respiration[a]	Inhibition (%)	Respiration	Inhibition (%)	Respiration	Inhibition (%)
Cu^{2+}	0	432	—	1344	—	2124	—
	363	318	26	1020	24	1584	25
Ni^{2+}	0	312	—	1362	—	2016	—
	425	216	31	882	35	1398	31
Zn^{2+}	0	282	—	1452	—	2358	—
	382	282	0	1068	26	1716	27
UO_2^{2+}	0	474	—	378	—	852	—
	150	318	33	354	6	792	7
MoO_4^{2-}	0	378	—	450	—	828	—
	5.2	60	84	318	29	474	43
	52	6	98	378	16	372	55

[a]μl O_2/mg protein·h

From Brierley (41)

to metals is similar to that of *T. ferrooxidans*, the well-studied leaching microbe. Thus, both types of microbes are sufficiently resistant to metals to function in leaching systems.

Several reports have been published which deal specifically with mineral leaching by the facultative, thermophilic, iron-oxidizing microbes (42–44). LeRoux and Wakerly (42), who were the first to report leaching using the facultative, thermophilic bacterium TH-1, found the facultative thermophile growing at 50°C more effective than *T. ferrooxidans* in leaching nickel from violarite (Ni_2FeS_4) as shown in Table 8. The rate of nickel extraction was

TABLE 8.   Comparative leaching of nickel from violarite by a facultative thermophile (TH-1) and *Thiobacillus ferrooxidans* with respect to pulp density

Microorganism	Temperature (°C)	Pulp density (%)	Residence time (days)	Extraction (%)	Nickel extraction rate (g/l·day)
T. ferrooxidans	30	3.00	5	67	0.16
Facultative thermophile	50	3.00	3	97.5	0.39
T. ferrooxidans	30	9.21	10	67	0.25
Facultative thermophile	50	9.21	5	81	0.60

increased 2.4 times. Increasing the pulp density had no apparent effect on extraction rate by TH-1. A soluble nickel concentration of over 3.0 g/l was achieved during leaching; this concentration was not inhibitory to the organism. TH-1 did not function well in leaching copper from a chalcopyrite concentrate; presumably this was a consequence of copper toxicity. The TH-1 strain was able to grow on pyrite with the medium supplemented with uranium (UO_2^{2+}) at a concentration from 0.05 to 2.0 g/l, suggesting that this facultative thermophile may be effective in uranium leaching systems.

Marsh and Norris (43) demonstrated chemolithoautotrophic growth of several strains of facultative thermophiles using ferrous iron and mineral sulfides. The yield of soluble iron from pyrite oxidation differed among strains. This difference may have been, in part, a result of the organisms' sensitivity to acidity. In general, pyrite dissolution declines as the pH approaches 1.15 to 1.25, although some strains have increased the acidity to values below pH 1.1 (J. A. Brierley, unpublished observation). Contrary to the work of LeRoux and Wakerly (42), Marsh and Norris (43) found that at least one strain of the facultative thermophiles could leach copper from chalcopyrite. After a three-week leach period at 50°C, the Lake Myvam strain leached chalcopyrite, achieving a soluble copper concentration of about 1.9 g/l. In contrast, *T. ferrooxidans*, growing at 37°C, leached 1.25 g/l soluble copper from chalcopyrite. These data provide increasing evidence for the ability and potential for use of facultative thermophiles in minerals leaching.

A recent study (44), involving the leaching of Chilean copper ores, suggested that the facultative thermophile TH-3 was the most effective microbe for solubilizing copper. This strain was compared with *Sulfolobus* sp., *Thiobacillus thiooxidans*, which would not be expected to catalyze leaching, and *T. ferrooxidans*. Unfortunately, the mineralogy of the copper ore was not described. The ore was characterized only as consisting largely of copper sulfides; presumably this meant chalcocite and chalcopyrite.

5. COAL DESULFURIZATION

The iron- and mineral sulfide-oxidizing bacteria have been proposed for removing sulfur, primarily pyritic sulfur, from coal. Although several studies have been conducted in which *T. ferrooxidans* was used to leach pyrite from coal, thermophilic bacteria can also be applied for the same purpose. The thermophiles may also have application in removal of organic sulfur from coal.

An example (Table 9) of the effect of microbial desulfurization using *T. ferrooxidans* indicates the effectiveness of these microbes in removing inorganic (pyritic) sulfur from coal. These organisms are not effective in extracting organic sulfur. This microbial process does result in a beneficial loss of ash content with no apparent loss of heat content (45).

TABLE 9. Effect of *T. ferrooxidans* on coal containing both pyritic and organic sulfur

	Content before leaching	Content after leaching	Removal (%)
Pyritic sulfur	1.89	0.26	86
Organic sulfur	1.10	1.24	—
Total sulfur	4.44	1.97	56
Ash	8.00	3.38	58
Btu/lb	11,670	12,939	—

From Detz and Barvinchak (45)

5.1. *Sulfolobus*

Detz and Barvinchak (45) compared the efficacy of *T. ferrooxidans* with *S. acidocaldarius* for desulfurization of a pyritic coal and evaluated the economics of microbial desulfurization processes. Their conditions for thermophilic leaching are listed in Table 10. Using *S. acidocaldarius*, greater than 90% removal of pyrite was achieved in 4 to 6 days, whereas, using *T. ferrooxidans* at 28°C under the same conditions (see Table 10), a similar degree of desulfurization required 15 to 20 days. The use of thermophilic microbes resulted in a more rapid rate of desulfurization and a savings in terms of reactor size. The cost for operating an 8,000 ton/day microbial desulfurization plant was calculated to be $9.48/ton (1979 dollars) using *T. ferroxidans* and $10.46/ton (1979 dollars) with *S. acidocaldarius*. The somewhat greater expense resulted from heating costs. The estimated cost for microbial desulfurization was 50% of competitive physiochemical treatment processes.

Desulfurization, using *S. acidocaldarius*, was studied using an airlift-external recycle fermenter to establish rate of sulfur removal (46). It was sug-

TABLE 10. Conditions for leaching pyrite from coal using *Sulfolobus acidocaldarius*

Coal particle size	−200 mesh (<0.05 mm)
Coal concentration	20% (w/v) slurry
Temperature	60°C
pH	1.5 to 2.0
Initial concentration of *S. acidocaldarius*	10^8 cells/ml (viable count)
Dissolved O_2	6 mg/l
Dissolved CO_2	17 mg/l

From Detz and Barvinchak (45)

gested that airlift fermenters are superior for kinetic studies for the following reasons: a) particle attrition is minimized; b) lower shear force decreases cell desorption from particle surfaces; c) oxygen mass transfer efficiency is higher than in mechanical agitated fermenters. The latter reason may not be valid if leaching is a function of ferric iron oxidation of pyrite and not direct attach by the organism. The experimental system used by Kargi and Cervoni (46) used a 5% coal slurry of ~125 μm particle size, containing 4% initial total sulfur (~2.1% pyritic sulfur) and a temperature of 70°C. The maximum rate of sulfur removal achieved after a 4-day lag was 1.8 mg S/ l·h. The maximum amount of sulfur removed was 30%, representing 1.2% of the initial total sulfur. Correlation of the increase of soluble sulfur with increase of soluble iron indicated that only pyritic sulfur was removed. Only about 57% of the pyritic sulfur was removed over a 16-day leach period. These results suggest that not all coal samples will have equal susceptibility to microbial desulfurization.

Kargi and Robinson (47) evaluated *S. acidocaldarius* as a means for oxidation of the refractory organic sulfur compounds present in coal and oil. A model compound, dibenzothiophene, was used in their study. The compound was inhibitory to *S. acidocaldarius* when concentrations exceeded 500 mg/l. Sulfur release from dibenzothiophene was determined by measuring the increase in sulfate concentration in solution. These investigators noted a 65% sulfur release in 28 days from a 300 mg/l suspension of dibenzothiophene. Although it appears that *S. acidocaldarius* is capable of oxidizing dibenzothiophene, it cannot be concluded that the microbe would be equally effective in removal of other thiophenes bound within a coal matrix.

The *S. acidocaldarius* desulfurization process has operational characteristics similar to those of metals leaching. Sulfur removal was dependent upon the sulfur content and particle surface area (48). The coal slurry pulp density also determined the amount of sulfur removal because the extent of sulfur removal decreased with increasing pulp density (49).

5.2. Facultative, Thermophilic, Iron-Oxidizing Bacteria

Few studies have been published on the use of the moderately thermophilic, iron-oxidizing bacteria for coal desulfurization. Murr and Mehta (50) attempted desulfurization of a coal sample with a total sulfur content of 5.5% using a strain of iron-oxidizing microbe growing at 55°C. Actual sulfur removal was not determined and desulfurization was implied by following the increase in soluble iron in the culture medium; iron solubilization was attributed to pyrite dissolution. The thermophilic culture solubilized 1.4 times more iron (~11 g Fe/l) than *T. ferrooxidans* (~8 g Fe/l) growing at 35°C over a 220 hour leach period. These results suggest that the thermophilic microbes may be more effective than *T. ferrooxidans* at removing pyritic sulfur from coal.

A more thorough study (51) evaluated the removal of sulfur species from lignite coal with TH-1 incubated at 50°C. Following 25 days' incubation, about 50% of the organic sulfur and over 90% of the pyritic sulfur were solubilized. These results are quite encouraging; it appears that the facultative thermophiles have potential for removal of both organic and pyritic sulfur, a feat not possible with *T. ferrooxidans*.

Although the studies completed to date are indicative of potential use of facultative thermophiles for desulfurization, the results are by no means definitive. Much research is still required in this area.

6.   OTHER PHENOMENA RELATED TO THERMOPHILIC MICROBIAL LEACHING

6.1.   Metal Reduction

Thermophilic microbe-metal interactions involve phenomena other than solubilization of metals. Both *S. acidocaldarius* and *S. brierleyi* were found to reduce molybdenum(VI) to molybdenum(V) when oxidizing sulfur under anaerobic conditions (52). Apparently aluminum in the growth medium enhanced microbial reduction of molybdate. Molybdate was not reduced when the *Sulfolobus* sp. was grown heterotrophically using yeast extract as an energy source. It was suggested that the *Sulfolobus* used molybdate as an electron acceptor during sulfur oxidation (anaerobic respiration), but the yeast extract was used in a fermentative mode of metabolism. Although at present there is no application of the reductive phenomenon, one can project the use of thermophilic microbes to promote metal redox changes for recovery of metals from solutions and to reactivate catalysts.

6.2.   In Situ Metals Leaching

In situ solution mining of metals is a possible alternative to the conventional approach to mining, which entails removal of ore by means of underground or open pit mining, crushing, grinding, and extracting the metal values. In situ mining involves the fracture of an ore deposit in place followed by solubilization of the metal values with a leach solution and recovery of the solution and metals. Figure 6 illustrates the various aspects of the in situ solution mining process. The role of microorganisms in this type of mining would involve their ability to oxidize ferrous iron and metal sulfides. Because prevailing conditions within an ore body being leached in situ include elevated hydrostatic pressures and temperatures, the thermophilic microbes offer greater potential for use than the mesophilic *T. ferrooxidans*. The effects of elevated pressure on the facultative, thermophilic strain TH-3 have been evaluated (53).

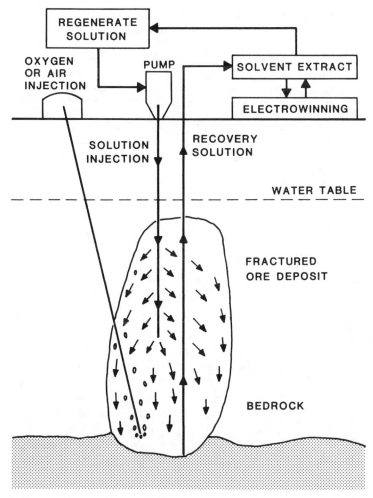

Figure 6. Schematic of a solution-mining operation employing solvent extraction and electrowinning for metal recovery from the leach solution.

Hydrostatic pressure in a water column increases with depth by about 0.10 MPa (1 atm)/10 m. The maximum pressure permitting growth of TH-3 with ferrous iron as a substrate was 25.3 MPa. However, the cell's iron oxidation system, evaluated after decompression, was neither affected by 68.9 MPa pressure nor the presence of 5 g/l copper or 1 g/l uranium at pressure. However, TH-3 was much more sensitive to hyperbaric compressed air. A pressure of 6.89 MPa inhibited microbial iron oxidation; this was attributed to the toxic effect of oxygen. These results indicate that TH-3 can function in a high-pressure, deep, in situ leaching operation; the microbe can facilitate metals extraction by maintaining iron in the oxidized

TABLE 11. Copper solubilized from chalcopyrite using TH-3 at 50°C under conditions of compressed air and hydrostatic pressure

Pressure (MPa)	Copper solubilized (mg/l[a])	
	Compressed air for 7 days	Hydrostatic pressure for 6 days
0.1	889	1040
0.34	813	—
0.69	651	—
1.03	382	—
1.38	224	—
13.8	—	1010
27.6	—	1040
41.4	—	880
55.2	—	980

[a]Values corrected for abiological copper solubilization in uninoculated controls
From Davidson (54)

state. However, the microbe cannot be used if the partial pressure of oxygen is increased.

The ability of TH-3 to solubilize copper from chalcopyrite under pressure was also evaluated (54). The inhibition of TH-3 by compressed air was reflected by a 75% decrease in copper dissolution at 1.38 MPa (Table 11). The microbes were, however, able to catalyze copper extraction under high hydrostatic pressure with only a small decrease in solubilization noted as the pressure increased above 41.4 MPa. Certainly, the TH-3 type microbes demonstrated a potential for metals leaching at elevated temperature and pressure.

7. CONCLUDING REMARKS

1. The research to date has clearly established the potential for use of thermophilic microorganisms for metals leaching.

2. Benefits to be gained using thermophilic bacteria for leaching are: a) The rate of reaction is accelerated by increase in temperature. b) These microbes can exist in "hot" areas of leaching systems, developing as a consequence of exothermic oxidation reactions involving sulfides. These hot areas are not compatible with the growth requirements of *T. ferrooxidans*, the microbe most commonly associated with metals leaching.

3. Problems of obtaining a more thorough understanding of the potential for use of thermophilic microbes in leaching include: a) The field

sites for study of the microbes are limited to large-scale leaching systems. These sites are often remote and may not be readily accessible. b) The research for field studies is quite expensive, because sampling requires drilling capability. Model systems would have to be set up in large containerized systems in which heating occurs (see reference 10 for a description of such a system).

4. There remain many unknowns regarding the use of the thermophilic bacteria for metals leaching. Both fundamental and applied research of the following nature is required: a) Extensive study of the ecology of the thermophiles in the leaching environment is necessary. b) As demonstrated for mesophilic conditions (26), mixed thermophilic culture systems (thermophilic microbial consortia) may be more effective for leaching in high temperature environments. Confirmation of this would greatly benefit the development of high-temperature bacterial leaching technology. c) Further benefit for technology development would be derived from study of the fundamentals of ecology, physiology and genetics of thermophiles.

REFERENCES

1. A. R. Colmer and M. E. Hinkle, *Science* **106**, 253 (1947).
2. L. C. Bryner and R. Anderson, *Ind. Eng. Chem.* **49**, 1721 (1957).
3. C. L. Brierley, *Crit. Rev. Microbiol.* **6**, 207 (1978).
4. D. P. Kelly, P. R. Norris, and C. L. Brierley, in A. T. Bull, D. C. Ellwood, and C. Ratledge, Eds., *Microbial Technology: Current State, Future Prospects*, Cambridge University Press, London, 1979, p. 263.
5. D. G. Lundgren and M. Silver, *Ann. Rev. Microbiol.* **34**, 263 (1980).
6. P. R. Norris and D. P. Kelly, in A. T. Bull and J. H. Slater, Eds., *Microbial Interactions and Communities*, Academic Press, London, 1981, p. 443.
7. A. P. Harrison, Jr., *Ann. Rev. Microbiol.* **38**, 265 (1984).
8. J. V. Beck, *Biotechnol. Bioeng.* **9**, 487 (1967).
9. J. R. Harries and A. I. M. Ritchie, in G. Rossi and A. E. Torma, Eds., *Recent Progress in Biohydrometallurgy*, Associazone Mineraria Sarda, Iglesias, 1983, p. 377.
10. L. E. Murr and J. A. Brierley, in L. E. Murr, A. E. Torma, and J. A. Brierley, Eds., *Metallurgical Applications of Bacterial Leaching and Related Microbiological Phenomena*, Academic Press, New York, 1978, p. 491.
11. C. L. Brierley, J. A. Brierley, P. R. Norris, and D. P. Kelly, in G. W. Gould and J. E. L. Corry, Eds., *Microbial Growth and Survival in Extremes of Environment*, Academic Press, London, 1980, p. 39.
12. T. D. Brock, K. M. Brock, R. T. Belly, and R. C. Weiss, *Arch. Microbiol.* **84**, 54 (1972).
13. C. L. Brierley and J. A. Brierley, *Can. J. Microbiol.* **19**, 183 (1973).
14. T. D. Brock, *Thermophilic Microorganisms and Life at High Temperatures*, Springer-Verlag, New York, 1978, p. 128.
15. R. M. Marsh and P. R. Norris, *FEMS Microbiol. Lett.*, **17**, 311 (1983).
16. W. Zillig, K. O. Stetter, S. Wunderl, W. Schulz, H. Priess, and I. Scholz, *Arch. Microbiol.* **125**, 259 (1980).

17. D. W. Shivvers and T. D. Brock, *J. Bacteriol.* **114**, 706 (1973).
18. O. Kandler and K. O. Stetter, *Zbl. Bakt. Hyg., I. Abt. Orig. C* **2**, 111 (1981).
19. N. W. LeRoux, D. S. Wakerly, and S. D. Hunt, *J. Gen. Microbiol.* **100**, 197 (1977).
20. J. A. Brierley and N. W. LeRoux, in W. Schwartz, Ed., *Conference Bacterial Leaching 1977*, Verlag Chemie, Weinheim, 1977, p. 55.
21. J. A. Brierley, *Appl. Environ. Microbiol.* **36**, 523 (1978).
22. J. A. Brierley and S. J. Lockwood, *FEMS Microbiol. Lett.* **2**, 163 (1977).
23. R. S. Golovacheva and G. I. Karavaiko, *Microbiologiya* **47**, 815 (1978).
24. J. A. Brierley, P. R. Norris, D. P. Kelly, and N. W. LeRoux, *Eur. J. Appl. Microbiol.* **5**, 291 (1978).
25. P. R. Norris, J. A. Brierley, and D. P. Kelly, *FEMS Microbiol. Lett.* **7**, 119 (1980).
26. P. R. Norris and D. P. Kelly, *FEMS Microbiol. Lett.* **4**, 143 (1978).
27. R. S. Golovacheva, *Microbiologiya* **48**, 528 (1979).
28. A. M.Schacklett, *Some Aspects of Autotrophic Growth by a Moderately Thermophilic, Iron-oxidizing Bacterium*, M. S. Thesis, New Mexico Institute of Mining and Technology, Socorro, 1983.
29. A. P. Wood and D. P. Kelly, *J. Gen. Microbiol.* **130**, 1337 (1984).
30. A. P. Wood and D. P. Kelly, *FEMS Microbiol. Lett.* **20**, 107 (1983).
31. D. J. Lane, D. A. Stahl, G. J. Olsen, D. J. Heller, and N. R. Pace, *J. Bacteriol.* **163**, 75 (1985).
32. C. L. Brierley and L. E. Murr, *Science* **179**, 488 (1973).
33. O. H. Tuovinen, S. I. Niemela, and H. G. Gyllenberg, *Antonie van Leeuwenhoek J. Microbiol.* **37**, 489 (1971).
34. C. L. Brierley, *Journal of the Less-Common Metals* **36**, 237 (1974).
35. R. W. G. Wyckoff and F. D. Davidson, in W. Schwartz, Ed., *Conference Bacterial Leaching 1977*, Verlag Chemie, Weinheim, 1977, p. 67.
36. C. L. Brierley, *Developments in Industrial Microbiology* **18**, 273 (1977).
37. C. L. Brierley, *Developments in Industrial Microbiology* **21**, 435 (1980).
38. B. W. Madsen and R. D. Groves, Bureau of Mines, Report of Investigations, 8827 (1983).
39. J. A. Brierley and C. L. Brierley, in L. E. Murr, A. E. Torma, and J. A. Brierley, Eds., *Metallurgical Applications of Bacterial Leaching and Related Microbiological Phenomena*, Academic Press, New York, 1978, p. 477.
40. R. M. Marsh, P. R. Norris, and N. W. LeRoux, in G. Rossi and A. E. Torma, Eds., *Recent Progress in Biohydrometallurgy*, Associazione Mineraria Sarda, Iglesias, 1983, p. 71.
41. J. A. Brierley, in P. A. Trudinger, M. R. Walter, and B. J. Ralph, Eds., *Biogeochemistry of Ancient and Modern Environments*, Austrialian Academy of Science, Canberra, 1980, p. 445.
42. N. W. LeRoux and D. S. Wakerley, in P. A. Trudinger, M. R. Walter, and B. J. Ralph, Eds., *Biogeochemistry of Ancient and Modern Environments*, Australian Academy of Science, Canberra, 1980, p. 451.
43. R. M. Marsh and P. R. Norris, *Biotechnol. Letters* **5**, 585 (1983).
44. F. Acevedo, J. C. Gentina, J. Retamal, A. M. Godoy, and L. Guerrero, in G. Rossi and A. E. Torma, Eds., *Recent Progress in Biohydrometallurgy*, Associazione Mineraria Sarda, Iglesias, 1983, p. 201.
45. C. M. Detz and G. Barvinchak, *Mining Congress Journal* **65**, 75 (1979).
46. F. Kargi and T. D. Cervoni, *Biotechnol. Letters* **5**, 33 (1983).
47. F. Kargi and J. M. Robinson, *Biotechnol. Bioeng.* **26**, 687 (1984).
48. F. Kargi and J. M. Robinson, *Appl. Environ. Microbiol.* **44**, 878 (1982).
49. F. Kargi, *Biotechnol. Bioeng.* **24**, 2115 (1982).
50. L. E. Murr and A. P. Mehta, *Biotechnol. Bioeng.* **24**, 743 (1982).
51. C. F. Gokay and R. N. Yurteri, *Fuel* **62**, 1223 (1983).
52. C. L. Brierley and J. A. Brierley, *Zbl. Bakt. Hyg., I. Abt. Orig. C* **3**, 289 (1982).

53. M. S. Davidson, A. E. Torma, J. A. Brierley, and C. L. Brierley, *Biotechnol. Bioeng. Symp.* **11**, 603 (1981).
54. M. S. Davidson, *The Effects of Simulated Deep Solution Mining Conditions on the Activity of Iron and Sulfur Oxidizing Bacteria*, Ph.D. Thesis, New Mexico Institute of Mining and Technology, Socorro, 1982.

INDEX